Managing Technology in Society

The approach of Constructive Technology Assessment

Managing Technology in Society

The approach of Constructive Technology Assessment

Edited by
Arie Rip, Thomas J. Misa and Johan Schot

PINTER
PUBLISHERS
LONDON AND NEW YORK
Distributed in the United States by St. Martin's Press

PINTER
An imprint of Cassell Publishers Limited
Wellington House, 125 Strand, London WC2R 0BB, England

First Published in 1995

© The editors and contributors, 1995

Apart from any fair dealing for the purposes of research or private study, or criticism or review, as permitted under the Copyright, Designs and Patent Act, 1988, this publication may not be reproduced, stored or transmitted, in any form or by any means or process without the prior permission in writing of the copyright holders or their agents. Except for reproduction in accordance with the terms of licences issued by the Copyright Licensing Agency, photocopying of whole or part of this publication without the prior written permission of the copyright holders or their agents in single or multiple copies whether for gain or not is illegal and expressly forbidden. Please direct all inquiries concerning copyright to the Publishers at the address above.

Distributed exclusively in the USA and Canada by St. Martin's Press, Inc., Room 400, 175 Fifth Avenue, New York, NY10010, USA

British Library Cataloguing in Publication Data

A CIP catalogue record for this book is available from the British Library

ISBN 1 85567 339 8 (hb)
      1 85567 340 1 (pb)

Library of Congress Cataloging-in-Publication Data

A CIP catalog record for this book is available from the Library of Congress

Typeset in Book Antiqua by Bob Kennedy, Enschede, The Netherlands
Printed and bound in Great Britain by Biddles Ltd., Guildford and King's Lynn

Contents

Preface viii
Christopher Freeman

Acknowledgements x

Introduction

1. Constructive Technology Assessment:
   A New Paradigm for Managing Technology in Society   1
   Arie Rip, Thomas J. Misa and Johan Schot

Part I: The Constructive Technology Assessment Discourse   15

2. Technology Assessment and Reflexive Social Learning:
   Observations from the Risk Field   19
   Bryan Wynne

3. (Constructive) Technology Assessment: An Economic Perspective   37
   Luc Soete

Part II: Steering Technology is Difficult but Possible 53

4. The Danish Wind-Turbine Story: Technical Solutions to Political Visions? 57
   Ulrik Jørgensen and Peter Karnøe

5. Steering Technology Development Through Computer-Aided Design 83
   Gary Lee Downey

6. Risk Analysis and Rival Technical Trajectories: Consumer Safety in Bread and Butter 111
   Fred Steward

Part III: Experiments with Social Learning 137

7. Learning About Learning in the Development of Biotechnology 141
   Jaap Jelsma

8. User Representations: Practices, Methods and Sociology 167
   Madeleine Akrich

9. Technologies as Social Experiments. The Construction and Implementation of a High-Tech Waste Disposal Site 185
   Ralf Herbold

10. Pollution Prevention, Cleaner Technologies and Industry 199
    Arne Remmen

Part IV: Constructive Technology Assessment: The Case of Medical Technologies 225

11. Why the Development Process Should Be Part of Medical Technology Assessment: Examples From the Development of Medical Ultrasound 231
    Ellen B. Koch

12. Social Criteria in the Commercialisation of Human Reproductive Technology 261
    Vivien Walsh

13. Decision Structures and Technology Diffusion: Technical and Therapeutic Trajectories for Diabetes Care 285
    Thea Weijers

Part V: Theoretical Analysis of Possibilities for Change 305

14. Technological Conception and Adoption Network: Lessons for the CTA Practitioner. 307
    Michel Callon

15. Firm Strategies and Technical Choices 331
    Rod Coombs

Epilogue 347

Index 355

Preface

Christopher Freeman

For many years economists have struggled with techniques of investment appraisal and cost/benefit analysis. For any project which stretches into the future such an appraisal is of course essential to avoid waste of resources or even bankruptcy. Yet there was always an uneasy awareness that uncertainty inevitably accompanied such estimates and that the *best laid schemes of mice and men gang oft aglay*. Especially is this the case where technical novelty is involved in a project. Where a new plant uses tried and tested techniques then of course much can be learnt from previous experience and relatively accurate estimates of construction costs can be made, although in a market economy estimates of future sales and hence of future rates of return must always be uncertain.

It was this awareness of these uncertainties that led Keynes to make his famous comments about the role of 'animal spirits' in investment decision-making. As little as in an expedition to the South Pole do the promoters of major new investments believe in the statements in their own prospectus. They set aside their doubts and fears as a healthy person set aside the fear of death. Keynes rightly stressed the benefits for posterity which derive from animal spirits in confronting these uncertainties but he was of course well aware that there were costs as well as benefits.

In the post-war world the necessity of taking into account social costs and benefits as well as private costs and benefits became even more apparent. Some famous attempts to make a thorough cost-benefit analysis of large public investment projects demonstrated the difficulties of attempting to place a monetary value on social costs affecting quality of life variables. At the same time, the extraordinary problems of forecasting the future consequences of such complex technologies as nuclear power also became more and more obvious. Conflicts of opinion between well-informed experts were by no means exceptional and the limitations of a purely economics-based assessment of social and environmental problems became clear.

It was in these circumstances that techniques of 'Technology Assessment' began to be used in an attempt to overcome the short-comings and limitations of cost-benefit analysis and to extend its range beyond the individual project to technologies affecting many products and processes.

TA has had a chequered history. Its formal adoption by the U.S. Congress and the establishment of TA procedures by several European governments and parliaments certainly marked a widespread recognition of the need to make some publicly available assessment of the potential risks, hazards, costs and benefits of developing some new technologies. It also marked recognition of the importance of parliamentary control of assessment procedures and the involvement of diverse actors.

However, many problems remained. Some of the most important ones are amply discussed in this volume. Its publication marks a major new stage in the development and application of TA techniques. Based especially on the experience of the Netherlands Organisation for Technology Assessment (NOTA) and the theoretical work of some of the authors of these papers, the concept of 'Constructive Technology Assessment' has gained increasingly widespread recognition. It marks especially the recognition that TA cannot be a one-off type of appraisal but must involve a continuous process, just as R&D project evaluation within the individual firm has to be a process. Secondly, CTA marks the recognition that TA must be in the nature of a continuous dialogue between potential or actual users of new products and processes, those who are affected by them and those who design, develop and promote them. Finally, it marks the recognition that in the end it is only a 'kritikfähige Öffentlichkeit' (a public opinion capable of informed critique of new technological developments) that can sustain democratic government. This book is essential reading for all those who wish to prevent the disenfranchisement of the technologically disadvantaged and the emergence of expertocracy.

Acknowledgements

The chapters in this book are based on papers prepared for the international workshop on Constructive Technology Assessment, held at the University of Twente, Enschede, The Netherlands, 20-22 September 1991. Authors and editors have profited from the discussion and the contributions of participants in the workshop, which gathered eminent scholars in economics of technical change, history and sociology of science and technology, technology assessment, as well as a small number of interested policy-makers. The financial support of the Netherlands Organization for Technology Assessment (NOTA, now The Rathenau Institute), the Dutch Ministry of Economic Affairs and the Illinois Institute of Technology's ERIF program is gratefully acknowledged.

Participants in the Workshop on Constructive Technology Assessment:

Madeleine Akrich
Ecole Nationale Supérieure des Mines
Paris, France

Håkon With Anderson
University of Trondheim
Norway

Joey van Boxsel
TNO
The Netherlands

Stuart Blume
University of Amsterdam
The Netherlands

Michel Callon
Ecole Nationale Supérieure des Mines
Paris, France

William Cannell
Commission of the European Communities
Brussels, Belgium

Norman Clark
University of Sussex
United Kingdom

Rod Coombs
University of Manchester
United Kingdom

Acknowledgements

Susan E. Cozzens
Rensselaer Polytechnic Institute
Troy, USA

Giovanni Dosi
University of Rome
Rome, Italy

Gary Downey
Virginia Technical University
USA

José van Eijndhoven
The Netherlands Organization for
Technology Assessment (NOTA)
The Hague, The Netherlands

Reinjer Grundman
Wissenschaftszentrum Berlin
Germany

Mikael Hård
Centre for Interdisciplinary Studies
Göteborg, Sweden

Ralf Herbold
Universität Bielefeld
Germany

Christian Holland
Ministry of Economic Affairs
The Hague, The Netherlands

Jaap Jelsma
University of Twente
Enschede, The Netherlands

Ron Johnston
University of Wollongong
Australia

Ulrik Jørgensen
Technical University Denmark
Denmark

Ellen B. Koch
School of Public Health
Houston, USA

John Law
University of Keele
United Kingdom

Thomas J. Misa
Illinois Institute of Technology
Chicago, USA

Arne Remmen
University of Aalborg
Denmark

Arie Rip
University of Twente
Enschede, The Netherlands

Johan Schot
University of Twente
Enschede, The Netherlands

Ruud Smits
TNO Centre for Technology and Policy
Studies
Apeldoorn, The Netherlands

Luc Soete
MERIT/University of Limburg
Maastricht, The Netherlands

Dirk Stemerding
University of Twente
Enschede, The Netherlands

Fred Steward
Aston University
Birmingham, United Kingdom

Michiel Schwarz
Cultural Cafe "The Balie"
Amsterdam, The Netherlands

Philip Vergragt
Ministry of Physical Planning,
Housing and Environmental Affairs
The Hague, The Netherlands

Vivien Walsh
University of Manchester
United Kingdom

Robin Williams
University of Edinburgh
United Kingdom

Thea Weijers
TNO Centre for Technology and
Policy Studies
Apeldoorn, The Netherlands

Brian Wynne
Lancaster University
United Kingdom

1 Constructive Technology Assessment: A New Paradigm for Managing Technology in Society

Arie Rip, Thomas J. Misa and Johan Schot

Any legitimate approach to managing technology in society must address the following dilemma. On the one hand, science and technology have become strategic political and economic resources for industry and government. For them, a rapid pace of technical innovation is the means for gaining competitive advantage in the short term as well as maintaining their survival in the long run. In addition, most people have been eager to take up the constant stream of new products and enjoy the living standards made possible by new technologies. On the other hand, technical change is deeply implicated in many of our most intractable problems, among others, environmental pollution, workplace safety, the invasion of privacy, and unsustainably expensive medical care. Cultural critics ranging from Lewis Mumford and Langdon Winner, to Jacques Ellul and Günther Anders have articulated the unease at being compelled to adapt to the inexorable demands of the machine, to the opaque forces behind technical change, but were ultimately unable to come up with viable resolutions to the dilemma.

As a society we seem to be addicted to new technology, but unable to build a constructive relationship with it. Phrased in this way, it becomes clear that the dilemma links up with long-term developments in society. The modernist adventure, one could say, is now confronted with its limits. And technology — one of the carriers of this adventure, and an icon of modernity — is part of the problem.

Need to bridge promotion and control activities

In response to the dilemma, our society has relied on a two-track approach that separates promotional activities from control and regulation. Our institutions are set up this way, with regulatory agencies separate from technology-promotional agencies, and the approach is embedded in our culture. The struggles within government, and those between environmental agencies and industry and technology agencies, while frustrating and sometimes counterproductive, can also be seen as an example of countervailing powers. A good example of the problems that can result when promotion and control responsibilities are vested in the same organization was the U.S. Atomic Energy Commission. Throughout the 1950s and 1960s, this agency actively promoted nuclear power but made only passive and ineffectual attempts to mitigate its dangers. Eventually, it became clear that the government's two activities in this area required two separate agencies (the Department of Energy and the Nuclear Regulatory Commission) with clear statutory responsibilities (Hewlett and Holl 1989). So, while there are valid political and administrative reasons to separate promotion and control, problems in the two-track approach have emerged as well, and increasingly so.

With the criticism and sometimes active resistance against new technology, and what is sometimes called a crisis of credibility afflicting business and government, the two-track approach has lost its earlier effectiveness. Take again the example of the nuclear complex, a focus of criticism in many countries, and beleaguered with credibility problems. The discussions and struggles have conjured up a storyline of proponents and opponents, which reinforces the two-track approach without resolving the problems at hand. Similar conflicts occur in areas like chemicals and transport. New technologies, such as biotechnology and information technology — despite their enormous promises — are also beset with resistance. The problem is not that there are opponents to new technology; criticism may be healthy. The problem is that our society is reorganizing itself, with the production and distribution of wealth being complemented, and sometimes overruled, by the production and distribution of risk (Beck 1992).

Since the late 1960s, technology assessment (TA) and risk assessment have been preached and sometimes practised as management tools to deal with the new situation. Such approaches have hardly been effective, however, in solving the underlying problem. While the aim of these approaches is to provide early warning and offer a perspective on future impacts of technologies, they serve primarily as after-the-fact gatekeepers. Even when not premised on two-track separation, these approaches fall victim of it. So the two-track approach has reached its limits; another track is necessary.

Resolving these problems, we believe, will require building bridges between promotion and control activities. This contention reflects a newly recognized, and essential, feature of technology in society, the co-production of technology and its effects. This feature has not been considered systematically in assessment activities and policy making, partly because the storyline of contesting parties has directed the attention to conflict-resolution mechanisms. Indeed, technology assessment has ignored this major dynamic of technology in society. Recent research has shown that social effects of any technology depend crucially on the way impacts are actively sought or avoided by actors involved in the development of technology (Daey Ouwens et al. 1987; Rip and Van den Belt 1988; Bijker and Law 1992). Social effects do not result from the technology alone (if such a thing as a 'technology by itself' exists at all). In the case of automation and of telecommunication and information technologies, this insight has been won through costly trial and error. The challenge is to reduce the errors, and implement the insights more broadly.

The notion of co-production of impacts goes beyond the limitations of the failed two-track approach by suggesting a shared responsibility of promoters and controllers, rather than presuming a strict division of labour between them. Approaches to managing technology in society that follow this insight must focus on the challenge of learning how to handle this shared responsibility. Whatever their ultimate promises, technologies confront major resistance and acceptance problems when and because their promoters fail to consider impacts and the impacted communities lack access to the pertinent decisionmaking processes. Citizen protests and regulatory challenges that come after-the-fact are symptomatic of this exclusion. What is missing are mechanisms and processes to facilitate societal learning about how to co-produce technology and its impacts, and how to achieve desirable outcomes. This book considers strategies that take up that challenge, and uses "constructive technology assessment" as a label for the new approach.

The new approach

Let us reformulate the TA philosophy to transcend the biases of the two-track approach. The goal is developing technologies with desired positive impacts and with few (or at least manageable) negative impacts. An intentional strategy to do so would integrate the anticipation of technological impacts with the articulation (and promotion) of technology development itself. The co-production of impacts must become reflexive, i.e. actors — whether they see themselves as 'promotion' actors or 'control' actors — must realize the nature of the co-production dynamics, and consciously shape their activities in terms of a shared responsibility.

At present, governments are trying to manage technologies in society in two main ways. First, many governments have lent financial and infrastructure support to the development of certain technologies deemed to be desirable. In the energy sector, for instance, the development of alternative technologies like solar and wind energy has been directly subsidized through R&D support and tax credits. Second, governments have tried to punish technologies deemed undesirable through strict regulations or punitive taxes. Both strategies of "technology forcing" are useful, though each has its limitations (for an analysis see Rip and Van den Belt 1988). In this book we start from the observation that there are always ongoing attempts to modulate and influence technological development. Consequently, alternative technology and technology forcing measures are particular cases of the new, and general, approach.

Without neglecting other options, those concerned with the management of technology in society should be especially interested in influencing technologies while they are undergoing development and taking on their durable forms. A logical first step is broadening design activities, which take place in all phases of technological development, to include (e.g.) societal and environmental impacts. It is obvious that policy instruments predicated on economic models will have insufficient influence on technologies before they enter the marketplace.

Stimulating the development of technologies with desirable impacts is not a clear-cut problem, of course. First of all, which impacts do we want to consider and in whose interest? This is a pressing problem, especially because in our postmodern time it is no longer possible to fall back on some overall rationality, a grand narrative which can legitimize action. Second, impacts are dynamic: Technical change is driven partly by the historical experience of actors, their views of the future, and their perceptions of the promise or threat of impacts; in turn, technical change generates impacts in a specific societal setting, and these impacts may or may not be aligned with developers' goals. In fact, goals evolve across the course of lengthy development and implementation projects. Therefore, even if clear values are present and shared, it is often impossible to identify an optimum strategy beforehand. This implies that experimentation and societal learning must be an integral part of management of technology in society.

While one cannot immediately and unambiguously identify "best-possible" management strategies, it is still possible to specify design criteria for such strategies. Realistic strategies for managing technology in society, we maintain, must consider impacts already during the development of technology, involve users and other impacted communities and contain an element of societal learning in how to co-produce technology and its impacts.

Technologies developed through such strategies will be socially more viable and accepted, which will enhance the economic viability of new products and processes. Presently, in the U.S. and increasingly also elsewhere, there are formidable legal and regulatory mechanisms available to those who might suffer from negative impacts, and cannot do much else than delay or harass a new technology after the fact. So it is already a matter of prudence to adopt such strategies. The design criteria stated above also include a normative aspect. They are based on the position that our current practice is morally inadequate because the full range of relevant impacts is not taken into account. Sometimes, it is the good fortune of a group not involved in the design process to deal with wholly desirable impacts; more often it is the grim predicament of an out-group to deal with negative impacts without gaining anything. John Staudenmaier (1985: 198-199) has coined the term "impact constituency" for these often-absent voices. Identification of positive and negative impacts of technology in order to work towards a more equitable distribution is an urgent problem. We add that the normative position sketched out here also implies a responsibility of the impact constituency to become active in the co-production process.

Constructive Technology Assessment

There are good reasons to shift the traditional two-track approaches to managing technology in society in new directions. So much will be clear by now. One could try out a variety of strategies, and keep whichever seem useful. Why speak of a new paradigm?

When Thomas Kuhn introduced his concept of a paradigm (1962, 1970), he was trying to understand the dynamics of scientific developments. Here, our object is much broader: strategies to manage technology in society. But Kuhn's basic idea still applies: a paradigm is an exemplary achievement that invites others to follow its example and articulate it further. In our case, the exemplary achievements refer to activities in the Netherlands and Denmark.

In the Netherlands, in the early 1980s, TA was broadened to what was called (in the title of a government policy memorandum of 1984): integration of science and technology in society. The Memorandum emphasized the importance of broadening the aspects, as well as the actors, that were involved and taken into account in the design and development of new technology. Thus, the construction of new technology would be broadened, and the anticipations and assessments necessary to realize this were labeled 'Constructive TA.'

The Memorandum also considered societal learning, but limited it mostly to an effort to improve public understanding of science and technology. When the Netherlands Organisation of Technology Assessment (NOTA) was established in 1987, it invested in Constructive TA studies, and took societal learning seriously (cf. Daey Ouwens et al. 1987). Broadening design and development was apparent in projects like PRISMA (learning about how to develop and implement cleaner technology). Broadening of interactions created new possibilities for developments in telecommunication and information technologies and especially in biotechnology (Fonk 1994). In Denmark, societal learning was foregrounded from the beginning, which relates to a political culture where public involvement was considered important. The "consensus conferences" organized by the Danish TA organization have become a model for other countries to follow (Cronberg forthcoming).

The exemplary achievements of the Netherlands and Denmark have been recognized widely, and CTA has become the label to refer to them. In an OECD report, *New Technologies in the 1990s: a socio-economic strategy*, for example, the emphasis on investment in new technologies is complemented by an argument for a more subtle approach to technology assessment. Under the heading "Towards A Broad-Based Consensus: The Role Of Constructive Technology Assessment," the OECD considers the role of the state in counteracting externalities of technological change, and recommends experimentation with new institutional structures and arrangements (OECD 1988, Section J of the Conclusions and Recommendations). For the OECD, the term "constructive" indicates the expectation of minimizing mismatches, wrong investments and possible social conflicts, which one can read as a version of our general formulation of constructive technology assessment. Also in scholarly analysis of these issues, CTA has become an approach to refer to and to articulate further (see Schwarz and Thompson 1990: Ch. 9 and Schot 1992). Thus, there are good reasons to refer to CTA as a new paradigm for managing technology in society.

Already, articulating the exemplary achievements has taken up both intellectual and practical aspects. Intellectually, the link with recent insights in the dynamics of sociotechnical developments has been important (Rip, Van den Belt and Schwarz in Daey Ouwens et al. 1987 and Schot 1991); it has become almost a defining characteristic of CTA.

Practically, the projects of NOTA in the Netherlands and of Technologinævnet (Danish Board of Technology) in Denmark were a way to experiment and learn about societal learning in managing technology in society.

Although developed for the public domain, CTA has recently been taken up by firms, non-profit bodies and social groups as well. Methods, instruments and opportunities will be at least partly different for these actors and need to be analysed and developed taking into account the special opportunities of non-governmental actors. To a certain extent new developments in industry would allow for the development of a "design for impacts." As part of a new concern with quality, some firms have integrated production and suppliers' and customers' concerns and activities into their design process. Recently the promotion of concurrent engineering within industry signals a shift towards an attempt to conduct all activities (design, development, purchasing, marketing, production) simultaneously in parallel tracks (Bloom 1989 and National Research Council 1991). Environmental pressures have led to so-called integral chain management with the ambition to close material cycles. Accordingly, waste handling activities need to be connected to design and production activities (Den Hond and Groenewegen 1993).

The aim of this book is to take the next step in the intellectual articulation of the paradigm, and also to draw out implications for working with CTA in practice.

The analytic focus

Our diagnosis is not that technology is out of control, but that the way our society handles its technology can and should be improved. Still, there is a question of control. Collingridge has argued (1980 and 1992) that efforts to control technology face two problems. First, an information problem: impacts cannot easily be predicted until the technology is extensively developed and widely used. Second, a power problem: control or change is difficult when the technology has become entrenched. Collingridge proposed overcoming these two problems by advocating flexible technology. However, entrenchment is necessary to implement a technology. To pursue full flexibility will make it impossible to develop any technology at all. Making technology inflexible is part of the domestication and appropriation of technology which involves control.

Collingridge's position is not so much wrong as conceptually flawed in that it assumes that there is a yes/no decision to be made at a certain time. Essays in this book amply demonstrate that this assumption is incorrect. Technical change is shot through with assessments by various actors. In fact, there are continual attempts to exert some influence, and it is out of the interplay of actors and their assessments that technology is shaped (Rip and Van den Belt 1988; Rip, Van den Belt and Schwarz in Daey Ouwens et al. 1987; and Schwarz and Thompson 1990, 148-151).

That entrenchment occurs, and certain paths will be followed, is inevitable. The point is that some paths are better than others, and that these should be actively sought and shaped. Thus, the problem is not that technology is out of control; it is that the present dynamics of control need some change.

To trace the existing dynamics of control requires an analytical project to conceptualize technology and social change, and to understand the effect of intervening in this process. Such an analytical project is well underway. A contemporary convergence in the sociology, history, and economics of technical change provides the outlines of a more robust model of technology and social change (see Coombs et al., 1992; Bijker and Law 1992). Some essays in this book explicitly build on this convergence (Callon, Coombs). The convergence offers exciting new possibilities for understanding technology in society. This book extends this recent work by focussing explicitly on the implications for broadening the processes of technological development, on opportunities for societal learning, and new ways to manage technology in society. In this way, it articulates the new paradigm of constructive technology assessment.

Content of the book

In the first part the concept of CTA is explored against the background of the failure of current directions within technology assessment (and risk analysis) and technology policy. Wynne and Soete coming from two different angles present an argument why CTA is needed as well as discuss problems and possibilities. Their arguments lead to the conclusion that both technology assessment and technology policy are two sides of the same coin: technology as a strategic factor in our society. CTA strategies are therefore part of a redefinition of their boundaries.

CTA tries to change technology in society for the better, and thus has to face existing entrenched technologies and societal regimes, and develop other entrenchment paths for these and new ones. Essays in the second part argue through case studies — food processing in the British dairy and baking industry (Steward), wind-turbine developments in Denmark (Jørgensen and Karnøe) and computer-aided design in a U.S. context (Downey) — that changing the course of entrenched technologies is difficult but possible. All three case studies show how technological trajectories are created, sustained and sometimes changed. These case studies suggest the range of factors and actors that block fundamental sociotechnical change. Conversely, these are entrance points for change efforts and opportunities to improve the management of technology in society. Of necessity, the possibilities for change will be situation-dependent and tentative. Learning how to handle technology in society is therefore perhaps a better phrase than "managing" it.

Since learning processes are so central to our topic, part three explicitly considers this theme. Discussion of conditions for learning as well as suggestions for some tools and strategies draw on case studies in environmental technology in the Danish fishing industry (Remmen), waste management in Germany (Herbold), biotechnology in the U.S. context (Jelsma) and information technology in France (Akrich).

Part four can be read as a demonstration that CTA policies ultimately will be sector specific. We have chosen to evaluate the options and possibilities for CTA in the medical sector. The essays in this section by Koch, Weijers and Walsh analyse how the decision-making and dynamics of technological development in this sector hamper the consideration of impacts and thus societal learning processes. Koch and Weijers reflect as well on actions to overcome obstacles for learning in this sector. Callon and Coombs in part five try to develop insights into productive management of technology in society on the basis of theory. Actor-network theory, evolutionary models and insights from strategic management are taken as starting points.

In the epilogue the editors wrap up the work done in this book by offering a series of suggestions for CTA action.

The wider issues

Especially from a no-nonsense management point of view, it may well seem too ambitious to talk of addressing a crisis of modernity. But our primary goal, to reform technology assessment philosophy and practice, is linked to the wider issues. Chapters of the present book consider issues of professional groups, of institutional and power structures that have evolved, of possibilities as well as limitations of societal learning. These issues lie at the heart of the problems and challenges of modern society.

While linking up with the broader *problématique*, the book aims at a practical result. It does not analyse the wider issues as such, but limits itself to the specific question of understanding technology in society and the co-production of impacts, and to drawing out strategies to do better.

Modern society has bound its fate to technology. The key is to replace the combination of modernist support and even glorification of technology, and the social, often distant and thus "outsider's" criticism of technology, by socio-technical criticism. This is constructive, not in the sense that technology can now be implemented and diffused more easily, but that the technology which does survive the socio-technical criticism is a better technology in society.

References

Anders, Günther. 1980. *Die Antiquiertheid der Menschen*. Teil I, and II, München.

Beck, Ulrich. 1992. *Risk Society. Towards a New Modernity*. London: Sage. (Translation of German original. 1986. *Risikogesellschaft*. Frankfurt: Suhrkampf.)

Bloom, Howard M. 1989. "Design for Manufacturing and the Life Cycle," in: S.L. Newsome, W.R. Spillers and S. Finger (eds.). *Design Theory '88 (Proceedings of the 1988 NSF Grantee Workshop on Design Theory and Methodology)*. New York: Springer Verlag, 302-312.

Bijker, Wiebe E., and John Law (eds.). 1992. *Shaping Technology, Building Society. Studies in Sociotechnical Change*. Cambridge, Mass.: MIT Press.

Collingridge, David. 1980. *The Social Control of Technology*. London: Frances Pinter.

Collingridge, David. 1992. *The Management of Scale. Big Organizations, Big Decisions, Big Mistakes*. London and New York: Routledge.

Coombs, Rod, Paolo Saviotti and Vivien Walsh (eds.). 1992. *Technological Change and Company Strategies*. London: Academic Press.

Cronberg, Tarja. forthcoming. "Technology Assessment in the Danish Socio-Political Context." *International Journal of Technology Management*.

Daey Ouwens, C., and P. van Hoogstraten et al. 1987. *Constructief Technologisch Aspectenonderzoek. Een verkenning*. The Hague: Staatsuitgeverij. NOTA Study V4.

Den Hond, Frank, and Peter Groenewegen. 1993. "Solving the Automobile Shredder Waste Problem: Cooperation Among Firms in the Automative Industry," in: Fischer, Kurt, and Johan Schot (eds.). *Environmental Strategies for Industry. International Perspectives on Research Needs and Policy Implications*. Washington D.C.: Island Press.

Ellul, J. 1954. *La technique ou l'enjeu du siècle*. Paris: Armand Collin.

Ellul, J. 1977. *Le système technicien*. Paris: Calmann-Lévy.

Fonk, Gertjan. 1994. Een constructieve rol van de consument in technologie ontwikkeling. CTA vanuit consumentenoptiek. (A constructive role for consumers in technical change. CTA from a consumer perspective. Enschede: University of Twente, Ph.D. thesis.

Hewlett, Richard G., and Jack M. Holl. 1989. *Atoms for Peace and War, 1953-1961: Eisenhower and the Atomic Energy Commission*. Berkeley: University of California Press.

Kuhn, Thomas S. 1962. *The Structure of Scientific Revolutions*. Chicago: University of Chicago Press. Second revised edition 1970.

Mumford, L. 1963. *Technics and Civilization* (1947). New York: Harcourt Brace Jovanovich, Inc. (originally published in 1934.)

Mumford, L. 1967. *The Myth of the Machine, I. Technics and Human Development*. New York: Harcourt, Brace and World, Inc.

Mumford, L. 1971. *The Myth of the Machine, II. The Pentagon of Power*. New York: Harcourt, Brace and World, Inc.

National Research Council. 1991. *Improving Engineering Design. Designing for Competitive Advantage*. Washington, D.C.: National Academy Press. Committee on Engineering Design Theory and Methodology, Manufacturing Studies Board, Commission on Engineering and Technical Systems, National Research Council.

OECD. 1988. *New Technologies in the 1990s: A Socio-Economic Strategy*. Paris.

Rip, Arie, and Henk van den Belt. 1988. *Constructive Technology Assessment: Possibilities and Constraints*. Enschede: University of Twente. Internal report, used as keynote paper for the international workshop on Constructive Technology Assessment, University of Twente, 20-22 September 1991.

Schot, Johan. 1991. *Technology Dynamics: An Inventory of Policy Implications for Constructive Technology Assessment*, The Hague: NOTA Working document no. 45.

Schot, Johan. 1992. "Constructive Technology Assessment and Technology Dynamics. The Case of Clean Technologies." *Science, Technology and Human Values*, 17 (1): 36-56.

Schwarz, Michiel, and Michael Thompson. 1990. *Divided We Stand. Redefining Politics, Technology and Social Choice*. New York: Harvester Wheatsheaf.

Staudenmaier, John M. 1985. *Technology's Storytellers. Reweaving the Human Fabric*. Cambridge, Mass.: MIT Press.

Winner, Langdon. 1977. *Autonomous Technology: Technics Out-of-Control as a Theme in Political Thought*. Cambridge, Mass.: MIT Press.

Part I

The Constructive Technology Assessment Discourse

The Constructive Technology Assessment Discourse

Introduction

CTA is an approach that contributes to bridging the gap between the divergent discourses of technology policy and technology assessment. Brian Wynne makes clear that CTA is a departure from the general philosophy behind TA practice, which holds that we should anticipate impacts of future technological developments, and accommodate such insights in post hoc decision making (Rip and Van den Belt 1988 and Remmen, this volume). This departure results from the recognition that impacts are not just passive effects of a given technology on its environment, but are actively sought (or avoided) by actors. Accordingly, CTA does not focus on predicting the impacts of a given technology, but on exerting leverage in its development and diffusion into practice.

This characteristic of CTA could also help to correct certain systematic weaknesses evident in technology policy making across the industrial world. The philosophy behind most technology policy activities is that some form of market failure leads to underinvestment in research and development, and that government through technology policies should stimulate such investment. But on the basis of industrial and organizational economics and recent empirical evidence, Luc Soete argues forcefully that this view is based on wrong economics. Existing market allocation already leads to overinvestment in many cases. This is especially true for new fields like information technology and biotechnology where research efforts tend to cluster around certain expected major technological breakthroughs. So-called critical technologies like robotics, composite materials, optoelectronics and HDTV are further examples (Branscomb 1993, chapter 2). Inevitably such efforts are poorly coordinated and consequently involve wasteful duplication. Technology policies resulting in subsidies for these clusters will exacerbate this misallocation of resources while exaggerating the lack of economic, social and societal integration of new technologies. Soete concludes that greater attention should be directed towards the capacity of society to incorporate technological changes, and we would like to add the need to anticipate demand and acceptability.

Such integration is possible in principle. Neoschumpeterian economics makes clear that technology is no exogenous manna from engineers and scientists (Dosi et al., 1988), and that new technologies do not originate outside the economic system and subsequently penetrate it. Rather, the view taken by Neoschumpeterian economics is that technological change is an endogenous process: technologies are conceived, developed and diffused by means of long and costly investments that are realized under economic and societal constraints. Accordingly, technology policies should be directed at guiding and stimulating the integration of technology in society (see also Branscomb 1993: 21-26, 167-201). Both Wynne and Soete are arguing that technology promotion and technology assessment are two sides of the same coin: a strategy for achieving goals of wealth, sustainability, safety etc. CTA can therefore be seen as part of a redefinition of the boundary between current technology policies and technology assessment: promoting those technologies that promise to have many desirable and few undesirable impacts.

Since a societal consensus on which impacts are desirable is rarely present and achievable, such a promotion policy needs to be conceptualized as one of social learning and experimentation. More is at stake than evaluating and discussing impacts. Wynne emphasizes that learning processes will involve deeply held values as well as tacit and unreflective models of social conditions assumed for — and perhaps required by the technology in question. Following Rip (1986) among others, Wynne discusses risk controversies as an important source of dialogue, learning and assessing of technology. Controversies may lead to productive learning but not when the social conditions presumed for the technology's viability remain hidden rather than explicit. In highlighting the clash of values, Wynne underscores that developing socially viable technology requires clarifying values and developing open mechanisms to relate divergent sets of values. Jelsma's analysis of biotechnology regulation and Remmen's discussion of dialogue workshops in part three make clear that if these social conditions are indeed discussed, controversies can lead to effective management of technology and (some) rationalization of the discourse between actors involved. Wynne concludes that such collective value learning will ultimately lead to enhanced safety, broader social legitimacy, and greater viability of technology. Could we identify criteria to decide if a learning process goes well? Once again, because of the indeterminate character of technical change, criteria are not static and will change over time. Wynne therefore argues that values are emergent. CTA should therefore emphasize the negotiated and reflexive element of social learning and experimentation. The point is very important, but one should add that empty pluralism is to be avoided. Substantive pluralism is to be aimed at, which allows for different values and points of view, but at the same time asks for real engagement which interferes with others' behaviour.

References

Branscomb, Lewis M. 1993. *Empowering Technology: Implementing a U.S. Strategy.* Cambridge, Mass.: MIT Press.

Dosi, Giovanni, Christopher Freeman, Richard Nelson, Gerald Silverberg and Luc Soete (eds.). 1988. *Technical Change and Economic Theory.* London: Pinter Publishers.

Rip, Arie. 1986. "Controversies as Informal Technology Assessment." *Knowledge* 8 (December): 349-371.

2 Technology Assessment and Reflexive Social Learning: Observations from the Risk Field

Brian Wynne

Introduction

Although the technology assessment (TA) literature has never explicitly referred to "social learning" to describe what it was about, its proponents believed that TA was a social learning process. It took form in the systematic identification and analysis of (hitherto neglected) "nth order" impacts of technologies and the transmission of such expert insight into policy decision making. That TA was flawed by an assumption of unilinear flow from 'objective' scientific discovery to 'necessary' social adjustment only reflected its correspondence with the broader field of science, technology and public policy, dominated by a similar weakness. Wynne's (1975) critique of the TA enterprise as a "technocratic rhetoric of consensus politics" implied that systematic obstruction of social learning was inadvertently occurring in the then-dominant discourses of TA and technology. The key commitments were the black boxing of technology and the conceiving of its development trajectories as essentially objective and non-social.

As several authors have suggested (Rip 1986; Cambrosio and Limoges 1991; Mazur 1981; Nelkin 1979) the suffocating conceptual, methodological and political confines of conventional TA could be fruitfully enlarged by regarding public controversies about technology as rich, engaged and grounded social processes of assessing technologies. Whereas TA assumes and in effect prescribes consensus about basic definitions, relevant types of impact, methods of analysis, relevant actors and knowledges, and boundaries of authority, controversy studies have highlighted the positive dimensions of adversary processes, vividly underlining that more is at issue than 'preferences' or even interests (Schwarz and Thompson 1990). Controversies expose neglected possibilities, clarify the limitations of accepted analyses, and identify the social values or interests concealed in existing 'objective' trajectories. They thus enlarge the scope of assessments when performed in the less orderly, but more discursive, pluralistic and open way typical of controversies.

Rip (1986) proposed that controversies about technology could be treated as informal social learning processes complementary to formal TA mechanisms. Cambrosio and Limoges (1991) have gone further, to argue that social controversy should be regarded as the general case, from which conventional TA is sometimes abstracted in protected social enclaves of consensus or supposed consensus: "The borderline case is not when controversy plays the role of an informal technology assessment. The actual borderline case occurs when controversy is at its lowest and when formal methodologies of technology assessment seem to be in charge." This is consistent with Wynne's (1975) view that TA methods and discourses not only reflected (non-typical) consensus, but attempted to *cultivate* consensus. Also consistent with this original critique are actor-network theoretical terms (Callon 1987; Law and Callon 1982), in which *formal* TA can be treated as part of the *enrolment* process.

In this paper I shall try to further the specific discussion about learning processes and TA discourses by coming at it from the field of risk studies. This field contains several different disciplinary perspectives, and it is closely related to policy processes, patronage and demands (Otway and Peltu 1985; Krimsky and Golding 1992). Underlying substantial diversity and confusion is a distinction between approaches that tend to decontextualise risk and artificially objectify its intrinsic meaning, and approaches that attempt to identify and locate people's own risk definitions within their indigenous cultural and social relations (Douglas 1986; Krimsky and Golding 1992).

Drawing on the risk field I want to elaborate the observation that boundaries between technology and its social environment are themselves social conventions, and that operational models of technology reflect objectively different ('internal' or 'external') social positions with respect to technology and its controlling power structures. In actor-network language the 'network' is experienced differently from different positions within it. In actuality, each node in a network is an intersection of different multiple networks. Thus associated risks are experienced and defined in fundamentally different ways, even though it is ostensibly the same risk-system.

In adopting a reflexive standpoint, I go beyond an approach which still conceives of assessing impacts of actual or proposed projects (Rip 1986). By a reflexive approach I emphasise technology as a social vehicle that already represents, and tacitly reproduces, social commitments; not as a social entity which only has post-hoc social impacts. That is, we need to pay attention to what a technology embodies, reflects and reproduces by way of *prior* values, identities, cultural forms, and social relations. What assumptions or commitments of these kinds does it propagate and reinforce? This reflexive critique has existed since the early days of TA (Tribe 1973, Wynne 1975) and it is time to integrate it into developing concepts of CTA. Claims about risks, including scientific analyses, embody implicit representations of technology which are shaped by tacit models of social relations and processes. The essential open-endedness of these models (hence of the risks they harbour) is concealed by deterministic discourses of technology and scientific knowledge.

Recognition of the social nature of technologies immediately introduces a cogent new dimension into the interpretation of risk controversies and offers new implications for social learning and CTA. Furthermore this social dimension of risk bears directly upon analysis and argument over physical risks, since as I shall illustrate below, analysis of physical risks is always conditional upon an underlying model, usually unconsciously exercised, of (the relevant parts of) society. There are direct analogies here with the identification of designer constructions of user-situations as a critical issue (Wynne 1989; Woolgar 1991; Callon 1987; plus compare Akrich, this volume).

Thus the sociological critique of the dominant paradigm in the risk field has developed along very similar, but so far separate, lines to the sociological analysis of technology. One theme with strong implications for social learning and constructive technology assessment is the extent to which the dominant discourses of risk and technology assessment conceal the indeterminacy or lack of closure of technological systems (Wynne 1988). The emergent approaches to CTA have not fully escaped from this non-reflexive discourse, and need to pay more attention to what we might call 'values-learning' or articulation, as well as to 'impacts-learning'. That tacit and unreflective social models condition objective risk analyses can be shown at several levels.

Objective risk discourse and social models

In the 1977 Windscale Public Inquiry into the proposed thermal oxide nuclear fuels reprocessing plant (THORP) the dominant assumption of the nuclear and government experts, vigorously reinforced by the High Court Judge in the Chair, was that the technology for 'objective' assessment was limited to the reprocessing plant — nothing more (Wynne 1982). The organisational, technological, economic and political integration of the proposed plant into the nuclear fuel cycle, and indeed into centralised energy supply was removed from the 'rational' problem domain. Thus those who suggested the likely impacts of the plant should include the future demand for fast breeder reactors to use the extracted plutonium, for additional reprocessing plants and associated fuel-cycle facilities all of which were implied in the specific planning proposal in question — were dismissed as introducing "emotive" images of the "plutonium economy". Officially, the inquiry asserted that those further questions could be dealt with as completely separate matters at the appropriate future date.

Thus the risk assessments of the establishment experts reflected this circumscribed 'objective' definition of the risk-generating technology. In terms of underlying social models, this dominant framing discourse took for granted the trustworthiness and impartiality of decision making institutions (of which that discourse's authors were eminent members). Hence the questions about THORP itself could be divorced from any future wider decisions that might be implied by it (fast reactors, etc.). This social model — that existing decision-making structures were impartial, and trustworthy — therefore strongly (if obliquely) defined the technology boundary and the risk problem.

Objectors on the other hand regarded the decision-making institutions as inextricable parts of the risk problem. This stance was not articulated as such, but was evident from the difficulty they had in trying to keep to the narrow THORP-only agenda. They assumed that pressure for future expansion of the nuclear infrastructure would inevitably follow from the THORP plant. For them, the risk issue was an extrapolation forward of the existing political body language (unswerving, dogmatic commitment to expansion). In this regard, independent regulators, policy makers, and nuclear industry proponents were indistinguishable. In addition opponents addressed the 'purely' scientific risk questions of the THORP plant in terms of the past behaviour of the risk-controlling institutions: What had they promised and had they delivered? What was their track-record in terms of rigour, openness, impartiality and competence? If they could not be trusted, as objectors argued the record showed they could not, the physical risks were multiplied accordingly, not just the perceived risk. In other words, the indeterminate *social-conditional* character of the physical risks was ignored by the scientific discourse of objective risk.

In the inquiry's official report and the subsequent debate, the dominant discourse set by the government, the chair and the industry experts prevailed. Controversy raged not around what the dominant discourse left out but only around whether the chair had misunderstood or misrepresented the "facts". The prior framing — and the social model which it both reproduced and reinforced — was not even identified as an issue. In effect, the inquiry chair's rejection of objections as factually ungrounded was a means of imposing by fiat the establishment's framing social model, rather than have contending models recognised and debated as equally legitimate. Each model reflected the objectively different social positions of their authors in relation to power and decision making about nuclear technology.

If the analysts of the cultural conditions of post-modernity are correct (Giddens 1990, 1991; Beck 1992), then such mistrust of dominant institutions and the sense of risk it portends, is pervasive. This insight led to the general argument (Wynne 1988) that the logical basis of public definitions of technological risks was not the taken-for-granted framework of technical experts (namely physical risk magnitudes such as probable rates of death or injury), but the trustworthiness of the social actors supposedly in control of those risks. Physical risks always occur in some particular context where social actors can affect the situation, in various ways. Expert abstractions of physical risk magnitudes out of such contexts are artefacts which may ignore important situational variables about which lay people (such as workers) may have more specialist knowledge (or equally legitimate commonsense judgement of social practices) than the experts. Indeed the dominant ('objective versus perceived risk') discourse, in so designating public experiences and identities, inadvertently confirms and reinforces the sense of *social* risk.

The rationalist rhetoric of risk and technology assessment allowed the dominant discourse to pit 'objective' risk assessments of the experts against the 'perceived' risks of the (allegedly misinformed and irrational) public. Having identified a legitimate social logic that framed the public's definitions of risk, the sociological approach to risk then demonstrated the symmetrical point about scientific expertise (Wynne 1989, 1992). The experts also derive their 'objective' risk analyses from equivalent, unspoken and taken-for-granted social models.

Similar insights can be taken from the controversies surrounding biotechnology. The European Commission has encouraged the development of a European biotechnology industry, with appropriate regulatory safeguards and public information programmes. In 1985 the Commission wished to examine the safety of biotechnology-produced animal growth hormones. The Commission requested a risk assessment from a committee of veterinary experts which decided to measure the residual hormone levels after different forms of treatment and time intervals. Since these particular hormones are naturally found in the gut of cattle the experts decided that a reasonable criterion of safety would be that the residual levels of deliberately administered growth hormone should be less than the natural background levels. On this principle they established that if the hormones were used under certain specified conditions then they would presumably be safe, hence permissible for commercial production and use.

The specified conditions were that the biochemicals should be administered under veterinary supervision; that maximum doses and dosages should be defined and respected; that injection should be only into non-edible tissue; and that there should be a minimum waiting period between the last dose and slaughter for consumption. It made no comment upon the feasibility of these conditions in real life.

On the basis of this expert report the Commission officials drafted legislation to permit full-scale production of the growth hormones which they confidently expected to be confirmed by the Council of Ministers. In the meantime however, a widespread campaign against this permission emerged, and by the time the Council of Ministers met in December 1985 only one Minister remained willing to vote in favour. The Commission was forced to redraft the legislation to reflect the Council decision. After this volte-face, Commission staff involved in the risk assessment process were enraged by what they saw as the readiness of Ministers to bend to popular irrationality and hysteria against biotechnology, ignoring what the experts described as well-prepared, scientifically sound decisions. How could civilised society be defended against such antiscientific hysteria, they railed, if political leaders would not give the proper rational lead?

Despite their portrait of objective science against popular irrationality and superstition, the Commission officials conspicuously failed to address the social conditions which the scientific advisory committee had at least pointed to. Could not the opposition's apparently extreme risk assessment reflect the (implicit) judgement that these conditions could not be built into the practice and regulation of farming up and down the length and breadth of the whole European Community, from Greece to Denmark, and from Ireland to Italy? Implicitly, the officials had assumed that these social-technical conditions could be guaranteed in real-life practice. The difference between the sides was not science versus anti science. It was an unrecognised difference of social models from which the conflicting risk assessments were derived. In principle each was legitimate and could be debated using relevant evidence.

We are now talking not so much about overt conflicts over claimed impacts, as tacit conflicts over social conditions of validity of technologies and associated risk assessments. The questions are: Are such conditions feasible? Are they acceptable? What level of social policing would be required? Indeed the articulation of a risk assessment in real life becomes in effect a prescriptive framework for the technology, as the social assumptions underlying the risk assessment take on the role of tacit commitments that must prevail to validate the technology (and the assessment).

Thus instead of a learning process occuring through the systematic identification and evaluation of such implicit social models or conditions, existing discourses encourage inadvertent concealment, defensive reactions and rationalisations. In this case European Commission officials still appear to believe simply that the wrong decision was reached, and that there was no conceivable logic to the successful political opposition. In such cases (see also Wynne 1989) modernistic discourses and institutions themselves undermine their own social legitimation, and fuel the social dimension of risk now articulated widely as 'risk society' (Beck 1992). To indicate the wider relevance of this analysis, I offer two further examples — from what might be called the hardest case, nuclear power.

The 'hardest', most purely physical form of risk analysis is that performed for reactor pressure vessels. In such risk analysis there are both elaborate analytical models and operating data on complete reactor systems and their components. Thus failure probabilities and modes can be built up in quasi-mechanical fault or event-trees, and effects of failure and mitigating design measures given quantitative estimates. However the elaborate and impressive technical analytic work does not reduce its dependence upon some key social assumptions or models. For example, when applied to new circumstances, the validity of empirical data on component reliability is conditional (*inter alia*) upon the same or better quality of manufacture, inspection, maintenance and operation being maintained. The same applies to reactor or pressure vessel systems. That such assumptions may be generally expected to be realised in practice in no way lessens their social character. Again they are prior social conditions of the expected or hoped-for operation of the system to specified performance standards. These assumptions also have to hold for widespread commercial use, when the technology is ordinary and routine, as opposed to an object of special attention as a pilot or demonstration plant.

A significant dimension of nuclear safety controversies can be attributed to unclarified differences of view over the long-term stability and quality of such social institutions. An implicit counterargument sometimes heard from the nuclear industry is that only by placing the prescriptive demand on the institutions, by establishing such technologies, will we force the maintenance of that necessary stability in the face of incipient social indiscipline. The mutual construction of technology and social order is here clearly illustrated.

The final example — nuclear dread and the weapons connection — requires us to fill in some necessary background. Psychometric studies of public risk perceptions have identified various 'attributes' of risks which supposedly affect public perceptions of risks beyond the purely scientific assessment of their magnitudes (Slovic 1992). This influential work was initiated to explain the troubling deviation of public perceptions of risks from experts' assessments. Examples of attributes of risks are: voluntariness/involuntariness; reversibility/irreversibility; concentration or dispersion (over time and space) of the same aggregate expected harm; familiarity or unfamiliarity; and anonymity or personalisation of casualties. In addition, the dread imagery associated with different risks was also identified. The archetypal example was thought to be the imagery of the mushroom cloud from nuclear weapons, often associated with civil nuclear power. Another was the Frankenstein imagery said to be evoked by biotechnology and genetic engineering.

These psychometric attributes of risks presume the universal object of people's experience and perception is indeed, risk, as defined by the analyst (some physical index of probable harm, such as mortality or mortality/morbidity combined, per unit of time or activity), and in relation to an assumed given discrete unit of activity, such as a nuclear plant rather than the full nuclear fuel-cycle. The attributes then explained why lay people 'distort' risk evaluations from the 'objective' magnitudes given by the experts (from within the framing assumptions outlined above). That people might be responding to an altogether different agenda reflecting their social position and experience of technologies as systems of power and social control, has been posited by sociological critics (Otway and von Winterfeldt 1982; Wynne 1982). The dread imagery associated with nuclear power was said to be composed of the atomic mushroom cloud evoking the horrors of nuclear annihilation. Technical experts lost no time in lamenting the apparent ignorance and misunderstanding on which this association of nuclear power plants with nuclear weapons was based. They emphasised the accepted technical understanding that a nuclear reactor is physically incapable of a runaway nuclear chain reaction as in a nuclear weapon, and advocated correction of this misperception to 'correct' exaggerated public fears about nuclear power.

As with most of the other psychometric attributes of risks however, this dread association can be given a sociological grounding which introduces symmetrical logics into the conflicting positions. According to the expert framing, the dread imagery connecting nuclear energy and nuclear weapons is unfounded and illegitimate. Yet an alternative way of interpreting the connection is that it represents in symbolic form that there are well-recognised instances of horizontal proliferation into nuclear weapons from the diversion of sensitive materials, technology and know-how from civil nuclear programmes. Institutions such as the International Atomic Energy Agency's Safeguards programme in theory protect against such diversion, but they confront uncommon pressures with inadequate resources and limited powers. Thus the 'imaginary' dread association could be seen as an expression of the concrete social judgement that those institutions supposed to maintain a complete divorce between civil and military nuclear technologies are simply not achieving their formal objectives. The technicist's perspective which calls the dread imagery groundless and bemoans the public ignorance and emotive exaggeration is just an opposite social judgement: that those institutions are doing a competent job and can be trusted to do so into the future.

Either of these conflicting social judgements or models is legitimate and arguable; there is relevant evidence to deploy, and structure to the institutional debate which would need to be articulated amongst the different parties. However the expert framing is no more objectively grounded nor automatically superior, just because it commands the rhetoric of asocial objectivity. A resolution of this controversy would have to be negotiated, not imposed *a priori* by defining it as a dispute between real risk-knowledge and mere subjective perceptions. A learning process would have to encourage each party to recognise, articulate and debate the unrecognised social models, assumptions and commitments tacitly shaping its factual discourse.[1] These are not 'impacts' models, but become prescriptive ordering commitments.

Controversy, indeterminacy and learning

The above examples illustrate that controversy about technologies and risks can induce different kinds of learning in the sense of adding information relevant to decisions or commitments, forcing the recognition of hitherto-excluded actors, and elaborating the agenda of issues to be addressed. They also show however, that the dynamics of controversy can inadvertently obstruct learning. Discussion of controversy as constructive TA (e.g. Rip 1986; Cambrosio and Limoges 1991) has failed so far to recognise the importance of more reflexive forms of possible learning which go beyond the addition of information and criticism regarding possible impacts. The sociological analysis of risk issues emphasises the unstated commitments by different parties to social conditions of viability of technologies, commitments which underpin and shape their factual discourses about impacts and risks. This emphasis upon assumed social conditions connects with the recognition within sociology of the essentially unclosed, or underdetermined property of established knowledges or technologies. Whether or not the tacitly assumed social conditions are realised is still an open question, contingent upon a variety of factors (including the dominating discourse).

Even stable technological systems, which appear viable, and acceptably safe, and fully determined by the relevant "explanatory" factors, sometimes are invisibly indeterminate. They are surprisingly open to destabilisation by trivial changes of conditions whose importance had not been recognised or which were governed by dynamics and rationalities independent of the technological system. An example (Wynne 1988) was the reduction of demand for water in Lancashire which led to an underground water-transfer tunnel remaining stagnant for long periods, whereas designers had assumed it would be pumped out continuously. This change in operating conditions allowed methane gas to accumulate in the Abbeystead tunnel, and led to an explosion that killed sixteen visitors. With tragic irony they were being shown the tunnel to reassure them about their fears of flooding downstream from the scheme's outflow.

There is no guarantee that more self-conscious examination of tacit social models and conditions of viability would anticipate and avoid all such problems; but that is not the point. As the sociology of science and technology has shown, technological systems and scientific facts advance and grow in the world not by demonstrating their pre-existing truth, but by reorganising the social as well as the physical world so as to affirm them (all the while benefitting from the rhetoric of demonstrated pre-existing truth). To render these social processes more explicit and more avowedly negotiated seems to be both a reasonable prescriptive goal as well as a useful instrumental one, for reasons of safety (as the Abbeystead case shows) and for broader social legitimacy and viability. The reflexive sociological perspective is a necessary tool for this job.

Greater reflexivity would result in more difficulty in developing hegemonistic technological systems — more fundamental pluralism. However successful the strategies of knowledge and technology growth may be as measured in terms of conventional indicators of their reach and power, the corresponding tendency towards hegemonic forms may entail both an increase of vulnerability to dislocation (less robustness towards ignorance and uncertainty), and strong undercurrents of alienation and lack of legitimation with social groups whose passive acquiescence at least is needed. The dominant 'objective' versus perceived 'risk' framework exacerbates these tendencies by denigrating and patronising 'irrational' publics. Wynne (1989) has criticised the mainstream paradigm of risk perception for gratuitously assuming that lack of expressed public opposition or dissent is equal to public acceptance of a risk or technology. This is another example of a contingently stable, but underdetermined system (here including the social legitimation) being wrongly taken to be fully determined and hence more stable than it really is.

Cultural theory of risk and technology is one of the most developed perspectives available to explore the implications of monopolistic rationalities for the social robustness of technologies (Schwarz and Thompson 1990). According to this theoretical perspective, developed from the cultural theory of Mary Douglas, there are just four basic cultural types whose interaction makes up the complexity of organisations, political cultures, risk management regimes, and technologies. These cultural styles integrate characteristic forms of social relations with confirming systems of a particular style, of belief and interpretation.

Each of these cultural styles reproduces itself by integration between its favoured patterns of social relations and its systems of interpretation and belief about nature and the cosmos (including salient types of risk). Each could be seen as a particular selection environment not only for natural knowledge but also for favoured systems of technology, since these reflect and combine both beliefs about nature and favoured social relations. However, an important element of cultural theory is that each of the four basic styles is able to define itself only in contradistinction to the other three, in dynamic interaction in the give and take of social life. To the extent that any one becomes predominant to the exclusion of the others, it loses robustness and learning capability. It may become more efficient in some respect, optimised around just one tidy set of principles or convictions; but this renders it vulnerable in more complex environments. It may eventually find it has to muster impossible powers and resources to control or suppress the surrounding disorder.

I have suggested the importance of one particular kind of social learning — one involving reflexive processes — in defining and debating risk issues as part of constructive social assessments of technologies. By reflexivity I mean something akin to the notion of self-reflexivity as discussed by various social theorists of modernity/post-modernity; but with scepticism about the intrinsic capacity of any actor to sustain self-reflexivity without its being encouraged by criticism and challenge from others. The specific form of reflexivity I mean here is identification of the tacit social models of the assumed conditions of viability and (safe) operation of technologies. These may involve:

- implicit judgements of the institutions that supposedly control the risky activities, which judgement also affects the defined scope and nature of the risk problem;
- models of the social relations and processes involved in operating, maintaining and regulating the technology, which models include assumptions of deterministic rule-following that conceal indeterminacy in such systems; and
- experts' implicit constructs of the public as risk definer and political actor.

These kinds of framing social assumption carry strong but unstated value commitments. They take on the role of prescription precisely because conventional discourses of risk and technology lack recognition of the essentially open, socially constituted character of technologies, and the mutual construction between technology and social context of validation. They cannot be recognised and articulated in public debate without a more reflexive approach which transcends the language of impacts, even if this language takes us inside the black boxes.

Conclusions

In this paper I have argued that dominant risk discourses, whilst appearing to respond to the negative implications of technologies, do so in ways which further conceal the prescriptive social ordering processes embodied in technological commitments. They thus tend to undermine constructive learning processes.

To talk about social learning is potentially pretentious and certainly risky. Part of its appeal is its very ambiguity. It does not seem posssible to conceive of learning except relative to normative principles injected by the analyst, or already identified by the actors. Some references to social learning do indeed operate with clear and elaborated notions of learning, but these are often parochial and unreflective, aimed at better information flow within policy agencies (Parsons and Clark 1992). Others (Sabatier and Mazmanian 1984) offer a middle-range perspective on mutual learning between environmental policymaking institutions and implementation and enforcement actors. The latter actors significantly reshape the policies and become recognised as policymaking actors, not mere ciphers. Finally, approaches informed by critical theory suggest that any increase in criticism and debate is by definition equal to social learning[2] (Eder 1987).

Although the critical theory approach to social learning can be criticised as failing to distinguish between learning and noise, it does emphasise the *relational* element of social learning. This distinction is valuable; learning is essentially a *process*. However, the dilemma is that the product of this learning process is a shared culture embodying the consensus that decisions about such technologies can legitimately be based upon one abstract dimension, namely annual individual fatality probabilities. Establishment of the shared focus on just one dimension results in the neglect of other important issues. In this view, learning is based in the developing social relations between actors (thus it might be regarded as learning capacity) more than in what specific actors may be deemed to know. Part and parcel of the developing recognition of the legitimate role and standing of others in the generation of collective knowledge and commitment, perhaps a necessary condition of it, is a reflexive critical awareness of one's own commitments and assumptions. This is consistent with the recognition of the essentially indeteterminate and open-ended nature of knowledges and technologies. Thus this paper builds on the relational concepts of social learning, emphasising that there is more to play for in the 'social assessment of science and technology' than is apparent from prevailing languages of impacts. The latter imply, falsely, that models of the human subject, social values and identities are not expressed and imposed in the developing technology itself.

The reflexive version of social learning would involve the systematic exposure, investigation and debate of the implicit social models and assumptions that structure 'factual' analyses of technologies, impacts and risks. Although these social commitments influence debates about technologies and risks, their hidden role often leads to the degeneration of debate rather than its enlargement. The commitments play an ambiguous role, ranging from virtual social hypotheses on whether society can be (re)organised in some way, to hard-line prescriptions on how society must be organised so the technology will behave in accordance with established risk assessments. These underlying roles and their interplay in the 'grand social experiment' of negotiating technologies are rarely explicated, to the detriment of more robust learning, and perhaps of more socially robust technologies (Krohn and Weyer, 1989; and Herbold, this volume).

This explication of underlying social visions differs in interesting ways from other treatments of technology development. The works of Callon and Latour offers an obvious parallel. However, their analysis emphasises the deliberate and conscious social visions that engineers and others have as the social templates for their technical programmes. Mine highlights instead the unconscious social visions shaping different models of technology and risk. Whereas Callon and Latour have focussed on technological empires in the making, my cases have involved the retrospective analysis of technologies which have run into trouble. Both traditions of work nonetheless share the most important feature, namely, the centrality of the indeterminate nature of the processes and products.

It is also worth noting the distinction from Rip's treatment of technical expertise in the risk field. He rightly observes a tension between the intrinsic intellectual tendency of science towards standardisation, and respect for the particularities of actual contexts of social use. What are the important properties, and which properties can be neglected, is usually an unstated dimension of conflict between scientific knowledge and lay knowledge. Rip sees a dilemma between scientific purism and social paternalism ("deciding what is good for the people"), with the expert having to make "assessments of the sociopolitical landscape" to be effective. In my analysis experts are deciding what is good for the people, but it is often decision by default, as unreflective and concealed social assumptions and models are converted into *de facto* social prescriptions. Experts must not only look outward at the sociopolitical landscape, but inward at their own prior frameworks and their shaping social models. These often derive from cultural and institutional factors. They may require a nudge from social criticism and controversy to do this. But constructive nudging needs a framework which recognises the need for systematic and explicit treatment of these dimensions.

It follows that CTA is still too strongly conceived in terms of identifying impacts of technologies and newly identified steering points rather than asking about what values, identities and incipient social orders may be reflected and reproduced in such commitments. These reflexive questions can be asked at the micro level inside technological systems just as they were asked of conventional TA twenty years ago (Tribe 1973; Wynne 1975). The main rationale for CTA, of social analysis inside the black box, still appears to be focussed on anticipation of impacts, now available at a more detailed level.

Finally, we should avoid the pitfall of automatically equating social explanation with social choice and steering opportunities. Social factors may not necessarily be subject to social choice.[3] More importantly, conceiving of the marriage between CTA as the fine-grained identification and anticipation of positive and negative impacts, and thus possible steering and selection, misses the central point. It is not enough to identify and then select for the positive and root out the negative impacts, in the micro-selection environments now identified in their sociological richness. It is also necessary to channel the interpretive formulation of different experiences and identities more fully into the continual negotiation of what impacts might be positive and which might be negative, (and what the implications of different commitments are even if all the impacts are not identifiable).

This more interpretive frame gives due recognition to the indeterminacy dimension, so long as the negotiatory discourse also allows for debate of what values and social-cultural identities the possible technical options and commitments signify, reflect and reproduce, and not just what they may cause. These values, interests and identities also have an indeterminate and emergent character, and are further shaped and consolidated in the processes of negotiation (Law and Callon 1982; Wynne 1982). Collective 'values-learning' in this sense may be the most critical and difficult element of the whole challenge of more constructive social shaping of technology.[4]

Notes

1. The extent to which research is implicated in such political issues whether it intends it or not can be glimpsed in a study of child safety in Glasgow housing estates. The usual approach would be to ask, what factors cause the observed high rates of accidents and injury to children? However this 'natural' professional discourse was turned on its head by researchers from Glasgow University who asked instead: given the highly dangerous design of the housing on such estates, why are there relatively so *few* accidents (Helen Roberts, personal communication). This research question immediately highlights factors which were *a priori* deleted from the normal framing, such as the expertise and inputs of parents in defending children from the intrinsically dangerous conditions. Thus the reflexivity point encourages awareness of the partisan role of our own framing social assumptions in the endless struggles over power.

2. As with any other concept of social learning ours is relative to some normative principle. Just as controversy does not automatically entail learning, it is difficult to accept that the indiscriminate increase of criticism and debate automatically constitutes collective learning. The tension between entropy and organisation seems as relevant here as elsewhere. The ambiguity is well captured in discussion of the evolution of acceptable risk criteria for risky technologies. Rip (1986) has suggested that the U.S. Rasmussen report on the safety of nuclear reactors initiated an important learning process in that, regardless of the rights and wrongs of the report itself, through the debate which it structured a shared culture was created about broader criteria of acceptable levels of risk for technologies.

    Whether learning combined with closure exemplified by Rip's discussion can genuinely be called learning may depend upon whether it leads to further development to consider the extra dimensions or freezes on that focus. This is itself an indeterminate matter, and it may invite analysis of what pressures exist to encourage or discourage such freezing or closure at a given level.

3. Indeed this very confusion between sociological *explanation* and practical opportunities for control by social actors, may be symptomatic of the epistemology of control which afflicts social science as well as natural science. Within such an epistemology, only social variables which can be controlled are candidates for social and explanatory factors.

4. In conference discussion Dirk Stemerding has argued that although social models may remain unarticulated beneath the 'factual' discourses they shape, the same is not true of values. I am not convinced that debates about values in relation to technology and risk fully address the richness of the 'value' dimensions. They often reflect individualist, non-relational rational choice theories of social relations, in which 'value' is not recognised in social relationships themselves, since they are only a means to realise one's preferred ends or 'values'. I would suggest that explicit expressions of abstract values in TA obscure the very concrete value commitments embodied in social models of desired or assumed operating conditions of viability of technologies. It is in these tacit social models that concrete values such as individualism or collectivism, hierarchy or egalitarianism, centralised control or decentralism, formality or flexibility are expressed.

References

Beck, Ulrich. 1992. *Risk Society: Towards a New Modernity.* London: Sage. (Translation of German original. 1986. *Risikogesellschaft.* Frankfurt: Suhrkampf.)

Callon, Michel. 1987. "Society in the Making," in: W.E. Bijker, T.P. Hughes and T. Pinch (eds.). *The Social Construction of Technical Systems.* Cambridge, Mass.: MIT Press.

Cambrosio, Alberto, and Camille Limoges. 1991. "Controversies as Governing Processes in Technology Assessment." *Technology Analysis & Strategic Management* 3(4): 377-396.

Douglas, Mary. 1986. *Risk Acceptability According to the Social Sciences.* New York: Russell Sage Foundation.

Eder, Klaus. 1987. "Learning and the Evolution of Social Systems: An Epigenetic Perspective," in: M. Schmid and F.M. Wuketits (eds.). *Evolutionary Theory in Social Science.* Dordrecht/Boston: Reidel, 101-125.

Giddens, Anthony. 1990. *The Consequences of Modernity.* London: Polity.

Giddens, Anthony. 1991. *Modernity and Self-Identity.* London: Polity.

Krimsky, Sheldon, and Dominic Golding (eds.). 1992. *Social Theories of Risk.* New York: Praeger.

Krohn, Wolfgang, and Johannes Weyer. 1989. "Gesellschaft als Labor: Die Erzeugung Sozialer Risiken durch Experimentaler Forschung." *Soziale Welt* 3: 349-373.

Law, John, and Michel Callon. 1982. "On interests and their transformation." *Social Studies of Science* 12: 187-202.

Mazmanian, Daniel, and Paul Sabatier (eds). 1984. *Effective Policy Implementation.* Lexington, Mass.: D.C. Heath.

Mazur, Allan. 1981. *The Dynamics of Technical Controversy*, Washington DC: Communications Press.

Nelkin, Dorothy. 1979. *Controversy: The Politics of Technical Decisions.* London/Beverly Hills: Sage.

Otway, Harry, and Malcolm Peltu (eds.). 1985. *Regulating Industrial Risk.* London: Butterworth.

Otway, Harry, and Detlov von Winterfeldt. 1982. "Beyond Acceptable Risk: On the Social Acceptability of Technologies." *Policy Science* 14(3): 247-256.

Parsons, Ted, and Bill Clark. 1992. "A Review of Theories of Social Learning." Harvard University, Project on Social Learning and Global Environmental Management.

Rip, Arie. 1986. "Controversies as Informal Technology Assessment." *Knowledge* 8 (December): 349-371.

Royal Society. 1992. *Risk: Analysis, Perception, Management*. London, Royal Society.

Schwarz, Michiel, and Michael Thompson. 1990. *Divided We Stand*. Hassocks: Harvester Wheatsheaf Press.

Slovic, Paul. 1992. "Risk Perception: Reflections on the Psychometric Paradigm," in: Krimsky and Golding (eds.), 117-152.

Tribe, Lawrence, H. 1973. "Technology Assessment and the Fourth Discontinuity." *Southern California Law Review* 46 (June): 617-660.

Woolgar, Steve. 1991. "Configuring the User," in: J. Law (ed.). *A Sociology of Monsters: Essays on Power, Technology and Domination*. Sociological Review Monograph 38, London/New York: Routledge, 58-99.

Wynne, Brian. 1975. "The Rhetoric of Consensus Politics: A Critical Review of Technology Assessment." *Research Policy* 4, 1-53.

Wynne, Brian. 1982. *Rationality and Ritual: The Windscale Inquiry and Nuclear Decisions in the UK*. Chalfont St. Giles: British Society for the History of Science.

Wynne, Brian. 1988. "Unruly Technology: Practical Rules, Impractical Discourses, and Public Understanding." *Social Studies of Science* 18: 147-167.

Wynne, Brian. 1989. "Frameworks of Rationality in Risk Management: Towards the Testing of Naive Sociology," in: Jennifer Brown (ed.). *Environmental Threats: Analysis, Perception, Management*. London: Belhaven, 119-132.

Wynne, Brian. 1992. "Risk and Social Learning: Reification to Engagement," in: Krimsky and Golding (eds.), 275-300.

3 (Constructive) Technology Assessment: An Economic Perspective

Luc Soete

This chapter puts forth an economist's perspective on reassessing the role and contribution of technology assessment in technology policy. Technology policy, I argue, should be centred around the societal integration of technology. An overview of traditional economic misconceptions of technology, including its measurement, sets the stage for discussion of the endogenous nature of technological change. Next, I argue that the traditional economic arguments for technology policy have resulted in unacknowledged supply biases. The last section advances an economist's agenda for CTA.

Economics and Technological "Progress": What is being assessed?

For most economists, assessing technological change appears something of a puzzle, far removed from economic reality. The main reason for this goes back to the traditional economic framework within which economists reduce technology to an "exogenous", external factor, whose impact on economic growth can be best described — just as in the case of population growth — in terms of a particular parametrical value: a "black box" variable, not to be opened except by scientists and engineers. Nonetheless, economists bring a particular vision and interpretation of the contribution of technology to economic development and growth. In this vision, technological change is (or practically by definition can only be) associated, thanks to the general allocative efficiency characteristics of the market, with welfare increasing aspects and with technological progress. Naturally, this vision resonates with ministries of economic affairs, and with spokespersons for technology. The danger is that the misguided view of allocative efficiency blinds one to the real opportunities offered by technology.

This view is reflected in the way technological change is measured. From an economic point of view, measurement of technological change is generally reduced to those new technologies having well-defined economic impact, either in productivity growth or new product demand. With respect to the latter, the methodological problems raised in effectively incorporating new products in the production function framework have generally led to a further reduction of the economic measurement of technological change to productivity growth – a weighted average of labour and capital productivity called total factor productivity. I raise this measurement issue because it illustrates the common perception in economic analyses that technological change and its societal impact can be correctly assessed only in economic terms. While this might seem an obvious bias to many audiences, it certainly is not obvious to most economists.

To economists it often comes as a surprise that many innovations have very widespread societal effects, but whose measurable economic effects are small or at best indirect in terms of macro-economic growth and efficiency. Oral contraceptives had a major impact on sexual behaviour in the 1960s and 1970s in most Western countries, and raised some fundamental issues about medical and social ethics (see Walsh, this volume). Yet the birth control Pill's economic impact was at best indirect through greater participation of women in the labour market. Genetic fingerprinting – a recent advance in bio-technology – is of great importance in forensic medicine, crime detection, and the judicial process, especially in cases of rape, assault and murder. It could also have major implications for medical prognosis and life insurance, again raising some ethical questions. Again though, the economic significance of this technical advance is difficult to predict, but probably rather small. For many innovations the societal impact may be very great even though the direct measurable economic impact is insignificant.

Another distinction that is insufficiently developed in economic analyses but has implications for technology assessment is the difference between innovations finding applications in only one sector and those affecting many or all sectors of the economy. In the several technological taxonomies suggested by Freeman, Nelson, Pavitt, and Rosenberg, technological advances are often identified as either 'localized' or 'pervasive' in impact. For example, the "float glass" process introduced by Pilkington's in the 1960s was of enormous economic importance for that firm and for the glass industry generally. Within a few years, it was licensed to almost all the major glass manufacturers in the world. However, it has no applications outside the glass industry and its *macro*-economic significance is therefore relatively small. By contrast, the microprocessor has found applications in practically every single sector of the economy, with (one suspects) major economic impact on the efficiency and growth performance of the economy.

Economists, in other words, are not only rarely aware of the societal impact of technological change. They are also insufficiently aware of the wide variance in economic impact of technological change. The problem has become increasingly severe. The tendency of most empirical economic studies to limit analysis to the industrial sector entirely fails to deal with the increasing number of service sectors that initiate technological change. In analysing service activities one is confronted with especially thorny questions about the actual direct economic impact of such technological advances.

Technological Change is Endogenous

Policy makers are increasingly recognizing that economists' traditional "exogenous" approach to technology and the related measurement biases are becoming a hindrance to conceptual frameworks for policy concerns. A recent report for the OECD stated:

> Technological developments, particularly the so-called "new" information and communication technologies, are not contributing to a satisfactory degree to growth in economic and social standards — the term "productivity puzzle" or "Solow paradox" has often been used to summarise this problem. Second, there are concerns about a division emerging, both within and among nations, between technology-haves and those with limited access to new technologies. Nations are concerned they may be left behind in the pursuit of new markets if they are not on or near the technology frontier; groups within nations worry that unless the opportunity to master new technologies is available the broad sweep of technological change may make their skills obsolete. Finally, rapid changes in technology together with the increasing integration of the international economy have brought new pressures to bear on the rules and institutions that regulate international co-operation. There are concerns that the burden and fruits of co-operation may be unequally shared among nations or the required timely co-operation in addressing common global problems may not be forthcoming. (Soete, 1991)

The productivity puzzle raises the general issue about assessing economic progress. It could be argued that in the post-war period up to the mid seventies — Jean Fourastié's "Les Trente Années Glorieuses" — there was general agreement that the quantitative economic data did provide a consistent picture of the growth in "progress" over this same period.[1] As Fourastié put it, "the great hope of the 20th century" has been fulfilled, based on "the facts of production, consumption, length of working hours, health care and life expectancy." More recently, however, and in line with the productivity puzzle evidence, increasing discrepancies are emerging amongst such facts and indicators, and in the perception of, and weights allocated to, material growth indicators and the many material and non-material "externalities" of such growth. There is far greater awareness that economic indicators do not correctly measure many of the social and environmental costs of economic growth.

Such paradoxes, puzzles, and discrepancies have altered how technology is being analysed at the economic policy level. The OECD Sundqvist Report (OECD 1988) was probably the first economic policy report to emphasize technological change as a wider process of social change. The report defined technology as a social process which, by meeting real or imagined needs, changes those needs even as it is changed by them. Thus technical innovation originates within the economic and social system (OECD 1988: 9). In other words, technological change, if it has to have beneficial effects on society will need to be "embedded" or integrated in society.

From such a perspective technological change is of course not so much an exogenous, "manna from heaven" factor, but rather an *endogenous* process whereby the technological changes are continuously being adapted and selected to the broad needs and requirements of society. The nature of the adaption and selection process then becomes of central importance if progress (in the broad sense) is to be realized. Consider the lack of development and growth in most East European countries over the last twenty years, despite massive investments in science and technology and higher education.[2] While the lack of economic integration, more specifically the lack of a market separating the technically from the economically feasible, pushed the science and technology system into isolation, this market failure is clearly only one aspect of this problem. In my view, one of the greatest paradoxes of development in the so-called socialist countries was the absence of *social* integration of technological change (lack of safety standards, higher health risks, lack of ergonomic considerations, etc.). Thus, in comparison to the so-called capitalist societies, science and technology were far more *imposed* on society and workers in particular, with the resulting lack of efficiency improvements at the shop floor.

The *endogeneity* of technology arises at all levels of technological development. At the level of technology "creation," technological innovation is not only impelled by scientific discovery, but is also induced by demand. The development of a potential economic idea into new products and processes requires many stages of experimentation in which market possibilities interact with the original idea. The acceptability of a product or process will also be conditioned by societal attitudes and norms. Thus, in broader terms, technological change stems from within the economic and social system and is not merely an adjustment to transformations brought about by causes outside that system. In other words, societies have a say in the shape technology is likely to take. Hence, the importance for technology assessment in technology. Yet the role of technology assessment, in the U.S. since the early 1970s, and in other countries more recently, has been to provide information and analysis, rather than input into technology policy. The reason lies with the conception of technology policy itself.

"Market Failure" Technology Policy as a Failing Policy

From a traditional *economic* point of view, science and technology policy have been guided by such minimalistic questions as: are there cases of market failure, or suboptimality in science or technological effort? There is general agreement that market failure is one of the intrinsic characteristics of science and technological activities and that *under*investment in research will logically result.[3] The fact that technological advances can be readily copied will deter companies from investing in them, even though a significant advance would lead to enhanced efficiency or performance. In cases where technological advances are not well protected by patents and can easily be copied, examples abound of such underinvestment in research. A typical example is agriculture: "Before the advent of hybrid corn seeds, which cannot be reproduced by farmers, seed companies had little incentive to do R&D on new seeds, since the farmers, after buying a batch, simply could reproduce them themselves. The farmers themselves had little incentive to do such work since each was small and had limited opportunities to gain by having a better crop than a neighbour" (Nelson 1987: 34). Similar arguments obtain for industries where scientists and engineers are mobile or where it is hard to keep secret information about the operating characteristics of particular generic designs, or about the properties of certain materials.

On the other hand, once the framework of the perfect atomistic market — as used in the case of farmers — is dropped, and some of the more common features of imperfect market competition are introduced,[4] it becomes obvious that market allocation can also lead to *over*investment in research. As illustrated in many major technological breakthroughs (superconductivity, HDTV, etc.), there is often a clustering of research effort involving duplication, which is inherently wasteful. Economies of scale and/or scope that could be achieved through coordination will be missed.

Many authors in the science and technology policy field have specified when market failure is more or less likely to occur.[5] It has always been tempting for policy makers to regard these market failures as justifying governmental actions to complement, substitute for, or guide private initiatives. In practice these arguments have led to active government technology support policies in most OECD countries centering on government support for research in the public and/or private sector. Large-scale research and development projects financed and planned directly by the state have emerged, but their modest achievements along with the difficulties of coordinating R&D as well as linking it to market information (if that is available) make it impossible to see any compensation of market failure. Indeed, some are dramatic illustrations of government failure in that public money was not allocated optimally. In recent times, acknowledgement of these coordination problems has led to support for *generic* technologies and for so-called *pre-competitive* R&D support.[6] Following a perception of Japanese policy, the US and particularly Europe now have major programmes of such pre-competitive, coordinated R&D support.[7]

This narrow, market failure approach has led to a massive *supply bias* in technology policy, exacerbating the lack of economic, social and societal integration of new technologies. It could be argued that the concerns of policy makers not to interfere with the market process have led them to support "pre-competitive" research that is identical with non-applicable research! Ironically, in cases of so-called generic technologies, there is a greater need for support at the application end. Generic technologies are *flexible in use*: improvements and learning are intrinsically related to the particular application. Often, research on generic technologies will involve uncertainties similar to those encountered in basic research.[8] Here, too, governments have a particular responsibility including public procurement, standard setting and other more application-oriented support. To summarize: the concern not to interfere with the market or with anything involving possible "competitive" applications has led to technology support in the *"precompetitive"* phase of research. Consequently many policy makers have become "over-concerned" with the creation of new, potentially pervasive scientific and technological knowledge, at the expense of possible applications falling outside government's immediate concern.

Thus, a market failure ideology leads to "supply bias" in many science and technology policies, particularly in countries at the technological leading edge. Such policies have paid insufficient attention to the capacity of the economic and social system to incorporate technological changes and transformations already in market (and industry-structure) terms. At some stage new technologies developed with the support of government may *further outrun the capacity of the economic system to adapt or generate new demand.* An illustration has been the European-sponsored megabyte project. It successfully led to the production by Philips and Siemens of so-called megabyte chips, but the demand for them was overestimated. The actual demand for such chips appears limited to the relatively limited and fragmented European mobile telephone market.

Endogenous Technology Policy: or How to Assess Technology "Constructively"

The previous policy discussion has focused primarily on the economic context of technology. However, the economic feasibility of a new process or product is only one part of the "societal" integration of technology. Other contexts — social, ethical and socio-political — play important roles. Given that the creation of technological capabilities involves a complex, endogenous process of change, negotiated and mediated by society at large, it is obvious that science and technology policies cannot be limited to the economic "integration" of technological change. They must include the broader societal "integration" of such change. It is in its broader interaction with society that technological change is adapted and selected and that the further realization of technological *progress* is enhanced. From this perspective, technological renewal encompasses research and development, diffusion and imitation of new technologies, as well as the associated social and organizational changes and innovations. The traditional notion of 'market' loses its meaning, and government intervention cannot be predicated on market failure alone. Policy making should take incorporation and integration of technology as its starting point, and attempt to redress failures there. How it should do so, and especially how it can do so without new types of government failure, is a difficult question.

Within this broad conceptual framework, constructive technology assessment has emerged as an instrument of government policy aimed at improving the societal integration of technology and influencing the use and further development of technology. Whereas technology assessment developed in isolation from economics and in particular economic policy, the present recognition of *both* the economic and societal context conditions opens up the possibility of a far more coherent and complete conceptual framework for technology policy. This framework in essence is and should be nothing less than (constructive) technology assessment.

The traditional, static TA issues are in economic language aspects of the unequal distribution of positive and negative externalities of technological change. As emphasized by Harvey Brooks (Brooks 1986) and others there is an apparent paradox in the distributional impact of technology. The costs or risks of a new technology frequently fall on a limited group of the population, whereas the benefits are widely diffused. Often the benefits to any restricted group are barely perceptible even though the aggregate benefit exceeds the total cost to the (limited) adversely affected group. Examples abound. "Automation" benefited consumers of a product by lowering its relative price, but the costs in worker displacement were borne by a small number of people, and may well be traumatic. A large electrical generating station may adversely affect the local environment, while providing widely diffused benefits to the population using the electricity produced. Workers in an unusually dangerous occupation such as mining carry a disproportionate share of the costs associated with the resultant materials, which may have wide benefits throughout a national economy. This disproportion between costs and benefits can also work the other way, as in many cases of environmental pollution and emissions. The effluents from a concentrated industrial area such as the Ruhr Valley or the American Great Lakes industrial complex may diffuse acid sulphates over a very large area which derives little benefit from the industrial activity, but may have its quality of life as well as agricultural productivity seriously degraded.

The issue of sharing costs and benefits of technological change shows how important it is, both from the national and international point of view, to draw up some "rules of the game". These rules should ensure that adverse effects are less harmful than they would be if everything were left to free competition, and they should be established fairly early on, before the vested interests acquire privilege and the fierceness of competition jeopardizes their compulsory application. The word static is of course inappropriate to describe all such distributional issues. At an early stage, most externalities have not occurred, nor can they easily be predicted. Still, some choices have to be made and measures taken. This becomes all the more pressing because, with the increase in the complexity of technology, possible risks threaten not only larger areas but also several generations.

Traditional static TA should therefore be complemented with a more dynamic approach, as argued above, focusing on the broad societal integration of technology as well as the adaptation of technology to societies' needs. From a practical economic policy perspective it seems useful to elaborate upon the externality terminology. As Nelson and Winter (1982: 413) put it: "technical change is continually tossing up new 'externalities' that must be dealt with in some manner or other. In a regime in which technical advance is occurring and organizational structure is evolving in response to changing patterns of demand and supply, new nonmarket interactions that are not contained adequately by prevailing laws and policies are almost certain to appear, and old ones may disappear. Long-lasting chemical insecticides were not a problem eighty years ago. Horse manure polluted the cities but automotive emissions did not. The canonical 'externality' problem of evolutionary theory is the generation by new technologies of benefits and costs that old institutional structures ignore." Thus, there are two levels of dynamic externalities: those that occur because of the dynamism of technical change, e.g. in creating networks; and those that relate to changes in overall patterns. The dynamics of externalities will not be susceptible to definitive once-and-for-all categorization, for they are intimately related to particular historical and institutional contexts.

Dynamic externalities are well known in the positive feedback effects of learning curves, which may enhance an original, and possibly fortuitous, advantage (Arthur 1988). At the end of the 19th century, the internal combustion engine became the dominant automobile engine, not necessarily because it was the best, but because it was taken up by mass manufacturers. Dynamic TA has to take such potential positive feedback into account in assessing different technological options. Similarly, path dependencies (see Callon's essay in this volume) cannot be avoided, but the goal should be avoiding lock-in to the worst paths. This point seems particularly applicable to network externalities. These have been discussed in terms of overcoming thresholds to expansion, e.g. when a telecommunications network has not enough subscribers and/or services to attract new subscribers or additional suppliers of services. With the imperative drive to overcome the threshold, paths of expansion may be set up that are less than ideal. Again, we have a case where TA should look not only at the distribution of externalities, but at their dynamics. To do this is a necessary condition for TA to serve the technology policy goal of societal integration. But it is not a sufficient condition: the patterns of technology and society may be evolving and this requires another form of dynamic TA.

Constructive TA or (as?) Experimentation and Feedback

When overall patterns change, the nature of benefits and costs changes. From this perspective the concept of a "socially optimal" way of assessing long term impacts of technological change loses much of its meaning. Accordingly, technology policy analysts have discussed the notion that society ought to be engaging in experimentation and that the resulting information and feedback will help guide the evolution of the economic and technological system. As in the case of "market failure," simple normative rules will not be very helpful in the design of "constructive" technology (assessment) policies. From this perspective, I would argue that such TA "policies should focus on problems of dealing with and adjusting to change. It involves in the first instance abandonment of the traditional normative goal of trying to define an *optimum* and the institutional structure that will achieve it, and an acceptance of the more modest objectives of identifying problems and possible improvements" (Nelson and Winter 1982: 394).

Three components appear necessary in the modest approach. First, there must be actual experimentation, and in such partially controlled ways that some societal learning will occur. The costs of experimentation, as well as the irreversibilities that may be introduced because of the real-life character of experimentation, must be balanced against the potential advantages. A second component derives from the possibilities of modeling non-linear dynamic developments. Models might allow virtual experimentation, provided they can model the relevant aspects with sufficient validly. The third component brings institutional structure back in. Exploration and feedback are more important than responsibilities to achieve an optimum situation. TA agencies in industrialized countries, provided they have enough room for manoeuvre, can fulfil such roles. Another possibility is increase of variety by institutionalizing access by more actors and/or acceptance of more aspects in the process of technical change (the original definition of constructive TA). For a society actually to engage in experimentation and feedback, the non-linear nature of the evolution of the economic and technological system should be recognized and taken into account. In that sense, technological change will become reflexive (which is not to say that it can be planned).

Notes

1. With some exceptions, particularly in the US. For one of the first economic analyses on the difference between material progress and happiness see Tibor Scitovsky's fascinating *The Joyless Economy* (1977).
2. See Gomulka (1990).
3. Such arguments were put forward more than thirty years ago in contributions of Nelson (1959) and Arrow (1962).
4. As emphasized in some of the more recent contributions in the field of industrial organization (see e.g. Dasgupta and Stiglitz, 1980).
5. See e.g. Ergas (1985), Justman and Teubal (1986), Mowery and Rosenberg (1989), Pavitt and Walker (1976), Freeman (1982, 1987), Nelson, Peck and Kalachek (1976), Nelson (1983, 1984, 1987), Rosenberg (1990).
6. For a particularly useful overview of some of the issues involved from a Dutch technology policy making perspective see van Dijk and van Hulst (1988).
7. For insight into the different meaning of "pre-competitive" research in the Japanese and US context, see Kodama (1988).
8. See also Ziman's (1990) contribution to the TEP Competitiveness Conference.

References

Arrow, K. 1962. "Economic Welfare and Allocation of Resources for Invention," in: R. Nelson (ed.). *The Rate and Direction of Inventive Activity*. Princeton: Princeton University Press.

Arthur, W.B. 1988. "Competing Technologies: An Overview," in: G. Dosi, C. Freeman, R. Nelson, G. Silverberg and L. Soete (eds.). *Technical Change and Economic Theory*, London and New York: Pinter Publishers.

Brooks, H. 1986. "The Typology of Surprises in Technology, Institutions and Development," in: W.C. Clark and R.E. Munn (eds.). *Sustainable Development of the Biosphere*. Cambridge, UK: Cambridge University Press, 325-350.

Dasgupta, P., and J. Stiglitz. 1980. "Industrial Structure and the Nature of Inventive Activity." *Economic Journal* 99: 266-293.

Dijk, J.W.A. van, and N. van Hulst. 1988. "Grondslagen van het Technologiebeleid." *Economische Statistische Berichten* 21, September.

Ergas, H. 1985. "Why do Some Countries Innovate More than Others?" Brussels: Centre for European Policy Studies.

Freeman, C. 1982. *The Economics of Industrial Innovation*. 2nd edition. London: Frances Pinter.

Freeman, C. 1987. *Technology Policy and Economic Performance: Lessons from Japan*. London: Frances Pinter.

Gomulka, S. 1990. *The Theory of Technological Change and Economic Growth*. London and New York: Routledge.

Justman, M., and M. Teubal. 1986. "Innovation Policy in an Open Economy: A Normative Framework for Strategic and Tactical Issues." *Research Policy* 15: 121-138.

Kodama, F. 1988. "Innovative Approach to Research Draws Conspiracy Cries from Abroad." *Japan Economic Journal*, November 26th.

Mowery, D., and N. Rosenberg. 1989. *Technology and the Pursuit of Economic Growth*. Cambridge: Cambridge University Press.

Nelson, R. 1959. "The Simple Economics of Basic Scientific Research." *Journal of Political Economy* 67: 297-306.

Nelson, R. 1983. "Government Support of Technical Progress: Lessons from History." *Journal of Policy Analysis and Management* 2 (4): 499-514.

Nelson, R. 1984. *High Technology Policy: A Five Nation Comparison.* Washington and London: American Enterprise Institute for Public Policy Research.

Nelson, R. 1987. *Understanding Technical Change as an Evolutionary Process.* Amsterdam: North Holland.

Nelson, R., M. Peck and E. Kalachek. 1976. *Technology, Economic Growth and Public Policy.* Washington DC: Brookings Institution.

Nelson, R., and S. Winter. 1982. *An Evolutionary Theory of Economic Change.* Cambridge, Mass.: The Belknap Press of Harvard University Press.

OECD. 1988. *New Technologies in the 1990s: A Socio-economic Strategy.* Paris: OECD.

Pavitt, K., and W. Walker. 1976. "Government Policies Towards Industrial Innovation: An Overview." *Research Policy* 5: 11-97.

Rosenberg, N. 1990. "Why do Firms do Basic Research (With Their Own Money)?" *Research Policy* 19 (1990): 165-174.

Scitovsky, T. 1977. *The Joyless Economy: An Inquiry into Human Satisfaction and Consumer Dissatisfaction.* Oxford: Oxford University Press.

Soete, L. 1991. "Synthesis Report of The Technology Economy Programme," in: *Technology in a Changing World*, OECD, Paris, December.

Ziman, J. 1990. "The Restructuring of the Links between Fundamental and Applied Research." Paper prepared for the TEP Conference on Technology and Competitiveness, OECD, Paris, June.

Part II

Steering Technology is Difficult but Possible

Steering Technology is Difficult but Possible

Introduction

The proposition that many technologies are deeply entrenched and consequently that changing them is difficult runs counter to much of the received wisdom about technology. For one, neoclassical economics presupposes that technology is infinitely malleable to market forces and that technology can be taken off the shelf as needed. From a different perspective, certain sociological accounts have posited actors who bend the world — including people and things — to their will. Essays in this section find, however, that cognitive and institutional structuring make technological change much more complicated than neoclassical economists have presumed and some sociologists suggest. We must start with the insight that this very complexity results in social entrenchment which makes changing course difficult.

To analyse the process of entrenchment the historian Hughes developed the concept of technological momentum and the economist Dosi rephrased the concept of technological trajectory. Although both momentum and trajectory are physical metaphors, they are best understood as social concepts. The concepts point not to the internal logic of technique but to the organizations and people committed in various ways to the technological system. The very presence of trajectories and momentum is a continual and contingent achievement of actors working to stabilize technical systems. Indeed, if technical trajectories are conceived as akin to the path of projectiles moving through empty space, impelled by some prime mover, all efforts at shaping technical developments — including CTA — would be futile. Such a mistaken conception of Dosi's and our notion of trajectories is a prominent feature of recent social constructivist analyses of technical change. Three essays in Bijker and Law (1992, pp. 8-19, 45-50, 290-305; see MacKenzie 1992 as well) reject the concept of trajectories, which is apparently misunderstood as being restricted to the patently false notions of unilinear development, the simplistic unpacking of assumptions built into previous technologies, and even inexorability.

Given the insight that trajectories are active constructions, it is possible in principle to alter them and thereby effect change in even well-entrenched and high-momentum technologies. Path dependencies and irreversibilities will result. The QWERTY keyboard's function in slowing down the typist's speed in order to avoid the hammers cluttering up is not necessary any more with the advent of word processors. But attempts to introduce other keyboards have foundered. The present energy production and distribution complex is difficult to change because of entrenchment. The entrenched character of technology and subsequent patterns and dynamics at the collective level will render actor-centred change strategies insufficient. Concerted and coordinated action will be needed. Often a collusion of actors and factors is needed to effect more radical change. Hence possibilities for change will be highly situation-dependent. Although much additional work needs to be done, with the increased insight into the nature of the dynamics of technological development and the way technology becomes embedded, some assessment of developments to be expected is possible and crucial change factors can be identified. Each of the essays in this section suggests certain clues.

Steward's long-run historical analysis of rival food-processing methods in the British dairy and baking industries provides an exceptionally clear illustration both of the collusion of the forces necessary for maintaining technical trajectories as well as the substantial forces needed to alter them. In both his cases, scientific evidence about the risks of chemical methods was suggestive but not conclusive. Indeed clear-cut objective evidence about risk is rare, and, as Wynne's contribution has made clear, all estimates of risk are contingent on certain assumptions about the locally situated use of the technique at issue. Steward shows that chemical methods for bleaching wheat flour were in the last fifty years transformed, but not abandoned, owing something to the poor development of alternative physical methods but more importantly to the strong organization and effective lobbying of the domestic baking industry. Inconclusive scientific evidence as well as absence of a well-organized lobby for change therefore yielded incremental change in this case.

By comparison, even after a similarly long period of dominance, chemical methods for preserving milk and butter were entirely abandoned by 1925 in favor of physical methods, including pasteurization and refrigeration. The key factors which came together and facilitated fundamental change in this instance were shifts in the dairy industry's structure and the direct effect of social groups lobbying for change. The emergence in the mid 1920s of integrated dairy firms facilitated the adoption of refrigeration and pasteurization technology. Abandoning chemical preservatives not only was a response to the public's desire for 'fresh' milk, an example of the influence of credibility pressure; the move also permitted the integrated firms to adopt new marketing strategies comparable to the 'green' strategies adopted by many firms today.

The case of milk and butter shows the catalytic role played by independent, public-health professionals who articulated the public's call for fresh milk. The importance of linking practising professionals as agents of change with social movements is developed, albeit implicitly by the essays of Herbold, Walsh, and Koch, for the cases of waste management, reproductive technologies, and ultrasound, respectively, and explicitly by the essay of Jørgensen and Karnøe. They make clear that social movements are not only doing cognitive work in carving out new cultural spaces, but can be envisaged as creating their own technological practices and trajectories to reflect to some extent their values (compare how environmental movements can be seen as new knowledge generators in Jamison and Eyerman 1994). Jørgensen and Karnøe examine the Danish wind-turbine industry, which emerged in the 1980s as an export-driven world leader. Competing against traditional sources of electricity, including atomic power, the Danish wind-turbine industry was nurtured from the 1960s on by a network of 'enthusiastic' inventors, entrepreneurs, and engineers linked to social activists opposed to the centralising tendency of atomic power. The resulting alternative energy movement thus effectively combined social vision with technical expertise. This network deployed a grassroots, learning-by-doing approach to develop wind-turbine designs that now compete successfully in the international market.

As the enthusiastic wind-turbine experts attest, the professional and emotional commitments of technology practitioners are a critical variable in assessing possibilities for change. Gary Downey, by extending recent insights in cultural anthropology, helps to comprehend how such commitments affect technical change. He suggests that any group's acceptance of a strategy for change will be modulated by that group's 'positional identity,' and how change affects it. This identity consists of the configuration of any particular agent in relation to other agents; thus, positional identity combines agency and structure. He finds that some developments in computer-aided design (CAD) like automated drafting have been readily accepted because they reproduce the existing identities of engineers, while others such as integrating design and manufacturing as well as computer-integrated manufacturing have not, in large measure because the actions generate internal tensions in engineers' identities. To be successful, according to Downey, a strategy for change should work to reproduce rather than threaten existing identities of practitioners. Practically, this finding indicates that change strategies should be localized, taking into account how proposed changes will redistribute the agencies of interacting participants including the agency of CTA. This local sensibility will make strategies easy if they do not shift agents into unwanted positions. Note however that it is sometimes necessary to break through existing positional identities.

References

Hughes, Thomas. P. 1985. *Networks of Power*. Baltimore: Johns Hopkins University Press.

Dosi, Giovanni. 1982. "Technological Paradigms and Technological Trajectories: A Suggested Interpretation of the Determinants of Technical Change." *Research Policy* 11: 147-162.

Bijker, Wiebe E., and John Law (eds.). 1992. *Shaping Technology/Building Society. Studies in Sociotechnical Change*. Cambridge, Mass.: MIT Press.

Jamison, Andrew, and Ron Eyerman (eds.). 1994. *Seeds of the Sixties*. Berkeley: University of California Press.

MacKenzie, Donald, 1992. "Economic and sociological explanation of technical change," in: Coombs et al. (eds.). *Technological Change and Company Strategies*. London: Academic Press.

4 The Danish Wind-Turbine Story: Technical Solutions to Political Visions?

Ulrik Jørgensen and Peter Karnøe

In the later part of the 1970s and the 1980s a new industry producing wind-turbines for power production emerged in the United States, the Netherlands, Belgium, Germany, United Kingdom, Sweden, and Denmark. The Danish wind-turbine industry has been especially successful, growing to a significant international position. In the same period, wind-power-produced electricity in Denmark developed from being totally dependent on subsidy to being price competitive with coal-produced electricity. Today each wind turbine typically supplies electricity to about 250 houses. In Denmark 3,000 wind turbines supply 2.3 percent of the total electricity consumption.[1]

This successful technological and industrial development in Denmark is based on the work of enthusiastic grass-roots activists, entrepreneurs, engineers and highly skilled workers who utilized a practically oriented development strategy. The design of the typical Danish wind turbine has a history dating back to the folk high school movement and the electrification of the countryside at the beginning of the century. These early designs were developed, especially in the 1950s when many of the principles in use today were invented and studied. The energy crisis in 1973 and the fight against atomic power plants stimulated a grass-roots energy movement that in several ways played an essential role: It made the support of nuclear power politically impossible, it influenced Danish energy planning by establishing a "counter expertise", and it brought the use of renewable energy sources on the national agenda by demonstrating in practice how they could be utilized. The alternative energy movement gave birth to many entrepreneurial ideas and created an institutional framework that nurtured the development of the modern wind turbine and the wind-turbine industry in Denmark. This bottom-up development strategy had its strength in combining a political vision with practical experience from experiments with small-and medium-sized wind turbines (50-250kW).

The established political system also embraced wind power as a solution to the energy crisis, but its efforts were not successful. Wind technology was put on the energy agenda by experts, the electric utilities, and the newly established Ministry of Energy. Their idea was to develop and build large-scale wind turbines (1 to 3 mW). The project was put in the hands of a scientist at the Technical University and a utility project group with Danish firms as subcontractors. This top-down strategy was based on a science-push logic. Its lack of practical results can be seen in Sweden, Germany, and the United States where the investment of billions of U.S. dollars did not bring about commercially viable wind technology, or even a usable concept for further development.

The development of modern Danish wind technology breaks some well established propositions about the development of energy production technology. First, that the most efficient trajectory for power production is to centralize in ever-larger units. Second, that scientific R&D is the best means of developing a new technology. Third, that the utility companies and private industry are the best organizational frame of development. Rather, Danish wind-turbine technology has been a surprising combination of smaller units, new industrial expertise, low costs and high performance.

This chapter illustrates how social and political interests play an important role in the construction of a technology. Such "external" support may be necessary for an upstart technology to break through the well established economic and technological positions of existing institutions. The case also shows that the process of construction, although developing unevenly, never ends and thus never permits the actors to rest on their laurels. But these facts at the same time pinpoint a theoretical question about the distinction between the technical and the social. Is it possible to identify a genuine, core type of wind-turbine technology undergoing the process of shaping? Or is it more likely that the social visions and the technical solutions cannot be separated and are equally engraved in the nuts and bolts of the resulting wind turbines?

The perspectives presented in this paper are not new in the Danish (and international) discussion. They have been an important part of the on-going self-reflections of the alternative energy movement. None of the participants in the Danish discussions can fairly be accused of having made simplistic conceptions of the interweaving of social and technical ideas. For analytical clarity, however, we will simplify the social visions behind many of the discussions and point to their importance in establishing the institutional setting for the Danish wind-turbine industry and (more broadly) as an integral part of the construction of this technology.[2] We try to locate this story of social visions materialized in technical artifacts in a broader analytical context of social and political change.

The Heritage of Modern Wind Power

Even before the striking developments beginning in the mid-1970s the utilization of wind power had already been on the Danish agenda three times, during 1891-1920, 1940-45 and 1947-1962. While the motivation behind wind power differed, all three periods shared the idea of decentralized power production by small- and medium-sized wind turbines (30-200 kW).

The development of wind turbines started in Denmark in 1891 with the work of Poul La Cour. His wind turbine in 1895 made it possible to illuminate Askov Folk High School (højskole) using electricity. A professor of physics, La Cour was engaged in several projects using electricity to modernize and improve farming methods and living conditions in the countryside. He was active in the farmers's association, several cooperative undertakings, and many activities for public enlightenment through the Danish folk high school movement (folkehøjskolebevægelsen). In a 1906 article entitled "Electricity in the Service of Agriculture and Smaller Industries" La Cour described his vision: "19th century cheap labour offered by the steam engine has cramped people in factory centres and increased the populations of major cities on a large scale. ... Electrical technology carries with it a contrary movement. It has started spreading and will further spread the all-too-dense population, and it will again transform a large number of wage-earners to independent business people."[3]

La Cour and his allies saw electricity as a progressive force in the restructuring of society. The desire for restructuring had a strong local political base in Denmark, which was further provoked by the conservative, anti-parliamentarian provisional Government in the years 1901-1904 that both the peasant movement and the social democrats fought. Wind power thus became aligned with Denmark's progressive, democratic, and rural political interests.

La Cour creatively combined wind turbines with the storing of electrical energy by batteries as well as by separating oxygen and hydrogen, and he supplemented his systems with petroleum engines. This combination of energy production and energy storing was obviously a necessity because of the intermittent nature of wind power. A local energy producing system must combine different production facilities or integrate energy storage. With financial support from the Danish government, he improved the aerodynamic design of the blades and designed a wind turbine with two, three or four narrow blades. In 1903 he helped found the Danish Wind Electricity Company, which delivered several hundred turbines (3-30 kW) to farms, engine shops and villages.

Wind turbines were introduced very early in the history of electricity, in parallel with other power-producing technologies like water turbines and steam engines. Because of his work electricity came to the countryside earlier (or at least at the same time) than to the cities, where gas had been the most widely distributed source of energy. La Cour's work was continued into the 1920s by the engineer Lykkegård who built bigger turbines (30-75 kW) of a slightly different design. Inspired by the development of aerodynamics in the 1920s researchers and inventors were attracted to wind turbines, and many designs were made including the Darrieu design in France from 1929. However, these inventive efforts were not put into practice, since they were not linked to groups with a practical interest in decentralized or rural electricity.

In the 1920s wind turbines were ousted by the establishing of electricity networks, the reduction in prices first of diesel generators and then of coal-fired power stations. Yet wind technology was not totally forgotten. Due to the lack of oil, both during and after the Second World War experiments were carried out to meet the need for alternative energy sources. During the war the Danish cement-producing company, F.L. Smith, developed and constructed 60 wind turbines. They were used despite their high maintenance costs, which were traced to the failure-prone wooden blades produced by Skandinavisk Aeroindustri. Problems with energy supply during the war stimulated research and experiments with wind turbines in many other countries. By the late 1940s thousands of wind turbines were in use in the United States, Germany and the Soviet Union.

Juul's Experiments in the 1950s

By all accounts, Johannes Juul was the leading force in practical designs of wind turbines in the third period of practical Danish wind power.

From 1947 to 1958 — starting when he was 60 years old — Juul was engaged in an experimental programme financed partly by his employer, the power producing company SEAS in south Sealand. His early inspiration came from Askov Højskole, where he had seen La Cour's wind turbines. But rather than local energy producing units, Juul's objective was to construct a wind turbine that could be connected to the power network.

SEAS let Juul start these experiments because it faced a shortage of energy in the late 1940s. In the early 1950s his experiments were also funded by an OEEC programme to meet the energy needs of postwar Europe. His work produced the basic concept that underlay the success of the Danish wind-turbine industry two decades later. He combined an efficient and reliable design with a simple set of principles for calculation. His carefully prepared articles — with a stepwise description of the way he established his insights — served almost as detailed building manuals when the first wind turbines were built in the mid-1970s. In many ways his design was not surpassed until the 1980s.

He started with a 8-metre rotor with blades shaped like those of an aeroplane, not like a propeller. The next turbine, with a 13-metre rotor, was used to test the connecting of the turbine to the power network. Then came the design of a 100 kW turbine, later enlarged into a 200 kW turbine with 24-metre rotor and built in 1957 in Gedser with Dkr (Danish kroner) 300,000 (approx $50,000 US) support from the Marshall Plan. This turbine worked regularly for ten years. It had blades built of steel-reinforced wood covered with aluminum plates. The rotor was relatively slow, stall-regulated, and featured a braking system that combined a turnable blade tip and a mechanical brake.

Despite considerable technical success Juul's wind turbines were never put into industrial production. His programme was stopped in 1962 when the Windpower Commission (Vindkraftudvalget) concluded that the cost-effectiveness of wind turbines should be calculated not from the value of their produced power (user-prices) but rather from the amount of fuel saved in the power stations. On this basis the cost-effectiveness had to be doubled. The argument was that the uneven power production from wind turbines would merely reduce fuel consumption but not reduce the necessary capacity of the traditional power plants. Juul did not agree: "first now I know how wind turbines should not be constructed." At the same time atomic power plants came into the arena as the strategic alternative, which for a period eliminated the last few supporters of windpower technology.

In wind power the 1950s showed a very complex scope of experiences based on the great variation in design concepts (at least 14 different inventors in the industrialized countries). They ranged in energy from 15 kW to 14,000 kW, had two to four blades, a regulation based on pitch, flaps or variable speed, DC or AC, and other fundamentally different characteristics (Golding, 1955). Only two of the wind-turbine designs from the 1950s produced power without serious breakdown for 8 to 10 years. One was Juul's wind turbine in Gedser; the other was constructed by a German, Professor Hütter, a fast-turning two-bladed turbine of 100 kW with a 34-metre rotor. These designs framed the vague contours of two technological trajectories (Sahal 1985; Dosi 1982).

Networking and Centralized Power Production

Power plants based on diesel engines made it possible for each town to establish its own central power production unit, and many such companies were founded in the 1920s. During the 1930s and 1940s a power network slowly covered the whole country. Danish energy production and distribution were not thoroughly centralized, however, until the 1960s. This had the very important consequence that Danish industry became interested in atomic power plants much earlier than the electric utilities, which were still locally organized. During the 1960s the power producing companies were reorganized, and centralized organizational bodies governing the power network were established.

Along with the construction and erection of power plants, a new technological perspective grew in the minds of both engineers and politicians. Each new generation of power producing units turned out to be more efficient and cost-effective than the earlier ones; this was especially the case with steam turbines. Even the added costs of the distribution networks did not negate this trend. The idea of locally-owned and -governed power production units was left behind, while the myth of the efficiency of ever-larger units and steady improvements based on scale, size and design experience was consolidated. One consequence was the possibility (and necessity) of locating power plants at the seaside with easy access to cooling water and where ships could deliver oil or coal directly. This perception of energy technologies being a showcase for economies of scale prevailed, and not only concerning the traditional power production technologies based on coal, oil and water. The centralized power plant also was the prerequisite for atomic power.

The Atomic Power Conflict

Before we turn to the development of today's wind turbines it is necessary to describe briefly the atomic power conflict in Denmark. It is difficult to understand the later developments without recognizing the ensuing confrontations between grass-roots and official experts. The self-consciousness that grew out of this battle gave the grass-roots movement an imperative to demonstrate a working alternative. From this background wind power emerged.

Like many countries in the 1950s Denmark established an Atomic Energy Commission (Atomenergikommissionen), and in 1958 set up its research centre at Risø near Copenhagen. Risø's aim was to study the peaceful uses of atomic energy in power production, industry and science. The broader societal view behind this research- and expert-based strategy was very clearly stated by the atomic physicist Niels Bohr in 1950 in his open letter addressing the United Nations. "As there is hardly any question for humanity to denounce the prospect of improved material conditions of civilization through nuclear energy sources, a profound adaption of international relations is obviously necessary if civilization is to continue its existence."[4] Technology, for Bohr, was the fact to which societies must adapt by changing their political attitudes, organization and interests.

In 1962, the Danish parliament gave the Minister of Education the sovereign right to decide on the erection of atomic power plants. Nine years later one of the Danish power distribution companies, ELSAM, decided to build atomic power plants and in 1974, when the energy crisis was at its most unpredictable stage, they proposed an atomic power plant in Jutland. At a meeting with inhabitants from the region where the power plant was planned, the director of ELSAM said: "You can discuss as much as you want, but atomic power you will get anyway." This blunt statement typified the technocratic outlook of top management which dismissed public reactions as merely local self-interest. Until this time no one had challenged atomic power in Denmark. For the experts in charge, atomic power was seen as the "clean" energy technology that could fulfil the necessary and unavoidable growth in energy consumption.

The principal grass-roots reaction to atomic power in Denmark was the establishment of the Organization for Information about Atomic Power (Organisationen til oplysning om atomkraft, OOA) in 1974.[5] OOA's organization and public activities were prepared during a period of seven months in 1973, and it grew very rapidly to be one of the biggest and most influential grass-roots movements in Denmark.

OOA succeeded in enrolling many other social groups by gaining access to the discussions in the Parliament through some of the political parties, establishing contacts with researchers, having members in the official steering committees for public R&D programmes, and being able to influence the energy policy planning processes from their very beginnings. Through these efforts OOA was able to position renewable energy sources as a necessary element in Denmark's future energy supply.

Many of the activists in the alternative energy movement had experience from the campaigns against the atomic bomb, and consequently one of the basic principles of OOA was non-violent action. It was a conscious strategy for OOA also to show a non-radical image in all its published material. OOA's published political ideas focused on "neutral" phenomena like growth, centralization, and a concern for the environment.[6] There were no direct links to leftist criticism or other socialist ideas. On the contrary, a vision of a self-sufficient local community with the idyllic village was the typical motif.

OOA saw a direct and insidious connection between the centralization of power in society and introduction of atomic power. The growing need for energy was directly connected with the narrow focus on economic growth. OOA criticized the side effects of the welfare society based on specialized mass-production and consumption predicated on continuous growth in energy demand. "Atomic power, coal and oil are of the same kind, because they will influence the development of society in the same way," stated the OOA in 1975. "They are based on complicated high technology and big production units, which demand central regulation and planning. And they all have a negative impact on the balance of trade, because the fuel has to be purchased outside Denmark."[7]

OOA very clearly set the agenda favouring the alternative. Renewable energy sources, by contrast, "have the opposite impact on development. The renewable types of energy like sun, wind, biogas, etc. are simple technologies, based on small units which give room for local planning and local democracy. At the same time they are based on renewable and inexhaustible sources, which need not be purchased outside Denmark and cannot be monopolized and create dependency." The linkage between a social vision of society and the character of its technology is fully articulated.[8]

After a period of aggressive reactions — for example portraying OOA as an subversive organization conspiring with the enemies of the country — the power companies and the Board of Industrialists (Industrirådet) changed their attitude and became actively engaged in campaigns both for the safety and cleanliness of atomic power and for the restriction of energy consumption in industry and private households. To oppose OOA, a counter-movement called Reliable Energy Information was established by some of the right-wing parties in parliament, but they failed to recruit broad support.

In the first phase OOA successfully argued that Parliament (not the Minister of Education) should decide on the establishment of atomic power plants. In 1975 in accord with the Danish tradition for public information and debate (a legacy of the folk high schools) an official committee, The Energy Information Board (Energioplysningsudvalget) was established to organize a public information campaign in which all interested parties were entitled to make their points.

Although opinion polls showed a majority of the Danish people were against nuclear power, no decisions were made in Parliament, and not until 1984 were atomic power plants removed from the government's strategic energy plans. (It has still not been explicitly decided in Parliament not to build atomic power plants in Denmark.) The non-violent strategy was kept as the central idea in the OOA, although confrontations especially in Germany showed how fundamental a threat the antiatomic movement seemed to be to established interests. The OOA campaigns were supported by the increasing unpredictability of cost estimates for atomic power plants, which led the Board of Industrialists to drop atomic power as their main priority for future energy sources. At the same time, in the aftermath of the Three Mile Island disaster, pointed questions were raised about the safety of atomic power plants and the environmental damage that might come from atomic waste deposits. Although an explicit decision never was made, Denmark is one of the few countries without atomic power plants as part of its present or future energy supply.

Institutionalization of the Alternative Concept

OOA aimed not only to fight atomic power but also to present realistic alternatives. The energy movement engaged in an important cooperation with established physics researchers.[9] At first the researchers provided expert information about possible environmental damage and the risks connected with atomic power. But soon they also became actively engaged in research on alternative energy, energy saving, and energy planning in society. The alternative energy-plan for Denmark, for example, was written by a group of researchers and published in 1976.[10] This publication marked OOA's overall policy-orientation in this period. The major points made in this plan have become part of today's official policy in Energy Plan 2000 (Energiplan 2000).

A subgroup in the energy movement pointed to the need for realistic and visible alternatives. It was not sufficient, these activists held, just to point to possible alternative energy sources and rely on public research and energy policy. They wanted the energy movement to become actively engaged in developing technological alternatives. The result was the establishment of the Organization for Renewable Energy (Organisationen for Vedvarende Energi, OVE) in 1975. Its purpose was to set up a framework for exchanging information and knowledge on the most promising technologies.

Apart from the political scene, the most important activities were the practical actions by the alternative movement. Many different types of actors together formed a vital entrepreneurial network. These included grass-roots entrepreneurs building wind turbines on their own, the first industrial entrepreneurs transforming prototypes into produced series of turbines, the idealistic buyers of the functioning but still unstable wind turbines, and the anti-nuclear power engineers starting the Test Station for Small Wind Turbines.

In the first phase of innovations during 1974-78 the grass-roots entrepreneurs and do-it-yourself builders were the most important groups. In this period a great variety of different wind-turbine designs were constructed and tested. The basic experience already available from earlier wind-turbine constructions was rediscovered. The wind-turbine builders found Johannes Juul's advice to be both farsighted and practically valuable.

OVE was of great importance in this period organizing Wind Meetings (Vindtræf). Up to four times a year these meetings brought together all kinds of people engaged in the construction of wind turbines to exchange information and ideas. This informal network played a crucial role as a forum for the diffusion of knowledge, experience and new ideas. Most of the wind turbines from this period were small (with generators from 20 to 50 kW). The only exception was the wind turbine built by the Tvind schools, an alternative folk high school and youth school in Jutland with many activities in the third world. The 2 mW Tvindturbine, at that time the largest in the world, was built by volunteers from all over the country. The basic design, the blades and the electrical equipment were designed and constructed by engineering experts from different companies and research institutions.[11]

Later developments drew specifically on Tvind's experience in constructing fibreglass blades. They were designed by an engineer who had previously worked for Skandinavisk Aeroteknik as well as with the blade construction for F.L. Smith's turbines in the 1940s. Before 1977 turbine blades were not produced as a separate component on a market basis. After 1977, together with standardized electrical switching equipment, industry-produced blades formed the basis of several new wind-turbine companies assembling standard components for the technologically difficult parts of their own turbine designs. From this time on industrial entrepreneurs became the dominant group, and about 20 small companies (most of which had produced machinery for agriculture) started their own production or assembly of wind turbines. The initial market for wind turbines was created by idealistic buyers, whose early purchases made it possible to build up expertise. Without the initial purchases of around 100 demonstration turbines, it would never have been possible to develop feasible wind turbines or the wind-turbine industry.

The years 1978 and 1979 were critical ones in institutionalizing wind-turbine technology. From the political system, significant developments included the start of a interlinked system of research, stimulation programmes, and collective safety and reliability standards. Also in 1978 two new interest groups emerged, the Association of Danish Wind Power Owners, and the Association of Danish Wind Mill Manufacturers.

The most important technological institution was established in 1978, too. The Test Station for Smaller Turbines (Prøvestationen for mindre vindmøller) was funded as part of the government's wind energy research programme with Dkr 5 million (around $1 million) over three years. Ironically it was set up at Risø, the atomic research installation.[12] The test station saw its aim as helping to establish a technologically competitive industry and to expand the energy produced by wind power. The first rupture in the energy movement was visible, however. Risø embodied a different social perspective focusing on the diffusion of a new technology and on a strategy that reinforced the traditional mechanisms of industrial production, learning, and knowledge accumulation.

The test station was essential in establishing a new industry based on small companies with limited experience and limited resources. The test station worked as the common R&D department for the whole industry. It played an important part in fostering cooperation among the competing, and sceptical companies, and made it possible for the wind-turbine industry to get a sign of approval from an independent authority. Also in this case grass-roots entrepreneurs helped transform Juul's principles into standards for the security system of a wind turbine based on the combination of air and mechanical brakes. Later this approval facilitated the process of penetrating and competing in the U.S. market.

The Ministry of Energy, established in 1979, supported wind technology by initiating specific state regulations and market subsidies for wind technology.[13] The buyers of wind turbines were heavily subsidized with up to 30 percent public funding. Moreover an agreement between the small wind-turbine owners and the utilities regulated the prices that were to be paid for power that exceeded the wind-turbine owners' own consumption. Beginning at this time, wind-turbine generating stations were typically established by small, local cooperatives seeking economic returns. Demand stimulation with investment subsidies was a new instrument in Denmark's industrial policy, and this combination was influential in diffusing the technology and giving industry a niche in which to develop.[14]

Two Different Knowledge Strategies

The renewed interest in wind-turbine technology in the 1970s and 1980s illustrates the existence of two different developmental approaches. While the the wind turbine is easy to assemble, it forms a complex system. Building aerodynamic and reliable rotors as well as regulating the turbine's connection to the power network are difficult.[15] These technical characteristics, in combination with different traditions and styles in different branches of industry, account for the bottom-up and top-down approaches existing side-by-side.[16]

The bottom-up strategy is based on the principles of learning-by-experience and learning-by-using in which scaling-up is a slow and step-wise process (Rosenberg 1982). Scale effects are assured on each level by producing turbines in series for the market. In contrast the top-down strategy is based on planned research and development programmes from which results can be transformed into principles for technical construction. In the top-down strategy the fundamental problem was defined as how to determine the "optimum" design and size, heading directly towards the final wind turbine (Karnøe 1993). What rotor diameter, blade tip speed, and height would in commercial production result in the cheapest electricity price? This strategy has been very expensive and quite unsuccessful; it has not yielded cost-effective wind turbines.

Why was the bottom-up strategy so effective? One reason is that the scientific model from aerodynamics was not directly applicable to wind turbines. Aerodynamic theory has been established around the construction of airplanes, where both blades and propellers are working under forced air flows. The blades of wind turbines, however, run at a speed where the wind pressure is constantly changing, which consequently causes "reverse tensions" in the blade. Danish turbine blades were designed by experienced aerodynamic engineers, but they took into account the accumulated practical experience with wind turbines by introducing many small modifications and continually comparing these results with the theoretical design models. And they did not apply the design principles from aerospace industry, as was the case in the U.S. industry, whose designers sought the lightest type of construction and disregarded the structural dynamics of the blades. The result in the U.S. has been an excessive rate of fatigue fractures.[17]

Denmark had a good starting position based on its previous experience with wind-turbine production and use. And the process of learning-by-doing (and learning-by-interacting) was supported by the many small competitors working and creating variations in a relatively small market, and then having an early supply of information on successes and failures from the demanding selection environment. By contrast, it is difficult to obtain variation and learning effects in the top-down strategy where unsolved problems with blade design, structural dynamics, and the introduction of light materials must be solved through further technical research.

The bottom-up approach was based on experiences with relatively small turbines (around 100 kW) run as local power stations over long periods of time. Indeed, one of the arguments for the top-down approach was that only big turbines would deliver the energy needed (1 to 2 mW) and that they would be more economical both in construction and maintenance. These turbines were to be run by power-producing companies and would replace other types of centralized power stations.

Just as the bottom-up strategy has deep roots in Denmark, the top-down strategy has deep roots in the United States. In the period from 1934 to 1941 the world's biggest wind turbine (1250 kW) was designed in the United States by leading scientists and engineers. It was taken down in 1945 after a definitive disaster when one blade snapped and landed 250 metres away. Thirty years were to pass before the design was again taken up in the renewed American top-down programme for wind power. When wind-turbine technology was reintroduced in the U.S. the design concept developed by the German professor, Hütter, was chosen as the starting point. Although many technical problems with this design soon arose, the U.S. research is still based on this concept.[18]

When the Danish Ministry of Energy was established in 1979 it also devoted the main part of Danish research funding to the top-down approach. The new ministry was at first overwhelmed with its various tasks, the deploying of different energy sources, the saving of energy, and the launching of special energy research programmes with a yearly spending of at least Dkr 30 million (around $6 million). The power companies have had great influence on the research and development programmes in the Ministry of Energy, which has most heavily financed innovations in traditional energy technology and large-scale wind-turbine experiments. Over the years the power companies have received money for several test-turbines and test programmes, all relying on theoretical models for scaling-up. Simultaneously the Ministry of Energy, as mentioned, also launched the most influential programme on renewable energy sources that supported wind-turbine installations.

Taking into account all the subsidy and funding channeled into the wind-turbine industry in Denmark — but excluding the few not very successful and commercialized big wind turbines in the Danish top-down approach — the total government expenditure from 1978 to 1989 has been Dkr 326 million (around $65 million). Out of this amount Dkr 210 million were spent on subsidies to the wind-turbine owners and Dkr 116 million were spent on research support for innovation.[19]

These figures can be compared to the Dkr 65 million spent just on the large 2 mW wind turbine built by the Jutland power company (ELSAM) in Esbjerg. Or to the U.S. funding of Dkr 1.7 billion (around $340 million) and the German funding of Dkr 900 million (around $180 million) spent on R&D activities alone. It seems the Danish people have received their money worth. The Danish wind-turbine industry by 1990 had achieved total exports of around Dkr 7 billion ($1.4 billion).

The Success of Industrial Production

Notwithstanding the grass-roots designers, industry has taken over the further technological development of wind turbines. As we already have discussed, the accumulation of knowledge and learning-by-experience seem to be crucial factors in the refinement of technology. In this process industry has taken on the integrative role of the wind meetings.

With indirect support from the Danish state, the wind-turbine industry first developed its market in Denmark. Growth in production in 1982-1986, however, was the result of success in the export market, especially in California where local government subsidies furthered wind power. Danish companies competed with 15 American and 8 European companies but had the technological lead, along with one U.S. company which had also based its development on learning-by-experience. In 1985 Danish wind turbines held 50 percent of the Californian market (Stoddard 1986) and about 700 mW of the 1500 mW wind power installed in California wind farms were produced by Danish manufacturers (California Energy Commission 1990). The institutional context in California was fundamentally different from the Danish concept of local users financially supported by the state. In California wind power was installed by investment groups, often set up by industry itself. The user contact in the U.S. was oriented to financial entrepreneurs instead of energy enthusiasts.

While the wind industry succeeded in California, the remaining grass-roots entrepreneurs and enthusiasts tried to re-institutionalize their original vision. In 1983 a new test bed and research centre was established, the North West Jutland People's Centre for Renewable Energy (Nordvestjysk Folkecenter for Vedvarende Energi). This new institution was a reaction to the separation of the wind-turbine industry from the alternative energy movement and its social visions. With funds from the Danish Ministry of Energy and with support from OOA and especially OVE, the People's Centre has tried to strengthen the combination of the social vision of local production and responsibility with the design of wind turbines and other technologies based on recycling of waste and renewable energy sources.

The People's Centre introduced a modular design for "blacksmith" turbines of different sizes. The idea was to supply local producers with technical drawings and construction details, enabling them to assemble their own wind turbine. But this approach lacked the opportunity of learning-by-experience. Although the People's Centre has continually tried to update the technical data and designs of its wind turbines, its designs have not been able to compete with turbines produced by industry. Also, a mutual distrust has emerged between the wind-turbine industry and the grass-roots enthusiasts. Industry has claimed, for instance, that certain developments have been "stolen", copied and distributed by the People's Centre.

The latter part of the 1980s rapidly changed the competitive conditions for the wind industry and threw it into a deep crisis. The California market was already difficult to cope with. The collapse of exports in 1987 led to the closing of many wind-turbine companies. Simultaneously, the Danish direct subsidies were reduced continuously and entirely removed by 1991. Wind turbines, it was reasoned, had become cost-effective and competitive in free market conditions. Only an indirect subsidy remains since the government has refunded the general energy tax put on power produced by wind turbines.[20]

Since 1988 the domestic and export markets have been of the same size. But the prospect of growing exports to other countries in the EEC gives the well established competence and competitive position of Danish wind power industry a good starting position. By the end of 1992 460 mW was installed in Denmark and 765 mW was installed in the rest of Europe (VE Information 1992). The German, Dutch and United Kingdom markets are growing fast, and around 4,000 mW wind power is expected to be installed in Europe by the year 2000 (European Wind Energy Association 1991).

The competitive position of Danish wind turbines is closely connected to their growing cost-efficiency. From the very beginning the Danish wind-turbine industry faced difficult competitive conditions. The historic Danish dependence on fuel and coal imports has led to very cost-effective power production in Denmark based on imports of cheap coal. Given this background, making the wind turbine cost-efficient[21] in Denmark really was a challenge. At present the cost of power produced by wind turbines compares favorably with that from big coal-fired power stations.

In the Danish debate the question of cost-efficiency of wind turbines is still important. A report was published by the utilities in June 1991 on some of the strategic problems. Power companies and wind-turbine representatives switched roles. Surprisingly, it was the wind-turbine industry arguing that wind-turbine technology was still not mature and therefore needed continued public funding, while the power companies argued that wind-based energy production was mature and should compete on free market conditions.

A real breakthrough in wind power is today dependent on scaling-up the size of wind turbines. Official energy policy bodies, the power companies, and industry agree on this point. Differences remain only in the choice of a strategy for scaling-up. This has divided the field into two technological trajectories with different alliances between the social groups supporting them: the top-down and the bottom-up approaches.

As commercial wind turbines grow bigger, conceivably reaching the size of the large test turbines, successful breakthroughs might come from the large wind-turbine test fields. Our study, however, does not regard this possibility to be likely. The actual 400-500 kW designs worked out now (both the Danish three-blade and the American two- and three-blade turbines) are very different from the earlier large-scale projects. Also, tomorrow's 750-1,000 kW turbines will not be based on radical new knowledge from the large-scale test fields.

The use of still bigger turbines will, on the other hand, reinforce the power companies' role in the development of wind turbines shunting aside the alternative energy movement even more. Their role will be strengthened if power companies combine buying and installing wind turbines as part of the power network with financing the development costs. Such a development will also represent a trajectory based on the scale-effects of growing size and power.

Table I: Development of the wind turbine

Year	Phase of development	Turbine power / rotor diameter
1975-1979	Grass-root entrepreneurs and energy movement	15-30 kW / 10-13 metre
1980-1983	Early industrialization and home market	55-65 kW / 15-17 metre
1984-1985	Fast growth and export orientation	75-99 kW / 17-20 metre
1986-1988	Professional R&D, markets declining and concentration	130-160 kW / 21-23 metre
1988-	Scaling-up ?	200-400 kW / 25-32 metre

Source: Peter Karnøe, 1990.

Ending the story at this point might lead to a conclusion often implied in economic theory: Social visions and social movements can play a temporary role in technological change, but in the long run techno-economic rationality will unfold and determine the path for future developments. This way of analysing the case reduces the controversy between the two approaches, and the alternative social visions discussed earlier, to a question of simply choosing the most efficient knowledge strategy. It fails to appreciate the important differences in the social and institutional environment propagating the two strategies. From the viewpoint of economic theory, social visions are primarily connected to the social environment for setting the criteria for selection of the technologies but not engraved in the nuts and bolts of the technology.[22]

Variation and Selection - The Visible Hand

As the case of wind turbines demonstrates innovation can be fostered by the political and social motivations as well as institutional conditions facilitating the generation of many variations. One dimension of variation can be illustrated by the Danish bottom-up versus the U.S. top-down approach, each originally founded on designs by Juul and Hütter respectively. Another dimension of variation was connected to the experiments by the grass-roots entrepreneurs trying out every possible concept of wind turbines. Even now, on a smaller scale, there is some variation between the different designs produced by the industry.

In economic theory the selection environment leads essentially to some techno-economic criteria of the product. But the case of wind turbines shows that additional components can be important selection mechanisms, for instance the ability to make financial arrangements and establish social alliances. These abilities are closely connected to the understanding of the technology, which makes it difficult to distinguish between the selecting environment and the technology to be selected, just as it is difficult to make a clear-cut distinction between the social environment creating variation and the technology to be created.

The tendency to invoke an ideal "invisible hand," as well as to make artificial distinctions between variation, selection, and the technology are the outstanding conceptual problems in evolutionary economic theory. This case shows the necessity of making visible links between the technology and the institutions fostering it. For both the variation process and the selection environment are created in parallel with the technology; they are the outcome of the same social construction processes. These can be constructive in the sense that they lead to the creation of new working technologies, but they can also be destructive by shaping a selection environment that counteracts the possibility of collecting experiences and that rules out alternative technologies. In this wind-turbine case we have illustrated the dynamic and interactive co-evolution of institutions, technologies, organizational forms and social groups which leads beyond the traditional focus on techno-economic parameters in innovation studies.

Social Visions Left Behind?

The years since 1988 have been characterized by concentration in the wind industry, and consequently its need for an alliance with the alternative energy movement seemed to vanish. Industry has recently focused on keeping its experience in-house, getting access to export funds, and indeed altering its orientation from novel interests organized in wind cooperatives to traditional interests organized in industrial boards. Social enrolment was a crucial factor in the first phase of the development process and has functioned in favour of the new industry, but now traditional economic forms seem to (re)establish their dominance.

Nevertheless, the effort to develop the wind turbines has had wide-ranging and durable effects on energy policy and institutional arrangements. Alternative energy sources are now completely accepted as a supplement to oil and coal, a dramatic shift in Danish energy policy. As well, changes in the institutional set-up have been working in favour of a more decentralized and integrated energy production system, while weakening the traditional concession system of power production and distribution. The power companies are struggling to regain their former positions by adopting renewable sources of power production. Although their top-down strategy was not successful, they have the capital necessary to invest in wind-turbine installations concentrated in so-called turbine parks. Accordingly, their influence has been growing steadily since the late 1980s.

The alliance between the energy idealists, especially OVE, and the group of wind-turbine owners created a political pressure group with strong and direct connections to political parties in parliament. The lasting importance of this effort is illustrated by the fact that 160,000 Danish households own shares in at least one turbine. The wind-turbine owners are typically organized in wind power cooperatives and their interests are practical, based mostly on economics. Such an ownership pattern has promoted the connecting of wind turbines to the power network, de-emphasizing local responsibility and local energy supply.[23] By cancelling the agreement on payments for surplus power from wind turbines, however, the utilities have recently asserted their central bargaining position. The wind-turbine industry is now independent of direct subsidy, so the unclear future conditions for connecting wind turbines to the power network as well as the prices paid for surplus power have resulted in stagnation in the home market.

Despite setbacks in its technical strategy, the People's Centre has continued to play a central role in the wider debate on alternative, renewable energy sources. By the combination of different energy sources and the focus on possibilities for energy saving in local communities, they continue working from the vision of local systems and local responsibility. Nevertheless, the conflict between industry and the "blacksmith turbine" still indicates a fundamental weakness in one aspect of the vision and the strategy of the energy movement: combining local and decentralized responsibility in the production and consumption of energy with the idea of local and decentralized production of the turbines. While the People's Centre is presently an important test bed for biogas installations, it seems likely that this technology too will end up being developed in an industrial context. Knowledge accumulation seems to be one of the strengths of the industrial mode of production that leads to "industrial dominance" in technological development processes.[24]

The ability of established and organized interests to sustain a period of attack yet regain position illustrates how power is exercised through sociotechnical relations. These processes cannot simply be explained, however, in terms of power relations based on the institutionalization of one technological regime. Power is as much as other social elements the result rather than the precondition of such development. Still, it is important to recognize the existing power relations as they appear for instance as anticipated behaviour. Here, the utilities were transformed as well. The utilities' adoption of the ideology of energy saving and their acceptance of using integrated production facilities, by which they have regained their leading role in Danish electricity, did not follow from their former role in the established institutional structure.

The idea of local energy production satisfying local needs has been undercut not only by economic interests but also by the unstable production from wind turbines that necessitates external power supply or connection to a network. These practical considerations undermined one of the basic motivations of the alternative energy movement, reducing the great costs related to distributing centrally generated electricity and the political/social power founded on this basis. In this sense, it was not the wind turbine as defined by the energy movement that was successful; rather, it was the small power station connected to the network as defined by industry and turbine cooperatives. This also explains the new problems arising from turbines being spread around in the countryside. Their localization is typically less connected to local interests, and more attention is directed to, for instance, the noise inconvenience from turbines.[25]

The different interpretations of wind-turbine technology are not simply a question of different social environments or political views of the same artifact.[26] Although many components in wind turbines might be the same, the two different visions put priority on different development paths for both the technology and the supporting social institutions, and consequently lead to material differences. One example is the use of asynchronous generators and the design of the electronic control equipment closely related to the turbines' connection to the power network. These technical choices are the direct consequences of viewing the wind turbine as a small power station producing stable and coherent power for a network. They are solid examples of how the nuts and bolts of wind technology have been formed by social interests.

The success of wind technology has divided it from the other alternative energy technologies, and the energy movement has not been successful in integrating these different technologies in a combined and multifaceted strategy. The energy movement's vision for wind turbines is still a possibility if they could be combined with energy storage as well as with other means of energy production that could provide power when the wind is not blowing. This combination could solve some of the problems of local electricity supply, and thereby establish the background for local governance and responsibility.

The most promising technologies for completing wind turbines are being developed outside of the alternative energy movement. At present, the Ministry of Energy plays an important role in promoting energy technologies based on biomass as well as the co-production of power and heat in local gas and "halm" fired plants. A restructuring of the alternative energy movement's alliances to align itself with the promoters of these complementary technologies seems necessary if the movement is to sustain its social vision.

Table II: Danish wind-turbine industry production, home market sales and exports

Year	1976-78	1979	1980	1981	1982	1983	1984	1985	1986	1987	1988	1989	1990	1991
Number of companies	6	8	12	14	10	10	10	15	20	20	-	14	12	10
Turnover mill.Dkr.	-	-	18	44	94	343	895	2275	1460	570	630	840	1000	-
Employment	-	-	50	70	200	500	1100	3300	2000	900	1200	1200	-	-
Exports mill.Dkr.	-	-	-	-	30	300	800	2100	1300	400	210	440	600	-
Home market mill.Dkr.	-	-	-	44	64	43	95	175	160	170	420	400	400	-
Turbines installed in Denmark	100	120	200	220	150	100	150	330	320	211	290	280	-	-
Installed new mW in Denmark	1,5	2	5	7	7	5	8	25	30	22	40	45	60	60
Effect in kW of new turbine generations	20	-	-	-	55	-	75	-	-	160	-	400	-	-
Cost-effectiveness Dkr./mWh	-	-	690	-	-	-	550	430	470	340	310	300	-	-
Number of households supplied by newest turbine type	-	-	8	-	-	-	45	-	-	-	-	-	275	-

Source: Karnøe, 1991.

Notes

1. This supply rate would grow to 5-7% by replacing all older and smaller wind turbines (30-100 kW) with the ones produced today (400-500 kW). The estimated lifetime for the essential parts in a turbine is 20 years.

2. The structure of our presentation of the case may lead the reader to overemphasize the distinction between the two strategies.

3. In *Tidsskrift for Industri* (Magazine for Industry) 1906, pp. 212-214.

4. Cited from the newspaper *Information*, 1950.

5. The organization came into existence in the back garden of a vicar in Lyngby, who had been actively engaged in environmental questions and was one of the organizers of the alternative environmental conference in Stockholm in 1972.

6. We are only able to point to a few written sources of inspiration: Amory B. Lovins, *Soft Energy Paths - Towards a Durable Peace*; Barry Commoner, *The Poverty of Power*, and (later) Robert Jungk, *The Atomic State*.

7. From the study circle material published in OOA's member magazine: "Atomkraft 12-13, studiekredsnummer", Copenhagen 1975.

8. In this sense the report *The Limits to Growth* by Meadows et al., although it was an important source of inspiration, did not satisfy the need for constructive alternatives. Here the earlier Danish experience from wind-turbine technology and ideas presented by Barry Commoner and Amory Lovins have played an influential role as advice. But we are still seeking material to explain how "sun and wind" (and later biogas) after only a few years were taken for granted as the socially acceptable alternative energy sources and furthermore that practical activities could be started immediately.

9. For instance Professor Niels I. Meyer and Professor Bent Sørensen. Meyer, especially, could be seen as a Danish variant of Barry Commoner.

10. This plan contained a complete programme for both the energy production and energy savings in Denmark. Wind turbines were never seen as the only alternative energy source and estimates never let wind energy account for more than 15 percent of the total energy supply.

11. The Tvindturbine is really a paradox; it was a low-cost (6 million Danish kroner) and well-functioning wind turbine, although it never reached the planned 2 mW (it typically produces around 1 mW for local use - not connected to the power network). In the U.S., expensive constructions were set up by NASA-Lockheed, Boeing, and Standard Hamilton based on a top-down programme (costing *hundreds* of millions of Danish kroner).

12. The young engineers who started the test station together with an experienced engineer from F.L. Smith, as the director, sympathized with the anti-nuclear power movement.

13. This had some very strong industrial policy effects, market pull, and worked like an "infant industry policy."

14. The Danish principles for market subsidy for the buyers investing in wind turbines are being copied in Germany and Holland.

15. Contrary to the common impression of wind turbines as being an old and well-established technology mostly to be characterized as low-tech. The development cost related to the latest Danish 450 kW design from the Bonus company was $3-4 million, while the American company U.S. Windpower invested $15-20 million in their newest 300 kW design.

16. This has been discussed in the innovation economic literature: "Technology often develops without a scientific understanding of how it works" (Rosenberg 1982:122); "Communities of practitioners developing technology (building knowledge) with their own (idiosyncratic) methods

(culture/tradition)" (Laudan 1984); and "Analytically and practically science and technology are two different bodies of knowledge developed by scientists and technologists, but with a very complicated interaction" (Layton 1971). An alternative to the linear science-push model is the chain-link model developed by Kline and Rosenberg (1986).

17. One could characterize the "Danish design" as being heavy-weight and over-dimensioned, at least according to the design principles from the aerospace industry. In the beginning, the Danish design was a result of playing safe by applying design criteria from traditional mechanical productions, rather than thorough investigations in structural dynamics.

18. Although they got access to a test programme carried out on the Gedser turbine in Denmark, which made it possible to compare the two designs.

19. In the Danish debate, opponents often add a further amount of 248 million Danish kroner in "returned energy taxes" - but it is questionable if this simply is a subsidy since the objective of this energy tax was to reduce the amount of imported oil, coal and gas.

20. The aim of the energy tax was to reduce the import of fuel by giving an economic incentive to save energy. But due to EEC regulations it has been difficult to tax imported fuels and electricity without taxing power produced in Denmark. In this case energy taxes were placed on power consumption and not on the real target: the imported fuel. A recent study financed by the EEC Commission showed that wind technology should receive subsidy from the state if market prices for power were corrected by the indirect costs put on society by oil, coal and nuclear power (Hohmeyer 1988). Taking this into account wind technology would have been price competitive already in 1982.

21. In 1980 the price for power produced by wind turbines was 0.70 Dkr/kW, while in 1988 it was below 0.30 Dkr/kW (Morthorst and Hjuler Jensen 1987).

22. These notions tend to neutralize the specific and very different types of social processes in action, which (like biological connotations in evolutionary economics) might easily mislead one to identify the outcome of the selection processes as satisfying one optimal choice — a determined path of development defined by the "survival of the fittest."

23. This changing interest can be illustrated in the criteria used when allocating state subsidy to wind-turbine buyers. In the beginning the turbine should be placed not more than 3 kilometres from the owner's house, but this criteria has been changed and was finally removed.

24. This structural phenomenon questions one of the basic ideas in the SCOT approach, where power and strength are co-produced by the actors in the same process in which technology is constructed (See Bijker et al. 1987; Bijker and Law 1992).

25. The noise level from the newest turbines is lower than a highway's noise level, and it is often the case that turbines are placed at a distance from houses because the wind regime is better on free spots in the landscape. The discussion about noise is part of a political fight concerning the "green" character of wind technology.

26. The notion of "interpretative flexibility" in the SCOT approach could easily lead to an understanding of social construction only assigning meaning to the technology, but not influencing its nuts and bolts.

References

Bijker, Wiebe E., Thomas P. Hughes and Trevor Pinch (eds.). 1987. *The Social Construction of Technological Systems*. Cambridge, Mass.: MIT Press.

Bijker, Wiebe E., and John Law (eds.). 1992. *Shaping Technology/Building Society*. Cambridge, Mass.: MIT Press.

Blegaa, Susanne et al. 1976. *Draft of an Alternative Energyplan for Denmark. (Skitse til alternativ energiplan for Danmark.)* Copenhagen: OOA & OVE. (In Danish)

California Energy Commision. 1990. *Report from The Wind Performance Reporting System*. Sacramento.

Commoner, Barry. 1976. *The Poverty of Power*. New York: Alfred E. Knopf.

Danielsen, Oluf et al. 1975. *Alternative Energy Sources - Energy policy. (Alternative energikilder - energipolitik.)* Copenhagen: The Energy Information Committee. (In Danish)

Dosi, Giovanni. 1982. "Technological Paradigms and Technological Trajectories: A Suggested Interpretation of the Determinants and Directions of Technical Change." *Research Policy* 11: 147-62.

European Wind Energy Association. 1991. *Wind Energy in Europe*. Time for Action. (Internal Report): Copenhagen.

Golding, E.W. 1955. *The Generation of Electricity by Wind Power*. London: SPON.

Hohmeyer, Olav. 1988. *Macroeconomic Impact of Wind Energy*. Proceedings of European Community Wind Energy Conference. Herning, Denmark.

Jungk, Robert. 1979. *The Atomic State. (Der Atom-Staat.)* München: Kindler Verlag. (In German)

Karnøe, Peter. 1993. *Approaches to Innovation in Modern Wind Technology: Technology Policies, Science, Engineers and Craft Traditions in the United States and Denmark 1973-90*. Discussion Paper no. 337. Center for Economic Policy Research, Department of Economics, Stanford University.

Karnøe, Peter. 1991. *Danish Wind Turbine Industry - a Surprising International Success. (Dansk vindmølleindustri - en overraskende international succes.)* Copenhagen: Samfundslitteratur. (In Danish)

Karnøe, Peter. 1990. "Technological Innovation and Industrial Organization in the Danish Wind Industry." *Entrepreneurship & Regional Development* 2: 105-123.

Kline, S., and N. Rosenberg. 1986. "An Overview of Innovation," in: Ralph Landau, and Nathan Rosenberg (eds.). *The Positive Sum Strategy*. Washington: National Academy Press.

Laudan, Rachel. 1984. "Cognitive Change in Technology and Science," in: Rachel Laudan (ed.). *The Nature of Technological Knowledge*. Dordrecht: D. Reidel.

Layton, E. 1971. "Mirror-Image Twins: The Communities of Science and Technology in 19th century America." *Technology and Culture* 12: 562-580.

Lovins, Amory B. 1977. *Soft Energy Paths - Towards a Durable Peace*. Penguin Books.

Meadows, Dennis L. et al. 1972. *The Limits to Growth. Report to the Club of Rome*. New York: Universe Books.

Morthorst, P.E., and P. Hjuler Jensen. 1987. *The Economics of Wind Turbines. (Samfundsøkonomi og vindmøller.)* Risø: The Test Station for Wind Turbines. (In Danish)

Münster, Ebbe. 1984. *Village energy - local production. (Landsbyenergi - lokal produktion.)* Copenhagen: OVE. (In Danish)

OEEC. 1953. *Wind Power: Technical papers presented to the wind power working party*. Paris/London.

Rosenberg, Nathan. 1982. *Inside the Black Box: Technology and Economics*. Cambridge: Cambridge University Press.

Sahal, Devendre. 1985. "Technological Guideposts and Innovation Avenues." *Research Policy* 14: 61-82.

Stoddard, W. 1986. *The California Experience*. Paper presented at the Danish Wind Energy Association Conference.

VE Information. 1992. October issue. Copenhagen.

5 Steering Technology Development Through Computer-Aided Design

Gary Lee Downey

Beyond the CAD/CAM Fix

At first glance, computer-aided design and computer-aided manufacturing (CAD/CAM) appear to provide a technological fix for constructive technology assessment (CTA). CTA strategies combine an analytical project with a normative project. The script for the analytical project involves synthesizing sociological, historical, and economic analyses of technology into integrated theories of the dynamics of technology. A central intellectual achievement of recent research on the dynamics of technology has been to reconceptualize technology from an independent force that acts upon society from the outside to a social activity in itself (Noble 1978:318). An implication of this claim is that technology development within corporations, for example, is shaped by interests and considerations that extend well beyond both the organizational boundaries of the firm and the narrow economic logic of profit maximization (compare Coombs, this volume). The script for the normative project is to formulate strategies for steering technology development in socially desirable ways, where "socially desirable" generally means the reduction of social inequity. This dual orientation provides both intellectual and political considerations for evaluating CTA accounts and modulation strategies.

CAD/CAM technology looks exciting at first because it appears to be a technology designed for CTA purposes. Do you not already believe, for example, that integrating computer technology into engineering design will somehow improve the development of new products and increase the range of variations that engineers consider in designing new products? In principle, many new types of considerations can be factored into design decision making using CAD/CAM technology. Following this logic, if CTA advocates could only convince engineering designers to integrate social equity considerations into design decision making, then perhaps the CTA movement could become entrenched politically through relatively minor modulations. CTA activity might even then focus on writing CAD/CAM software to integrate the socially desirable design criteria. In short, CTA would have a technological fix.

But our analytical project suggests that technological fixes are always more complex than they first appear. Using a technological fix to solve a social problem *appears* attractive when inserting that technology into society requires little, if any, social adjustment. But if technological development is a social activity shaped by heterogeneous considerations, then achieving a technological fix is a genuine, and probably rare, social accomplishment. Getting caught up in the enthusiasm of CAD/CAM technology without examining its heterogeneous developments and implementations is to fall victim to precisely the form of technological determinism that research in technology dynamics is designed to overcome.

In the account below, I present a picture of CAD/CAM implementation that differs from its more usual image as a technological fix. Focusing on the United States, I describe CAD/CAM implementation as the production of three distinct technologies — two dimensional drafting automation, three dimensional wireframe and surface modeling, and solid modeling — that are endowed with the agencies of three different types of users. Although a nationalist script has positioned CAD/CAM as a technological fix that will unite design and manufacturing activities in a coordinated, integrated, and flexible manufacturing enterprise, none of the three technologies is oriented toward uniting design and manufacturing.

In varying ways, the implementation of CAD/CAM technologies challenge previously stabilized design activities and serve as resources to empower some groups while marginalizing others. The implications for CTA concerns could be significant. Most important, accepting CAD/CAM technologies means accepting the increased mathematization of design. That is, CAD/CAM use grants new importance to mathematically based design activities, increasingly demanding engineers and managers to place confidence in mathematical methods they are unlikely to master themselves. This is particularly true for 3D wireframe and surface modeling, the technologies that could be modulated most easily to pursue CTA objectives. I illustrate both the opportunities and the likely problems involved in steering technology development through CAD/CAM technology by means of a brief case study: an ongoing attempt of aircraft designers to minimize the sonic boom produced by the High Speed Civil Transport (HSCT), a proposed commercial aircraft that would fly at supersonic speeds.

Localizing CTA Strategies

This paper constitutes a theoretical argument for localizing strategies to realize the social equity objectives of constructive technology assessment. The dual orientation of the CTA movement provides a mix of audiences for CTA theories and strategies. To have influence, these theories and strategies must be meaningful not only to CTA analysts but also to those involved in implementing CTA policies. Granting policy managers and recipients some authority over the content of CTA theories and strategies is an innovative move that generates new conceptual considerations. I would argue that CTA theories should be relatively easy for nonspecialists to understand and should have culturally positive significance. For example, the term "constructive technology assessment" indicates that the CTA movement wants to construct rather than destruct, and the term "modulation" suggests strategies for change in society that do not alter its fundamental character. In similar fashion, the conceptual objectives of CTA theories should include not only descriptive plausibility and explanatory power but also accessibility to nonspecialists and a supportive disposition toward stakeholders in areas of proposed change. If the implications were otherwise, constructive technology assessment might marginalize itself sufficiently to be ignored as irrelevant.

I do not consider developing a general theory of technology dynamics a necessary CTA objective. A theoretical implication of viewing technology as a social activity is that technology indeed becomes a social activity. That is, the value of technology-specific theory reduces as the focus shifts to account for heterogeneous social processes. This analytical trend appears to cut across the normative project of promoting equity, for the more usual strategy for cultivating allies in policy positions is by advancing theory that appears to be both systematic and predictive. It is difficult to posit heterogeneity and be systematically predictive at the same time.

The many models of technology dynamics that economists, sociologists, and historians have devised are enormously useful, for these provide taxonomies that help one to categorize and interpret case material. But models are context specific; they abstract bounded structures that do not apply equally to all cases. Developing a relatively shared set of taxonomic categories can be important for integrating contrasting academic communities into the CTA movement, as has been the case for the rhetoric of selection and variation in the evolutionary model. But not only should such models be shaped so that nonspecialists can understand them easily, there is also no reason to expect an accumulated mega-model of technology dynamics, or even a set of sector-specific models, to generate more than some general categories of modulation strategies. The more difficult task, and the area in which CTA theory is least developed and most vulnerable, is to make these strategies work by sensitively overcoming local opposition and fitting them into local scenes.

Cambrosio and Limoges (1991) offer the interesting insight about technical controversies that each one establishes a unique "controversist space" within which decisions must be made in order to prove acceptable. In other words, the form and content of acceptable solutions vary from controversy to controversy. In similar fashion, I believe that implementation of CTA strategies is likely to produce many mini-controversies with controversy-specific solutions. Viewing CTA from the perspective of controversy theory, the types of theories that the CTA movement needs most to guide its strategies are "theories of acceptance," i.e., accounts of people and groups agreeing to follow CTA policies. Theories of acceptance are theories of actor interactions that can link models of technology dynamics to modulation strategies.

Several candidate theories are already in use, each with its special strengths and limitations. The Dutch hosts for the CTA workshop locate acceptance in a willingness to follow the rules and participate in strategic games (e.g., Rip and van den Belt 1988; Van der Meer 1983). Actor network theory portrays actants in power terms, locating acceptance in submission to occupy a place in another's network (e.g., Callon 1986). Social constructivism locates acceptance in a group becoming convinced by another's problem definition (e.g., Bijker 1987). Participating in recent developments in cultural anthropology, I view acceptance as varying according to the manner and extent to which some action reproduces or transforms the "positional identities" of participants in an interaction (e.g. Rosaldo 1989; Downey 1992a, 1992b, 1992c).

I use the concept of positional identity to characterize how cultural objects endowed with agency move themselves around in relation to other cultural objects by positioning and repositioning themselves. The term identity accordingly refers to the positional meanings and powers of cultural agents in relation to one another. Agents can be human or non-human actors, including groups, organizations, and even technologies. The identity of any particular agent consists of its configuration of positions in relation to other agents, and the production of identity for an agent is simultaneously an attribution of meaning and an act of empowerment. For example, the CTA workshop and this volume produce a distinct identity for constructive technology assessment by positioning it in relationship to technology assessment, social impact assessment, science and technology studies, Netherlands Organization for Technology Assessment (which provided financial support), etc. In other words, CTA seeks meaning and power as an agent that extends far beyond the academic community. One pathway toward success would consist of agents from these other realms reproducing the same identity for CTA in their actions.

Each event of action raises a key question about positional identity: How does the event reproduce or transform existing positions and, therefore, power relations? Four different types of process occur regularly. A particular action may remake an existing identity by (1) fulfilling or reproducing some positions; (2) transforming some positions; (3) generating internal tension by reproducing some positions while transforming others; and (4) having no relevance to some positions.

A theory of acceptance built on the concept of positional identity focuses on the relation between who the agent *is* and who the agent *seeks to be*. From this perspective, accepting or rejecting a particular position is a choice about who an agent seeks to be, but the content of that choice also depends upon who the agent is. For example, gaining access to CAD/CAM technologies transforms users by empowering them with new agency, but the precise changes in meaning and power that occur depend upon whether the user already occupied the position of draftsman, design engineer, or manufacturing engineer.

But the implications of accepting or rejecting particular positions can often be unclear, for positional changes frequently produce tensions and ambiguities among the constituent positions of an agent's identity. The experience of change becomes an exercise in the management of tension and ambiguity. I show below that some CAD/CAM developments have been accepted readily because they empower design positions without restructuring the relation between design and manufacturing, while others have received varying levels of acceptance because they empower design positions at the expense of manufacturing positions or they produce tensions and ambiguities among the positions of design personnel.

Positional identity theory can be used in conjunction with other theories of acceptance to develop CTA strategies, for other theories tend to pay attention more to agents' objectives—who they want to be—than to who they are when they define their objectives. But CTA theory must link the two in order to ground acceptable strategies, or risk offering recommendations for change that could prove unattainable because irrelevant. At the same time, a theory of acceptance built on positional identity is not simply a restatement of interest theory, which holds that who agents *are* causally determines who agents *want to be*, i.e., their interests. Interests indeed occupy the space between who agents want to be and who agents are (cf. Latour 1987:108), but who an agent wants to be must be determined empirically and locally rather than be predicted blindly from an analysis of who the agent is. Accounting for agency as a product of interests is a *post hoc* rationalization.

A Nationalist Script for CAD/CAM

CAD/CAM technologies produce new agency in design by linking together previously distinct design activities and concentrating them in fewer locations at earlier points in product development. Each CAD/CAM technology is produced by identifying the 'informational'[1] content in various engineering activities, transcribing that information into binary code, and then reinserting the resulting technology back into those activities. For engineers involved in product design, accepting the agency of CAD/CAM technologies into their working lives generally does not involve simply a shift "from board to scope,"[2] i.e., from drawing board to computer scope. It also means bringing together such activities as drawing, checking, redrawing, doing analysis, calculating sensitivities, building prototypes, and planning manufacturing operations. In the process, it frequently means transforming engineers' career identities and pathways in varying ways.

In the United States, one component of this identity change has clear national significance and legitimacy. Both insiders in CAD/CAM development and representatives of American government and industry have stabilized an image of CAD/CAM technologies as a key agent in solving a national identity crisis. Since about 1980, many Americans have felt themselves under attack by outsiders in a new way. The dominant image has been a nation put at risk by economic defeats at the hands of international competitors, especially Japan. Since the late 1980s, the belief that the military threat from the Soviet Union has been reduced dramatically has intensified attention to the economic dimensions of national identity.[3]

Reproducing a cultural tradition of defining and solving social problems in technological terms (Downey 1992b), Americans have turned to technology and technology-driven industry as strategic agents for achieving national regeneration and resurgence rather than re-evaluating and restructuring institutional forms more directly. From this perspective, other social adjustments, such as modulating relationships among government, industry, and universities, have then been proposed and accepted as necessary to fulfill the technological fix. That is, technological developments have themselves served as the source of legitimacy for social policies needed to adapt to those developments.

The concept 'productivity' is now linked inextricably with the empowerment of national identity. That is, economic productivity now serves as a major vehicle for national redefinition, granting power and authority to all those individuals and groups who successfully incorporate into their own identities the national quest for increased levels of production with improved quality at competitive costs. Americans' cultural understanding of their nation has been tied to their understanding of productivity, changing the meanings of both at the same time.

In the context of national crisis, CAD/CAM technologies gain agency as "productivity tools" and CAD/CAM vendors sell productivity. For example, the leading vendor in 1980, Computervision Corporation, published a 300-page handbook (Machover and Blauth 1980) detailing the potential links between the technology and increased industrial productivity. This highly popular book begins with a warning that "U.S. Productivity [is] Slipping," and an announcement from the Chairman of the Board that "a new technology has evolved which ... will benefit all by improving mankind's standard of living and quality of life." "The technology is CAD/CAM," he proclaims, "and the benefit is increased productivity." CAD/CAM vendors have since repeated this message thousands of times.

If one extrapolates from vendor data, CAD/CAM technology appears to be enormously successful. For example, the major market research firm for CAD/CAM reported that between 1985 and 1988 the total number of computers worldwide using mechanical CAD/CAM, which is the major area of engineering application, increased by more than a factor of five (50,000 to 280,000 computers) (Dataquest Incorporated 1990). The company further projected that this number would nearly triple again inside four years. Clearly, something significant is taking place.

But the focus of nationalist agency also masks some important internal differences among CAD/CAM technologies. The technology that promised to save the American nation in 1980 was the integration of computer-aided design and computer-aided manufacturing, but CAD/CAM integration has fared poorly. As the nationalist script has been translated into localized searches for productivity, the determinist dream of CAD/CAM-induced integration has lost some of its rhetorical power. Rather, the development of CAD/CAM has produced technologies endowed with the agencies of different types of users, none of whom is oriented to uniting design and manufacturing. I believe that preoccupation with the image of a unified technological fix has inhibited insight into some of the more difficult power and identity issues raised by endowing CAD/CAM technologies with agency.

Waves of New Design Activities

By inquiring into how CAD/CAM technologies structure design activities,[4] I have discerned three different waves of development: (1) 2D drafting automation, (2) 3D wireframe and surface modeling, and (3) solid modeling. These waves have appeared and grown in roughly historical sequence, but they now travel concurrently. Not all organizations have experienced all three, nor necessarily in this sequence.

The first wave, drafting automation, shifts the engineering drawing process from board to scope. Engineers understand drafting as the process of producing detailed engineering drawings, which typically represent product parts in terms of 'views' in 'two dimensions', following mathematical rules of descriptive geometry. For example, an engineering drawing of a machine part might present how it looks from the 'top view', 'front view', and 'right side view' (e.g. Dent et al. 1983). The automation of drafting has been constructed on the image of a draftsman working at a drawing board, positioning 2D technology as a 'drafting tool.'

To the draftsman, designers, and engineers who do engineering drawing, automating the drafting process means replacing T-squares, triangles, compasses, French curves, and pencils with 'input devices' (e.g., keyboards, mouses), 'output devices' (e.g., printers, plotters), manuals for 'hardware' and 'software', and small screens for projecting images. Manually drawing points, lines, circles, and curves, becomes the manipulation of graphical 'primitives' and 'attributes' by combinations of programmed 'transformations' and 'control routines.'

By far the greatest proportion of CAD/CAM activities fall in this category.

Automated drafting is not positioned to fulfill the nationalist script of integrating design and manufacturing, however, even though it can increase dramatically the speed of repetitive tasks, such as making changes to drawings. Because 2D technologies are endowed only with the agency of drawing, their implementation does not reposition design and manufacturing activities in relation to one another. Rather, drafting automation tends to restructure relations on the design side alone between draftsmen and design engineers in ways that depend upon its local positioning. For example, 2D technologies can empower draftsmen by enabling them to appropriate some of the activities of engineers (cf. Hacker 1990:175-94), but also disempower draftsmen by forcing them to work nights and weekends and thus separating drawing activities from the weekday activities of design personnel (cf. Badham 1991). Finally, for CTA purposes, drafting automation is the least significant wave of CAD/CAM development, because shifting the agency of drawing to computer technologies does not position users better to introduce new design considerations.

The second wave of CAD/CAM development, 3D wireframe and surface modeling, transforms design activities in ways that open up the possibility of new design criteria, including CTA modulation. Key to this step is the shift from transcribing the agency of drawing to transcribing the production of geometric 'models' of discrete objects in three dimensions. What makes 3D graphical representations so significant is that these can be linked to other engineering activities that make up the design process beyond drafting.

A wireframe representation constructs an object as a collection of lines depicting the object's 'edges' (cf. Groover and Zimmers 1984:59-61). Picture, for example, a visual image of an automobile portrayed only by all the edges of its many components. A surface model represents an object as a set of curved surfaces. Picture the automobile now portrayed as a set of curved surfaces, perhaps with shading to give the exterior a sculptured look.

The major benefit of 3D models is that they add a great deal of engineering information to the representation (cf. Lynch 1988). Adding these different kinds of information is called 'doing analysis', which involves characterizing the object from the perspectives of different engineering sciences. For example, with a wireframe, engineers can view the object from any perspective and can use the point and line data to calculate the object's 'mass properties', e.g., volume, weight, center of gravity (location of the balance point) and moments of inertia (a measure of how easy it is to rotate the object in different directions, e.g., it is easier to roll a car over sideways than end over end). The point and line data can also be used to inquire into whether particular components interfere with one another.

The surface model is much more complicated mathematically because in order to represent surfaces it translates geometric data about points into differential equations about curves and then links these differential equations together. The surface model requires far more calculating time on a computer, but it intersects with a large number of analysis activities that build on information about surfaces. For example, such engineering sciences as heat transfer (how objects respond to heating or cooling), kinematics (how moving parts interact with one another), and fluid dynamics (how air, water, or other fluids behave when moving) all depend upon differential equations representing surfaces.

CAD/CAM surface models provide a common judgement site for the different groups of people who generate drawings, produce engineering calculations, and make larger design decisions. Concentrating these activities in one place can thus have the effect of blending very different identities. Repositioning agents in design, however, also rearranges power relations, which in turn defines the implications of differing levels of acceptance. For example, Kenneth Reinschmidt (1991:5), an industry leader and keynote speaker at a vendor's annual meeting, argued optimistically that linking drawing and analysis is "consistent with the trends toward shallower organizational structures and matrix management," which attempt to reduce hierarchy by giving more independence and problem-solving authority to subordinate levels. But as he further points out without recognizing the irony, the typical decision-making process to take this step "is characterized by a desire to use CAD/CAM to effect change" and "using the CAD/CAM system to impose the design structure on the engineering process ... may imply some controlling function ... " (Reinschmidt 1991:6).

In parallel with 2D technology, 3D wireframe and surface modeling are also not empowered to integrate design and manufacturing activities. Rather, by concentrating activities at an early point in the design process, CAD/CAM technologies are positioned to increase the influence of engineering designers in product development. The power of an engineering designer increases in proportion with each engineering capability added to the graphical image. As product development activities move 'upstream', so the identity and concerns of engineering design are extended 'downstream' into other areas.

The third CAD/CAM technology has been a much smaller wave of new design activities. A solid model represents the object as a solid, using one of two methods. The first is called 'constructive solid geometry', which builds models by adding and subtracting 'primitive' solid forms, such as spheres, cubes, and rectangular solids. Picture, for example, a model of an automobile constructed of chunks of spheres and cubes. The second method, 'boundary representation', produces a surface by linking together surface models to produce a 'closed volume.' In this case, a model of an automobile might break it down into its many components, each represented as a closed volume.

Solid models are very useful for making sure that product parts have enough space after these have been designed, i.e., for 'interference checking.' However, solid models do not transcribe very extensively the activities of either design or manufacturing. On the design side, the geometric representations in solid models are very difficult to modify using the results from engineering analyses. On the manufacturing side, engineers who turn to computers generally seek help in monitoring, controlling, and supporting manufacturing processes, which involves relating objects to their changing environments rather than simply picturing and manipulating them. Thus far, since solid models have not been very useful in either design or manufacturing, they provide poor candidates for CTA modulation.

CAD/CAM and Aircraft Design

Design engineers understand 'design synthesis' as the process of conducting different forms of analysis simultaneously on a proposed design. Although design synthesis antedates CAD/CAM development, the capabilities of 3D surface modeling are giving it greater prominence. Design synthesis has its longest and most involved history in the aircraft industry, mostly because of the close relationship between the geometric form of an aircraft and its performance in different categories of aeronautical engineering analysis. As the design engineer Richard Boyles (1968:486-7) put it, "The influence of the geometric definition of the aircraft on the analysis conducted to ascertain its performance and, conversely, the influence of the analysis upon the geometry of the aircraft are so great that the interaction between the man, the graphic interface, and the analytical capability of the computer are maximized." As a consequence, aircraft design may provide a good location for testing the use of CAD/CAM to steer technological development. At the very least, aircraft design offers well-developed cases for identifying CTA opportunities, strategies, and implications.

Consider the negotiation of ACSYNT, a computer program for the conceptual design of aircraft. ACSYNT, which stands for AirCraft

SYNThesis, was written over a twenty-year period by engineers at Ames Research Laboratory of the U.S. National Aeronautical and Space Administration. During the middle to late 1960s, a number of aircraft companies, including Boeing, Grumman Aerospace, Lockheed California, McDonnell Aircraft, and North American Rockwell, developed their own synthesis programs to aid in the early stages of design. These programs and their successors are proprietary, however, and are not open to public scrutiny. Not only is ACSYNT more available, its development has become the object of a cooperative venture involving NASA, five aircraft companies (Boeing, Lockheed, McDonnell Douglas, General Electric Aircraft Engines, and Northrop), and CAD/CAM researchers at Virginia Tech. By observing and participating in the activities of this venture, the ACSYNT Institute, I acquired a fairly detailed understanding of the program and the groups linked to it.

NASA's statutory responsibilities in aerodynamics include examining advanced aircraft technologies and evaluating proposed designs for military aircraft that contractors submit to the Department of Defense. NASA evaluation teams are minuscule compared to the engineering staffs of contractors. Ames engineers initially produced ACSYNT during the 1970s as a resource to give themselves greater independence and control in examining technologies and comparing proposals.

The engineers categorize ACSYNT as an exercise in 'conceptual design', a phase of design activities that stabilized in the aircraft industry after World War II alongside 'preliminary design' and 'detailed design.' Conceptual design practices specify the vehicle's initial geometric configuration, size, weight, and performance characteristics. During this phase engineers consider a much wider range of alternative vehicle concepts than at any other point in aircraft development. In the aircraft industry, groups responsible for conceptual design are typically small and do not command a great deal of power and authority nor play a great role in making company decisions to build an aircraft.

The leaders of preliminary design groups have traditionally held the greatest power by far in configuring a design. As both aircraft company and NASA engineers explained to me in interviews, the companies subdivide the activities of preliminary design according to the major disciplines of aeronautical engineering, such as aerodynamics, propulsion, and structures, and a combination of organization-specific considerations. Each disciplinary area has teams of engineers that can number in the hundreds. Starting with a small number of alternative concepts, these teams conduct computer-intensive analyses of expected vehicle performance in each area and then negotiate a narrowing of alternatives down to a feasible design that group leaders find acceptable.

By the time the phase of detailed design is reached, the design concept is well entrenched and very difficult to influence further. Company engineers regularly joke about the build-up of 'momentum' behind a design. Activities in this phase provide detailed specifications of all vehicle components, plan and schedule construction activities, and set up relationships with contractors. Evaluations of the design shift from computer simulations to experimental efforts and wind-tunnel testing with mock-up prototypes.

The ACSYNT Institute includes 15 to 20 regular participants from industry, all engineers working in conceptual design. A primary goal of these engineers, both individually and collectively, is to increase the influence that conceptual design has on company decision making by appropriating for conceptual design some of the functions (i.e., the agency) of preliminary design. As one engineer said in an ACSYNT meeting, "We're trying to do with the computer what we can't do with our organizations."[5]

The conceptual designers are particularly interested in ACSYNT because in 1987-88 CAD/CAM researchers at Virginia Tech wrote a surface modeler and linked it to the analysis features of the program (Wampler et al. 1988). With this CAD/CAM interface or front-end, the conceptual designer can input a geometric configuration, ask what additions or changes might be necessary for the vehicle to meet some specified mission requirements, and then view a three-dimensional shaded image of that vehicle on the screen. Prior to having access to CAD/CAM visualization, engineers had to analyze large amounts of geometric data from each computer run in order to draw visual representations manually. Participants in the ACSYNT Institute believe that having the capability to quickly analyse and then visualize alternative designs will enable them to enhance their decision-making authority.

The ACSYNT program itself consists of approximately 50,000 lines of commands, or code, that divide calculations along disciplinary lines into 'modules.' For example, the aerodynamics module determines the minimum drag on the vehicle, while the propulsion module calculates the performance of different types of engines on the vehicle, and the trajectory module uses data from both the aerodynamics and propulsion modules to calculate the fuel weights needed during each phase of specified missions. Other modules include geometry, weights, stability, takeoff, cost, advanced aeromethods, and sonic boom. Each module consists of detailed mathematical routines whose outputs vary with a limited set of input variables, or parameters.

Using the ACSYNT program transforms an initial geometric configuration by conducting and synthesizing several different forms of analysis. The synthesis process transforms the design in three steps: convergence, optimization, and sensitivity. I describe these steps briefly both to show that they constitute new mathematical methods for design decision making and to be able to illustrate the practical difficulties raised by attempting to minimize sonic boom in the High Speed Civil Transport.

Convergence refers to the production of a point design, which is a geometric configuration and a calculation of total gross weight that meets all the design constraints that the user inputs at the start. Garret Vanderplaats, at the time a NASA engineer heavily involved in ACSYNT development, applied ACSYNT to a "typical design problem" in an early paper (Vanderplaats 1976). The objective in this problem included estimating the optimum gross weight of a tactical fighter intended to fly a specified mission and figuring out what effects reducing gross weight by using more advanced materials might have on the vehicle's performance. Significantly, a limitation of ACSYNT is that it is only capable of analysing aircraft configurations whose geometries fall within the boundaries that define conventional fighter, bomber, and transport aircraft. In Vanderplaats' case, for example, it is "predetermined" that the vehicle will have a "conventional wing-tail configuration," which means no fancy geometries. Otherwise, the design will fall outside the envelopes of experience and theory that define the analysis routines. Also, this case included only five design variables: 1) wing loading, the amount of weight per unit surface area of the wing; 2) sweep, the angle between the wing and the fuselage; 3) thickness-to-chord ratio, the thickness of the wing relative to its average width; 4) aspect ratio, the tip-to-tip length of the wing relative to its average width; and 5) engine thrust, the amount of thrust per unit total weight.

Achieving convergence is tricky because it necessarily involves circular reasoning and depends upon prior experience. The two major contributions to gross weight are the combined weights of major components and the fuel. In order to calculate the amount of fuel needed, one must analyse the vehicle's performance along the mission trajectory that is planned for it (e.g., how much thrust is needed). But calculating how the vehicle will perform on its trajectory depends upon knowing the gross weight first. Also, the weights of various components (fuselage, wings, etc.) are calculated as fractions of gross weight, so making absolute calculations of component weights also depends upon knowing gross weight first. As a result of this circularity, to calculate the gross weight of a particular geometric configuration that meets all constraints, one has to begin by estimating it. The program then assigns values to component and fuel weights, uses these to calculate how the vehicle performs, modifies the geometry to meet all design constraints, recalculates the weight based on the new geometry, and then modifies the estimated weight to start all over again. The iterative process continues until the calculated and estimated values of gross weight agree to one-hundredth of one percent.

Although proponents of ACSYNT claim that this level of tolerance is both good and sufficient, it has no particular meaning to aircraft designers. They must simply accept that the mathematics of convergence require it. In the case of the tactical fighter, locating a single geometric configuration that would meet all specifications, i.e., the 'converged point design', took 22 iterations and 40 seconds of computer time.

The second transformation, optimization, begins with the point design and then resizes the vehicle and its propulsion system to find the geometric configuration that both meets all specifications and has minimum total weight. As we shall see below, the vehicle could be sized to minimize or maximize other parameters as well, including perhaps CTA considerations. For aircraft designers, optimization involves even more opaque mathematics than calculations of convergence. Furthermore, many competing methods exist and the field of optimization studies appears to be changing rapidly.

ACSYNT uses the 'method of feasible directions', but no one in the ACSYNT Institute fully understands how it works. Rather most everyone invokes, both orally and textually, the authority of Garret Vanderplaats, who borrowed the method from a Dutch mathematician (G. Zoutendijk). Vanderplaats refers to Zoutendijk's work without presenting any of its details. Optimization methods are necessary because one cannot optimize a design by varying one parameter at a time. Parameters must be varied simultaneously, yet in doing so the process becomes opaque to intuitions based on graphical methods.

In fact, as Vanderplaats explicitly acknowledges, the most efficient optimization strategy directly challenges standard design practices. In transforming an acceptable point design to an optimized design, the method of feasible directions authorizes moves through intermediate steps that do not converge, i.e., that do not meet design specifications. That is, rather than moving step-by-step from an acceptable configuration to an optimized configuration, optimization routinely moves through unacceptable configurations before reaching an optimal one. This makes no sense to designers who are always mindful of initial constraints. As Vanderplaats (1976:7) puts it, "this design procedure, using numerical optimization, represents a major departure from tradition conceptual design procedures." In the fighter example, Vanderplaats located the minimum-weight configuration with 102 iterations through the discipline modules and approximately three minutes of computer time.

In sum, integrating optimization methods such as ACSYNT into conceptual design activities means granting the mathematics of optimization and the mathematicians a place of authority. At the date of this writing, only one aircraft company uses ACSYNT in routine design activities, having integrated it long before the CAD/CAM interface was completed. At an Institute meeting in May 1991, Ames engineers and Virginia Tech graduate students began training industry engineers to use it with the interface.

Repositioning conceptual design can also bring with it reorganizing relations with entrenched groups and activities in preliminary and detailed design. One company member of the ACSYNT Institute has invested tens of thousands of dollars to demystify the mathematics by locating documentation for every mathematical calculation in the 50,000 lines of ACSYNT code. However, a wider acceptance of ACSYNT by conceptual designers could put them in direct confrontation with members of preliminary design groups. By contrast, conceptual designers at NASA-Ames need worry less about their relations with preliminary design since preliminary design is entirely an industry activity. Nevertheless, as we shall see below, the empowerment of conceptual design also introduces ambiguities for conceptual designers within the NASA organization.

An additional concern is that the mathematics of optimization tends to exacerbate errors within any given area of engineering analysis by bringing different areas of analysis into relation with one another. Vanderplaats (1976:7), for example, shows by checking ACSYNT against another aircraft program design that if the trajectory module underestimated the fuel weight by 10 percent the optimization "capitalizes on this error" and resizes the vehicle to reduce total gross weight by 25 percent. Recognizing that the credibility of design synthesis is threatened by the prospect of errors multiplying to produce unacceptable conclusions, Vanderplaats repeatedly reminds readers that "every effort should be made to ensure accuracy in the discipline modules or at least to ensure that the module information is slightly conservative" (1976:7).

But how are conceptual designers to know when their discipline information is conservative if they are exploring novel vehicle concepts? The answer to this question is to remember that ACSYNT's identity as a computer program is itself inherently conservative: the range of variation it permits is radically limited by the knowledge available about existing designs and by the choices among analysis methods it makes for each area. The only way to test ACSYNT is by using it to predict the characteristics of existing aircraft. ACSYNT documentation routinely claims, notably without detailed support, that such tests are consistently accurate to within 10 percent, but there is no way to test its reliability for new concepts. Also, virtually every analysis method in ACSYNT represents a choice among competing alternatives. Vanderplaats (1976:10), for example, shows how six different equations available in published and proprietary literatures for the relationship between aspect ratio and wing weight significantly contrast with one another. Each company has its own favorite equation, and choosing among them directly affects the optimization calculations.

The third transformation, sensitivity analysis, is much more meaningful to design engineers because it translates the results of optimization into a graphical form that they use routinely. Sensitivity analysis systematically varies a single design parameter to determine its effect on the total gross weight. Changes in some parameters, such as vehicle range, might have dramatic effects on weight while changes in others, such as tail length, might have a minimal effect. The results are plotted as a series of curves whose intersection defines the range of total weights possible. Sensitivity analysis is a strategy for ranking design considerations according to how "sensitive" the vehicle design is to changes in them. It is the most time-consuming of the three activities. Vanderplaats (1976:8) points out that in a typical analysis 20 to 30 hours of computer time would likely be used to determine the design sensitivity of variations in both the mission design and the technology used.

Minimizing Sonic Boom as a Design Criterion

Minimizing sonic boom has never been a significant consideration in aircraft design. In the design of supersonic military aircraft, environmental considerations almost never play a role. According to one NASA interviewee, designers of the SR-71 (a supersonic cruise reconnaissance aircraft, or spy plane) "worried about it" because the SR-71 needed to fly over U.S. land "but they didn't do anything about it."[6] Also, all American commercial aircraft fly at subsonic speeds. U.S. law prohibits commercial supersonic flights over land, with limited exceptions granted to the British and French Concordes. Since 1985, however, negotiations among the White House, NASA, Congress, and the aircraft industry produced a research program to evaluate the feasibility of a supersonic commercial aircraft, the high-speed civil transport (HSCT).

An earlier American effort to build a commercial supersonic transport (SST) was abandoned in 1971 after a prolonged controversy. The "controversist space" for decision making on the HSCT has not yet been defined fully, but this space certainly includes contemporary interpretations of the earlier SST controversy. A recent NASA program plan, for example, explains that the earlier effort failed "because of environmental concerns [sonic boom], economic uncertainties, and objections to government-funded prototype development" (U.S. National Aeronautics and Space Administration 1989:3). Program documents and interviews typically extrapolate from the SST controversy to add three new considerations. Two are additional environmental concerns: degradation of stratospheric ozone and increases in airport noise. Ozone degradation looms as the major barrier to overcome. The third is the overriding motivation to build an HSCT: nationalist fervor stipulating "that the nation cannot and will not allow [aeronautical] leadership erosion" (Executive Office 1985:1).

The White House Office of Science and Technology Policy first granted nationalist agency to the proposed supersonic transport through reports in 1985 and 1987 that defined goals and established initial plans. These actions shifted the burden to Congress, which directed NASA in June 1987 to "prepare a multi-year technology development and validation plan that will help the United States retain its leadership in aeronautics research technology" (U.S. Congress 1987:61). Each report highlighted European and Japanese threats to the United States' large trade surplus in aeronautics. NASA produced its plan in March 1988, identifying the HSCT as necessary to serve the rapidly growing trans-Pacific market (U.S. National Aeronautics and Space Administration 1988).

Before acting on this plan, a Senate committee acknowledged the potential of renewed controversy and sought briefly to map it out by sponsoring a workshop in May 1988 through the Congressional Research Service. The workshop provided a better forum for proponents than opponents, however, for it included ten participants from industry, three from government, five from universities, and one from an environmental lobbying group. Workshop organizers outlined the major issues in a lengthy report in January 1989 (U.S. Congress 1989), after which Congress voted to support five years of research and technology validation. Over 70 percent of this support is devoted to ozone-related research while research on airport noise and sonic boom divide up the rest.

By seeking to integrate new environmental considerations into conceptual design, the high speed civil transport program faces a problem that closely parallels the normative project in constructive technology assessment. At first glance, an obvious local solution to this problem is to integrate the agency for calculating environmental effects into CAD/CAM-based activities during design synthesis. In other words, simply optimize HSCT designs for minimum ozone damage, minimum noise, and minimum sonic boom, and then do some sensitivity studies to minimize all three at the same time. However, following this course is not likely to occur, for empowering design synthesis in such a way would significantly transform the identities of conceptual designers within NASA.

ACSYNT provides the exception that illustrates the problem. The current caretakers and spokesmen for ACSYNT at Ames Research Center have recently added a new analysis module that makes sonic boom calculations. From the perspective of aerodynamicists, a sonic boom is the product of a pressure disturbance caused by an aircraft flying faster than the speed of sound. When the aircraft cruises at supersonic speeds, the disturbance propagates behind the aircraft, intersecting with the ground to produce a 'footprint' within which people will experience a sonic boom. Aerodynamicists describe the pressure disturbance as an 'N' wave, which consists of a sharp increase in pressure when the initial shock from the aircraft's nose arrives, followed by a linear decrease to below normal pressure, then another shock from the rear of the aircraft that restores normal pressure. The geometric configuration of the aircraft plays a large role in determining the actual magnitude and shape of this pressure wave. For aircraft designers, it thus appears possible to identify an acceptable pressure wave-form and then design a geometric configuration to produce it.

It is significant that ACSYNT engineers have not attempted to add analysis modules for calculating ozone depletion or take-off noise. In the first place, doing so would appear to other researchers as a strategy for restructuring power relations within the NASA organization. Engineers at NASA's Langley Research Center gained primary responsibility for overseeing the research by having positioned themselves as advocates of HSCT for over a decade. But granting Langley engineers complete control over HSCT research would have empowered Langley Research Center at the expense of two other research centers, Ames and Lewis, and transformed the egalitarian relations that had stabilized among them.

Since Ames Research Center had long been positioned to do research on aerodynamics, some research groups at Ames acquired funds for research on sonic boom and on some general problems in atmospheric modeling. At the same time, ozone depletion and take-off noise are both linked to engine design, the technological arena within which the Lewis Research Center had been positioned but without making significant use of design synthesis. For synthesis experts at Ames, gaining formal approval to include engine design in ACSYNT would indicate that NASA headquarters had decided to empower Ames at the expense of both Lewis and Langley, allowing design synthesis to annex intellectual territory beyond aerodynamics, and reconceptualizing engine design according to methods used by aerodynamicists. In other words, including engine design in ACSYNT would be viewed as an attempt by Ames synthesis experts to impose their understanding of design problems on everyone else, which could position them in paradoxical ways. While such a move could reposition them into a position of power in HSCT research, it could also disempower them by suggesting they were not good NASA citizens. The outcome could be more likely to shorten their careers than win them new resources.

Furthermore, modeling ozone depletion and take-off noise in ACSYNT appeared less likely to produce results that relate geometric parameters to engineering analysis, and thus less likely to contribute meaningfully to conceptual design of the aircraft. Although atmospheric scientists have stabilized chemical descriptions of the reactions through which engine emissions deplete ozone, for engineers to translate these reactions into design criteria involves inserting the chemistry equations into highly simplified mathematical models of complex atmospheric systems. Judging the reliability of these models for design purposes is impossible without an accumulated body of experience. Take-off noise can be modeled more reliably, but the code that Lewis Research Center was using to model it suggested that the link between engine geometry and noise may be so complex that including calculations of convergence and optimization in ACSYNT would be too time-consuming.

Sonic boom presents the best-case scenario for CTA purposes, yet it also illustrates why using CAD/CAM to steer technology development also amounts to a political action in support of the mathematization of design. As the sonic boom module for ACSYNT stands at present, it does not participate in the optimization process. It cannot translate an acceptable pressure wave, even if such could be identified, into an optimal geometric configuration. Rather, after determining a potential geometry-based configuration based on other constraints, the user can only calculate the type of pressure wave that configuration would generate. According to one interviewee, integrating the sonic boom module into the optimization process would be extremely difficult because it means solving "one of the most complex optimization problems that hasn't been done yet — shape optimization."[7]

Figure I: Two Sonic Boom Waves of Acceptable Loudness

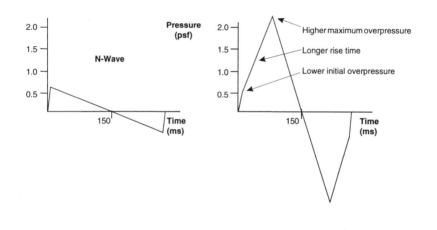

Source: Brown and Haglund (1988: 3)

The first difficulty in using the loudness of sonic booms as a design criterion lies external to the aircraft design process: how loud is too loud? After selecting some method of measuring noise from the more than ten alternatives available, design engineers must define a standard of acceptability. Langley researchers have been conducting 'human response tests' to develop proposed standards of acceptable loudness using dbA, or A-weighted decibels, which are calibrated according to the ear's changing sensitivities to different frequencies. A 1988 Boeing report used results from three such tests to identify 72 dBA as the likely highest acceptable loudness and its design goal (Brown and Haglund 1988:2). Researchers are acutely aware, however, that a new public controversy over the HSCT could affect significantly how different local groups might attribute acceptability. Four different variables in the pressure wave form contribute to its perceived loudness, three of which can be translated into design variables for the aircraft's geometric configuration. Different combinations of these variables can produce the same decibel level and be judged equally loud (see Figure 1). The 'maximum overpressure' is the highest level of extra pressure produced by the wave, which is linked to the gross weight of the aircraft. Keeping a sonic boom below 72 dBA by reducing the maximum overpressure alone would likely require it to be no higher than 1.3 psf (pounds per square foot). But the Boeing report argued that for any "large commercial transport it is unrealistic to look at anything below about 2.0 psf" (Brown and Haglund 1988:3). The Concorde produces an unacceptably high overpressure of 2.0-3.0 psf.

Since reducing gross weight is not an option, designers turn therefore to relations between the 'rise time', the amount of time it takes to reach maximum overpressure and 'initial overpressure', or the pressure caused by the aircraft's nose. By reducing the initial overpressure and lengthening the rise time, one can actually produce a wave with a high maximum overpressure that sounds like a much lower-pressure 'N' wave.

Although the desired pressure wave can be translated into a point design, it is very difficult to transform this point design into an optimized geometric description. The reason for this is that the geometric form of the aircraft affects the pressure wave in two ways. The first is through its volume: a slender aircraft produces less boom than a fat one. The second is by the distribution of 'lift area', or bottom surface area: an aircraft whose bottom surface area is evenly distributed from nose to tail produces less boom than one whose surface area is concentrated along a small portion of the fuselage. The desired pressure wave can be translated into a combination of volume plus lift area that can be used to identify a point design. However, optimizing a configuration would involve separating these two aerodynamic variables, correlating each with geometry variables describing the aircraft's components (wings, tail, fuselage, etc.), and then relating the two back together. This problem pushes the field of design optimization past its current limits. The Boeing report points to this problem in lamenting: "At this point the design process becomes difficult . . . because there are a great number of possibilities and there are very few design tools available to aid in this design process" (Brown and Haglund 1988:7).

In the absence of a new technological agent that could accept responsibility for design decisions (i.e., optimization), decision makers have adopted different strategies, each balancing choices to link who they are to who they seek to be. For example, Boeing design engineers offered a conservative solution: they simply selected a very conventional configuration and are perturbing it in small ways to reduce sonic boom. In this way, they can position themselves as leaders of a potentially successful project while posing a minimum financial risk to their company. NASA engineers are in some debate. Some endorse the industry strategy, thus positioning their organization as a research partner to industry and insuring its future stability. Others advocate a more radical concept, the oblique flying wing, which flies at an angle with the airflow rather than perpendicular to it, distributing lift evenly from the nose (one wing tip) to the tail (the other wing tip). Much more risky, this choice reproduces the dominant identity of NASA in earlier years as a producer of new ideas and technology through advanced scientific work.

Conclusion

The HSCT/ACSYNT example provides some clear insights into both the opportunities for steering technology through computer-aided design and the likely forms of resistance such efforts may encounter. Using CAD/CAM to achieve CTA objectives looks attractive because it can be applied at the earliest stages of the product design process. But even though the major features of a product may not yet be stabilized, the set and range of acceptable criteria for decision making typically are entrenched because design activities and design groups have become stabilized. Using CAD/CAM as a vehicle for introducing new agencies into design decision making does appear to allow the systematic introduction of some new criteria but only by restructuring local design activities and design groups. CAD/CAM technologies arrive not as a utopian technological fix but as waves of new, but structured, design activities.

It is important to keep in mind that CAD/CAM technologies possess the agency to affect design decision making most when they achieve reversible transcriptions between geometric models of objects and forms of engineering analysis. The first wave of CAD/CAM development, drafting automation, remakes two-dimensional drawing activities but not three dimensional design decision making. The third technology, solid modeling, fails to capture the activities of either design or manufacturing. Only in the second wave, 3D wireframe and surface modeling, is the technology achieving reversible transcriptions that restructure decision-making activities and permit the introduction of new design objectives.

Even wireframe and surface modeling, however, can only be applied to a subset of CTA objectives. Minimizing the environmental hazards of aircraft is a perfect case of using CAD/CAM to modulate decisions that affect the content of industrial products, i.e., of the objects modeled on the screen. But geometric modeling is not useful for formulating strategies to maximize levels of employment or quality of labor, objectives that pertain more to constructing manufacturing processes than producing designs.

The ACSYNT/sonic boom example shows that even though the agency in 3D CAD/CAM appears relevant to CTA product objectives, this strategy is likely to encounter at least two forms of resistance. In the first place, simply by introducing new design considerations, one injects an agency that has not been stabilized in design decision making. The design criteria for sonic boom have not stabilized, nor have those for ozone depletion and airport noise, for the controversist space for the HSCT has not yet been defined fully. Any attempt to integrate a specific criterion, such as Boeing's 72 dBA for sonic boom, becomes a politically controversial act of endorsement. The only solution for those representing new agency is to minimize the salience of this political act by considering alternative criteria acceptable to a range of participating groups.

Secondly, transcribing the agency of design activities into CAD/CAM technologies necessarily redistributes the agencies of existing human and nonhuman actors and, hence, the power relations among them. For example, not only does using CAD/CAM technologies empower design at the expense of manufacturing, it also redistributes the agencies of design engineers and draftsmen in ways that are highly variable and, hence, clear only at the local level. Thus, granting new agency to conceptual design through ACSYNT can have the effect of disempowering preliminary and detailed design. Members of the ACSYNT Institute are aware that some day they will no longer be able to characterize ACSYNT as simply a technology for conceptual design but will have to confront preliminary design directly.

Using 3D technologies for design synthesis and optimization is likely to prove most acceptable when applied to complex technological products, such as aircraft, automobiles, and ships. A wide variety of engineering disciplines inject their analyses into these product designs. A much more common occurrence, especially in smaller firms and less complex products, is the addition of information from only two or three different types of engineering analysis without any attempt at mathematically based optimization. A company might focus, for example, on fluid flow, heat transfer, or stress analysis. While the actual sequence of design steps might be open to change through CAD/CAM, the criteria for decision making have stabilized around these limited analyses and are less susceptible to change. In such cases, second-wave CAD/CAM is making some inroads through 'parametric design', i.e., correlating geometry with one parameter at a time. But then integrating CTA objectives requires convincing engineers to use entirely new parametric design programs. The close cultural linkage between CAD/CAM and national identity may keep this approach within the realm of possibility.

A major lesson of this case is that achieving any CTA objectives, whether through the agency of technology or not, requires highly localized strategies, and identifying such strategies requires a theory of acceptance. Using the concept of positional identity leads one to inquire: In what ways does the introduction of new agency redistribute the agencies of interacting participants, including the agency of constructive technology assessment? In other words, it attunes us to recognize the diverse agencies in technology and other non-human actors without assuming that such agency always includes a desire to expand power and control. It instructs us to investigate not only who actors *are* but also who they *seek to be* in order to identify strategies that do not shift them into unwanted positions. In short, the acceptance of CTA strategies need not take the form of acceptance of a CTA political identity such that everyone need have the same politics. Rather, the CTA movement can be as diverse as the range of identities that may find in CTA opportunities to achieve locally desirable positions.

Notes

1. I use single quotes to denote native cultural categories and double quotes to denote direct quotations and my own analytic concepts.

2. Interview with Timothy Bischof, Design Engineer Trainee, Boeing Helicopter, Inc., Blacksburg, Virginia, April 20, 1987.

3. For examples of reports detailing economic risks to the American nation, see National Academy of Engineering Task Force on Engineering Education (1980), National Commission on Excellence in Education (1983), Council on Competitiveness (1988), National Academy of Engineering (1985, 1987), Shapley and Roy (1985), Prestowitz (1988), and Dertouzos, Lester, and Solow (1989).

4. Over a three-year period, I conducted over 600 hours of participant observation research.

5. Statement made at closed members' meeting, ACSYNT Institute, Blacksburg, Virginia, May 3, 1991.

6. Confidential interview, NASA engineer, April 8, 1991.

7. Confidential interview, NASA engineer, April 8, 1991.

References

Badham, Richard. 1991. *Computers, Design, and Manufacturing: The Challenge.* Canberra: Australian Government Publishing Service.

Barlow, M. Andrew. 1990. "Benefits of Solid Modeling." *MicroCAD News* (May): 20-24.

Bijker, Wiebe E. 1987. "The Social Construction of Bakelite: Toward a Theory of Invention," in: Wiebe E. Bijker, Thomas P. Hughes and Trevor J. Pinch (eds.). *The Social Construction of Technological Systems: New Directions in the Sociology and History of Technology.* Cambridge, Mass.: The MIT Press, 159-190.

Boyles, Richard Q. 1968. "Aircraft Design Augmented by a Man-Computer Graphic System." *Journal of Aircraft* 5 (September-October): 486-497.

Brown, Jessica G. and George T. Haglund. 1988. "Sonic Boom Loudness Study and Airplane Configuration Development." AIAA-88-4467. AIAA/AHS/ASEE Aircraft Design, Systems & Operations, Atlanta, Georgia (September 7-9).

Callon, Michel. 1986. "The Sociology of an Actor Network: The Case of the Electric Vehicle," in: Michel Callon, John Law and Arie Rip. *Mapping the Dynamics of Science and Technology.* Basingstoke: Macmillan, 19-34.

Cambrosio, Alberto, and Camille Limoges. 1991. "Controversies as Governing Processes in Technology Assessment." *Technology Analysis & Strategic Management* 3: 377-396.

Council on Competitiveness. 1988. *Picking Up the Pace: The Commercial Challenge to American Innovation.* Washington, D.C.: Council on Competitiveness.

Dataquest Incorporated. 1990. *Forecasts* (Mechanical CAD/CAM). San Jose, California: Dataquest Incorporated.

Dent, Joseph B., W. George Devens, Frank F. Marvin, and Harold F. Trent. 1983. *Fundamentals of Engineering Graphics.* New York: Macmillan Publishing Co., Inc.

Dertouzos, Michael L., Richard I. Lester and Robert M. Solow. 1989. *Made in America: Regaining the Productive Edge.* Cambridge, Mass.: The MIT Press.

Downey, Gary Lee. 1992a. "Agency and Structure in Negotiating Knowledge," in: Mary Douglas and David Hull (eds.). *How Classification Works: Nelson Goodman among the Social Scientists.* Edinburgh: Edinburgh University Press, 66-98.

Downey, Gary Lee. 1992b. "CAD/CAM Saves the Nation?: Toward an Anthropology of Technology." *Knowledge and Society* 9: 143-168.

Downey, Gary Lee. 1992c. "Human Agency in CAD/CAM Technology." *Anthropology Today* 8 (October): 2-6.

Executive Office of the President. 1985. *National Aeronautical R&D Goals: Technology for America's Future*. Washington, D.C.: Office of Science and Technology Policy.

Groover, Mikell P., and Emory W. Zimmers Jr. 1984. *CAD/CAM: Computer-Aided Design and Manufacturing*. Englewood Cliffs, New Jersey: Prentice-Hall, Inc.

Hacker, Sally. 1990. *"Doing It the Hard Way": Investigations of Gender and Technology*. Boston: Unwin Hyman.

Latour, Bruno. 1987. *Science in Action*. Cambridge, Mass.: Harvard University Press.

Lynch, Michael. 1988. "The Externalized Retina: Selection and Mathematization in the Visual Documentation of Objects in the Life Sciences." *Human Studies* 11: 201-34.

Machover, Carl, and Robert E. Blauth. (eds.). 1980. *The CAD/CAM Handbook*. Bedford, Massachusetts: Computervision Corporation.

Meer, Frans-Bauke van der. 1983. *Organisatie als spel: Sociale simulatie als methode in onderzoek naar organiseren*. Enschede, De Boerderij, TH Twente.

National Academy of Engineering. 1985. *The Competitive Status of U.S. Industry – An Overview*. Washington, D.C.: National Academy Press.

National Academy of Engineering. 1987. *A Technology Agenda to Meet the Competitive Challenge*. Washington, D.C.: National Academy Press.

National Academy of Engineering Task Force on Engineering Education. 1980. *Issues in Engineering Education: A Framework for Analysis*. Washington, D.C.: National Academy of Engineering.

National Commission on Excellence in Education. 1983. *A Nation at Risk*. Washington, D.C.: U.S. Department of Education.

Noble, David F. 1978. "Social Choice in Machine Design: The Case of Automatically Controlled Machine Tools, and a Challenge for Labor." *Politics and Society* 8: 313-347.

Prestowitz, Clyde V. Jr. 1988. *Trading Places: How We Allowed Japan to Take the Lead*. New York: Basic Books, Inc.

Reinschmidt, Kenneth F. 1991. "Engineering in the Future: CAD." *CATIA Operators Exchange* (Winter): 4-9.

Rip, Arie, and Henk van den Belt. 1988. *Constructive Technology Assessment: Toward a Theory*. Enschede: University of Twente.

Rosaldo, Renato. 1989. *Culture and Truth: The Remaking of Social Analysis*. Boston: Beacon Press.

Schot, Johan. 1991. *Technology Dynamics: An Inventory of Policy Implications for Constructive Technology Assessment*. The Hague: Netherlands Organization for Technology Assessment (NOTA). Working document no. 45.

Shapley, Deborah, and Rustum Roy. 1985. *Lost at the Frontier: U.S. Science and Technology Policy Adrift*. Philadelphia, PA: ISI Press.

U.S. Congress. Senate. Committee on Commerce, Science, and Transportation. 1987. *NASA Authorization Act, 1988*. Report, 100th Congress, 1st Session, on S. 1164. Washington, D.C.: U.S. Government Printing Office.

U.S. Congress. Senate. Committee on Commerce, Science, and Transportation. 1989. *Commercial High Speed Aircraft Opportunities and Issues*. Washington, D.C.: U.S. Government Printing Office.

U.S. National Aeronautics and Space Administration. 1988. *Civil Aeronautics Technology Development and Validation Plan*. Office of Aeronautics and Space Technology. Washington, D.C.: National Aeronautics and Space Administration.

U.S. National Aeronautics and Space Administration. 1989. *High-Speed Research Program*. Office of Aeronautics and Space Technology. Washington, D.C.: National Aeronautics and Space Administration.

Vanderplaats, Garret N. 1976. "Automated Optimization Techniques for Aircraft Synthesis." AIAA 76-909. AIAA Aircraft Systems and Technology Meeting, Dallas, Texas (September 27-29).

Wampler, S.G., A. Myklebust, S. Jayaram and P. Gelhausen. 1988. "Improving Aircraft Conceptual Design: A PHIGS Interactive Graphics Interface for ACSYNT." AIAA-88-4481. AIAA/AHS/ASEE Aircraft Design, Systems and Operations, Atlanta, Georgia (September 7-9).

6 Risk Analysis and Rival Technical Trajectories: Consumer Safety in Bread and Butter

Fred Steward

This paper explores the assessment and choice of technological alternatives in food processing from the later part of the 19th century to the middle of the 20th century. Specifically, it focuses on the contrast between two technological trajectories: a chemical one based on additives and one relying on physical techniques. The use of chemical additives was criticised by scientific and public groups for their potential risks to consumer safety. Physical alternatives were promoted, with varied success, as a socially desirable alternative.

Two particular cases are investigated. Both concern staple foods where the production process was industrialised from traditional craft based techniques to large scale factory methods during the period. The industrialisation of flour milling was accompanied by new technologies for flour treatment to influence baking quality. Similarly, the industrialisation of the dairy industry was accompanied by new technologies for food preservation. The paper spotlights choices between chemical and physical trajectories in these two areas of flour treatment and dairy product preservation.

In both industries the chemical trajectory initially became dominant. Boron compounds were widely used in milk, butter, and cream from the 1870s. Chlorine related gases, particularly one called Agene, were used to treat flour from about 1920. Even after Agene was phased out after becoming the focus of health concerns, other chlorine related gases continued as the dominant technology for over 50 years into the 1970s in spite of physical alternatives. By contrast, after extensive use for 50 years, boron preservatives in dairy products were completely abandoned during the 1920s in favour of physical methods. In both cases the evidence of risk to consumers was suggestive rather than conclusive. Yet in one case the chemical trajectory retained its dominance, while in the other it was rejected in favour of a physical trajectory. The emergence of a chemical or physical trajectory was not therefore simply a product of risk assessment. It involved in both cases a process of social assessment and choice between competing technological trajectories. The study examines the role of industrial interests, technical opinion, and political influences to compare this process in the two cases.

The emergence of the technological trajectories discussed was associated with the decline of craft production and the industrialisation of UK food production during the 19th century. Two demographic changes — population increase and rapid urbanisation — had significant consequences for the technology of food production. These included a great increase in demand for food as well as a separation of the consumer from primary food production in the countryside. In the second half of the 19th century the growth in demand was paralleled by a massive increase in the importation of food from new overseas sources, inside and outside the Empire. Transportation of food became very important for this trade as well as for that between country and town within the UK.

Bread

Importation of wheat into Britain began with repeal of the Corn Laws in 1846, but its full impact was not really felt until the 1870s when the combination of poor harvests at home and cheap imports from abroad threw the traditional wheat-based agriculture into severe crisis (Fussell, 1958). This period was marked by the introduction of roller-milling, a major technological change in flour milling. First introduced in Hungary in 1840, this technology was rapidly adopted in the United States. The white flour produced by this process from the strong American wheats began to be extensively imported into the UK. Whiteness and strength were seen as desirable characteristics of baking quality compared with British flour. The British millers had stuck with the traditional stone-milling process, but such competition forced their changeover to roller-milling in the late 1870s (Morris, 1958). In 1878 the first roller-mill plant was established in Manchester by Henry Simon. The same year a new trade organisation was established for the milling industry, the National Association of British and Irish Millers (NABIM).

Within about 10 years a 'revolution' had occurred in the technological transformation to roller milling (Simon, 1893).

However, in spite of the efforts by UK millers, the imports of American wheat grain flour increased steadily. The quantity of imported flour continued at a high level until 1903, when imports declined significantly. The new UK milling industry survived the threats posed by the U.S. by establishing large mills in the port cities of London, Liverpool, Hull, and Bristol at the expense of the country mills inland. The industry became increasingly concentrated in a few giant companies. Three firms, Spillers Ltd., Rank Ltd., and the Cooperative Wholesale Society (CWS) accounted for nearly 40 percent of flour output by 1935 and upwards of 70 percent by 1944.

This formative experience of the threat of overseas millers led to an abiding preoccupation with the 'baking quality' of British milled flour. British flour millers concluded that their product needed the white colour and baking strength of American flour. From the 1880s onwards there was a concerted technical effort to find flour treatment processes that could produce such qualities in British milled flour. A leading consultant to the milling industry defined colour and strength as the two critical properties of flour that were of the highest importance to its commercial value (Jago 1911). A property of wheat protein, or gluten, strength was defined as: "a large relative yield of bread due to a high water absorbing capacity, the power of producing a large loaf, that of producing a bold loaf, a well-piled loaf." Influencing the strength of a given batch of flour was a key technical objective for the industry. At this time a physical trajectory was identified ("by simple mechanical attrition of the dough profound changes are made in the character of the gluten") as was a chemical trajectory ("additions to the flour are capable of effecting material improvements"). From early in the life of the new milling industry, then, flour treatment technology was practised and investigated.

Between 1890 and 1920 a number of chemicals, including ozone, nitrogen peroxide, and potassium persulphate, were used to bleach and improve flour. In 1921 J.C. Baker, a researcher at the American company of Wallace and Tiernan, discovered that the gas nitrogen trichloride could be used to treat flour as a combined bleacher and improver (Langrish 1972). Within 10 years after its introduction in 1923, nitrogen trichloride was used to treat 50 percent of British flour, rising to 80 percent after 10 more years. By the late 1940s over 95 percent of the flour used for baking bread in the UK was being treated with this chemical bleacher/improver. Nitrogen trichloride was given the proprietary name Agene because it reproduced the ageing effect obtained by letting flour stand.

When the Agene process was introduced in the 1920s there were two alternative flour treatments that did not involve the addition of chemical bleachers or improvers. One was natural ageing. Storage of flour for a period of time brought about bleaching and improving through oxidation on exposure to air. The transport of American flour across the Atlantic Ocean by ship permitted such natural ageing during the journey, which was one reason why the imported flours appeared whiter and stronger. The other was heat treatment. During the 1920s D. W. Kent-Jones introduced the KJ Continuous ThermoProcess which involved heat treatment of flour with "no chemicals whatsoever employed." It was advertised as allowing the miller to use a "weaker and cheaper wheat mixture" (Kent-Jones 1930). Several plants outside England used similar processes.

By the mid-1920s concern over the growing use of chemical flour treatment methods was evident. One government expert committee, the 1924-1927 Departmental Committee on the Treatment of Flour with Chemical Substances, assessed a variety of issues concerning the risk of and need for these chemicals. This committee adopted a cautious attitude towards the assessment of risks to human health. Its emphasis was on whether "in the light of present knowledge and the trend of research there is any ground for suspicion" rather than on demanding direct evidence of injury from the use of flour bleachers and improvers. The risks from the use of Agene were considered to arise from its reaction with the gluten. There was the possibility that products formed in the flour might actually "act injuriously" on the human organism. Although it acknowledged that the evidence was inconclusive the Committee pronounced its disapproval of Agene and concluded that its use was "undesirable" (Monro Committee 1927).

In response, the National Association of British and Irish Millers argued strongly in favour of chemical bleachers and improvers. In relation to improvers three commercial considerations were presented by the home flour millers: a greater uniformity of product, the ability to use any supply of sound wheat available, and greater utilisation of English wheat. The Departmental Committee doubted the validity of all of these claims, and other commercial considerations were identified such as the use of inferior wheats. The chemical trajectory was promoted by a powerful combination of the home millers trade association, NABIM, and the manufacturers of Agene, Wallace and Tiernan. The establishment of the Research Association of British Flour Millers at St. Albans in 1923 gave the millers an organised scientific base for the presentation of their case. Opposition to the use of chemicals was expressed by the National Association of Flour Importers and by the bakers through the National Association of Master Bakers, Confectioners and Caterers. Promotion of the physical trajectory was limited to one expert with a personal commercial interest in the process.

The Committee proposed that in view of the possible risks to the consumer it would be very desirable if "a purely mechanical or physical method" could supersede the use of chemical improvers. One obvious approach was "to store flour...so as to secure natural ageing." The recently patented heat treatment process, in spite of promising results was seen as an experimental option which required investigation and elucidation. The Committee did not refer at all to the technique of mechanical attrition of the dough which Jago had identified as an option many years before. It appears that the scientists on the committee were constructing their own arguments in favour of a physical trajectory but were not drawing on a coherent or well developed perspective from any external group of experts or section of the bread business.

The Committee did not pursue its own preference for the physical option with great conviction. Although it concluded that it would "like to see progress made" in the use of physical methods, it did not feel able to recommend "the complete elimination of the bleaching agents and improvers" then in use. This was certainly a desired long term goal but in the first instance the Committee confined itself to calling for a ban on the more objectionable chemicals notably Agene. In spite of its preference for the physical over the chemical trajectory, it had retreated to relative risk assessment between different chemicals.

It was apparent to the Committee that although NABIM represented 450 of the 650 millers in the country there were many millers who did not use chemical treatment. Most bakers and farmers were also against chemical methods. A deputation of MPs representing the National Farmers Union — which had not been able to give its view to the Committee — pressed the Department of Health during 1928 for action to control chemical bleachers/improvers. They were told that action was not to be taken because it might increase both the price of bread and the importation of flour (Hickinbotham 1938, 5 Dec.)

It was not until the late 1930s that the Ministry of Health undertook a review of the situation. Experts were again consulted as to the state the technology had reached. Agene had by then become the most widely used bleacher/improver. Heat treatment was commercially possible but more expensive. The millers claimed to have carried out experiments and produced workable methods, but heat treatment had not been developed in the UK "to the same extent as in a few continental countries."

A senior civil servant in the Ministry of Health was positive about the viability of both heat treatment and natural ageing as alternatives to chemicals. Regulation was seen as a way to stimulate the physical methods: "the total prohibition of chemical improvers might lead to an increased use of both alternative methods" (Hickinbotham 1938, 5 Dec.) However, subsequent discussions within the Department led to the view that the best course was to implement the recommendations of the 1924-27 Committee and permit a limited number of chemical methods as well as the physical ones.

There was particular concern that there would be strong pressure to allow Agene: "the Agene Combine is probably fairly powerful...[we may be]...unable to resist" (Hickinbotham 1939, 3 Apr). It was decided to test the possible opposition by making an unofficial approach to the Millers Association only, leaving any discussions with the National Farmers Union for a few weeks (Hickinbotham 1939, 6 Apr.) In response the Millers Association argued strongly for Agene stating that "such a treatment can have no injurious effect" and its widespread use had produced "no evidence of harm" (NABIM 1939, 12 May). A deputation from the Association said that prohibition of Agene would be disastrous and would "once again put the home miller at a disadvantage...the use of Agene has made it possible for the home miller to compete successfully with the foreign miller in producing a flour of good baking quality." After this predictable response, the Department gave way on the issue (Hickinbotham 1939, 24 May.) Following another meeting with the millers, draft regulations were drawn up permitting the use of Agene — despite the bakers' and flour importers' continued opposition to the chemical.

In 1946 the issue was dramatically brought to public attention by scientific studies implicating Agene-treated flour in a nervous disorder of dogs, canine hysteria, observed in the UK since 1924. Mellanby (1946) had fed dogs diets containing Agene-treated flour and untreated flour, finding only the former produced the abnormality. The flour treatment process used at the time by 95 percent of the home millers was clearly shown to have harmful consequences for animals, which naturally confirmed suspicions about possible risks for human consumers. Subsequently the toxic agent formed by the agene treatment of flour was positively identified as an amino-acid derivative, methionine sulphoximine. These results confirmed the 1927 Departmental Committee's suspicion that nitrogen trichloride could react with a constituent of flour to produce a toxic material.

The Agene crisis renewed interest in physical alternatives. The Ministry of Food agreed that the general question of flour improvement should be investigated and that "the scope of the researches should cover physical processes of improvement as well as chemical treatments." The strategy envisaged in early 1947 was that the Medical Research Council should pursue the research on Agene and human health while the Flour Millers Research Association investigated alternative improvers. However, this did not work out in practice. In April 1948 a conference between officials of the concerned ministries and NABIM was chaired by the Chief Medical Officer, Wilson Jameson, who attempted to get agreement to the desirability to "discontinue the use of chemical improvers, if possible, even if they were not harmful." The millers did not think there was an alternative: "a great deal of work had been done to find a practicable substitute, but without result yet." The NABIM scientists referred to two techniques, heat treatment and use of oxygen under pressure, but concluded that it was a slow job. While Jameson hoped that the search for alternative methods would be pursued, NABIM argued that it was simply not practicable at that time. The conclusion was to continue the use of Agene, and to search for alternatives, including chemicals.

Following this conference the Ministry of Health established an informal scientific committee that included scientific representatives of the milling industry. It first met in July 1948 and rather than pursuing physical alternatives agreed to concentrate on finding a chemical alternative to Agene. Chlorine dioxide was considered to be the main contender; there simply was no further discussion of physical methods. By September 1949 it had been agreed that the milling industry would shift over to the use of chlorine dioxide, a product also supplied by the Agene manufacturers Wallace and Tiernan. It was to be a voluntary switch without legal enforcement. A public statement of this policy was made in January 1950. Again, following an interest in a broader assessment of technological alternatives, the government and its advisers had retreated to a narrow risk assessment perspective.

Unfortunately, little work had been done on chlorine dioxide and in November 1950 Moran found that like Agene it also interacted with amino-acids in the gluten to form new substances. This finding threw the agreed-on policy into confusion, and a series of meetings with the millers' scientists was hurriedly called. At these meetings the possibility of using ascorbic acid (Vitamin C) as an improver was raised. Investigations as to its availability and cost were made with a promising outlook. By April 1951 it had been agreed that the changeover to chlorine dioxide was to be "temporarily suspended" because they were considering another and possibly preferable alternative. The use of Agene continued amid growing political pressure from the Housewives League, scientists such as Mellanby, and Members of Parliament. Ironically, the policy of continuing with chemical improvers resulted in greater pressure to remove chemicals altogether.

The continuing crisis over the use of chemical improvers reinvigorated interest in physical methods. The initiative this time came from within the flour milling industry itself. C.A. Loombe was a senior figure within one of the big three flour millers, Joseph Rank Ltd., as well as within NABIM. In May 1951 at another of the Ministry of Health/NABIM conferences he announced that his company had developed a new physical method for bleaching/improving. This so-called aeration process went back to the principle of mechanical dough development during the baking process as a way of achieving the necessary strength for bread making. The aeration process used a high-speed batter mixing machine, and had been patented by Rank in September 1950 as "An Improved Method of Making Bread with Untreated and Unbleached Flour." "No chemical improvers whatever" were added "at any stage in the process." The May 1951 conference examined loaves made by the two methods and declared that the loaf produced by the aeration process was "superior in appearance and texture." Closer examination of the process showed that while its running costs would be low it would require extensive capital investment by the bakers rather than the millers. By July 1951 the Scientific Adviser's Division of the Ministry of Food felt that "if the aeration process justifies the claims it would provide the ideal method, providing that simultaneous with its adoption we abandon or prohibit the use of all other improvers and bakers additives." Such a course would require consultation with and agreement by the representatives of the baking industry.

In August 1951 a meeting involving the bakers representatives was held for the first time. The baking industry had also established a Research Association, at Chorleywood in 1946, and was now in a stronger position to enter the technical debate. J.B.M. Coppock from the British Baking Industries Research Association maintained that the aeration process would not work on all flours and that more research work was necessary. Loombe said the millers were "ready and anxious to abandon all chemicals." By October Coppock thought one or two years would be needed before a definite decision could be made. The whole scientific staff of the bakery trade were said to be carrying out experimental work on the aeration process.

In December 1951 a critical point was reached. The Ministry of Health's informal committee concluded that a public statement should be made that the milling and baking industries were anxious in principle to abandon the use of chemical improvers of all kinds, that investigations were underway into promising alternatives "notably the use of a physical process such as aeration," and that in the meantime the use of gaseous improvers would be discontinued. However, the senior civil service view was to avoid any public statement. The opportunity to ban chemical agents was missed.

Evidence that the aeration process was a viable option mounted during 1953 and 1954. A number of bakers including Lyons, Beattie and Hill started to use it. By August 1953 agreement had been obtained from Joseph Rank that the patent would be made available royalty free to any user — provided that the government prohibited the use of any chemical bleacher/improver and recommended the use of the aeration process. Meanwhile, in the middle of 1952 NABIM had informed its members that a changeover to chlorine dioxide was in order. In 1954 a Medical Research Council investigation into the toxicity of chlorine dioxide had not found any definite evidence of harmfulness. It was an easier political option simply to confirm the position stated in 1950. In December 1954, the Government announced that "effect should now be given to the decision taken in 1950 to discontinue the use of Agene." By the end of 1955, the use of Agene drew to a close but it was simply replaced by chlorine dioxide. Chemical flour treatment continued as the main technological trajectory. The physical alternative continued to be developed and used, but in conjunction with the use of chemical improvers rather than as an alternative.

In 1960 the Preservatives Sub Committee of the Food Standards Committee reviewed the position on flour improvers. It defined the physical technologies as natural ageing, heat treatment, and aeration. The first two were considered inferior to chemical methods while aeration was said to have been "tried a few years ago by one or two firms who reported that it gave bread which compared favourably both as regards colour and texture with that made from chemically treated flour." Its drawback was the large initial capital outlay but it was worthy of further investigation. The use of a chemical improver was considered a "technological necessity." As late as 1974 80 percent of bread flour in the UK was still treated with chlorine dioxide. The chemical trajectory continued to dominate after more than 50 years.

The continuation of the chemical trajectory in the face of extensive public opposition and a major crisis over the safety of Agene is remarkable. The conception of risk assessment versus choice of alternative trajectories was clearly articulated by the 1927 Departmental Committee. Within the Ministry of Health civil servants believed that technology forcing of the physical trajectory could be achieved by prohibition of chemical bleachers and improvers. Yet no decisive steps were ever taken to this end.

It is apparent that one section of the bread industry, the home millers, had become very strongly organised early on. The flour millers established a strong technological basis to their industry in which cereal chemistry occupied the centre stage. Although only one section of an industry, in which farmers and bakers were also important, the flour millers exercised a hegemonic role by exploiting a notion of technological progressiveness and the "national" interest. Due to concentration in their industry, the flour millers were able to establish a representative body covering the majority of the baking industry. Their commitment to the use of chemical bleachers and improvers was established early on and consistently and forcefully maintained.

By contrast the promotion of the physical trajectory was never effectively taken up by any institutionally significant industrial or technical group. Support for physical methods was mainly expressed by an expert committee and by administrative staff within the Ministry of Health. The absence of any external constituency from which to reinforce the technology assessment perspective — that is, focusing on a choice between the two rival technical trajectories — meant that the Ministry of Health was constantly pressured by the highly organised milling interest to retreat to a comparative risk assessment between chemical flour treatment agents.

Even at the height of interest in the physical trajectory, when the Agene crisis had got out of control, the forces promoting it were unable to consolidate its position. The problem of dealing with specific company interests and industrial property issues emerged as serious problems when a government department tried to take the leading role in the promotion of the physical aeration process. Supporters of a physical alternative also were obliged to demonstrate its technical feasibility to skeptical and sometimes antagonistic technical and industrial opinion leaders. The failure to use regulatory prohibition of chemical agents as a forcing device left physical methods without sufficient leverage.

Broader political factors also played a role. During the war the flour millers had established intimate links within the ministry that had administered the bread and flour supply. After the war, the flour milling industry's technical experts were sometimes fulfilling dual roles as representatives of the milling industry lobbying government and also as "independent" advisers to government. The government's unwillingness to be open about the Agene crisis and its suspicion of public pressure groups meant that it was unable to bring in any countervailing force to the millers. Instead government perpetuated a special status for the millers within the consultation process, perpetuating the technological trajectory to which its industrial association, NABIM, was committed.

Butter

In response to intense competitive pressures on its traditional focus in wheat, British agriculture reoriented itself toward dairy products. In 1876, the first annual Dairy Show was held in London and the British Dairy Farmers Association was founded. Yet the international market also began to affect this sector as well, with increased attention to dairy goods by the new world producers. Imports concentrated primarily on butter since overseas producers could not compete in the market for fresh liquid milk (Fussell 1958). As a consequence the UK dairy farms became very strongly oriented to expanding production of liquid milk and distributing it to the industrial towns. This trend was boosted by the cattle plague (rinderpest) of 1865 which wiped out the herds of cows kept in urban centres and "compelled dairymen of London and other large towns to seek milk all over the country" (Morris 1958). The railways provided a convenient vehicle. The records of the Great Western Railway showed a spectacular 15-fold increase in the quantity of milk carried in 1866 compared with 1865. This had increased 200 times by 1900. From the 1860s onwards new retail companies, led by George Barham's Express Dairy Company, developed in London (Jenkins 1970).

The dairy industry during the period 1870-1920 was a fragmented one. The emergence of the new wholesale and retail companies added new sectors to the continuing farm-based producers. All of these sectors contained small and large companies. A tripartite structure existed throughout this period with competing wholesalers buying from farmers and selling to retailers. This was accompanied by a large number of producer-retailers, yielding a pattern of small and large firms linked by imperfect competition. The absence of any developed representative organisation for the industry led to the continued expression of a diversity of trade interests.

It was not until 1928 that a national and broadly representative organisation was established with the National Dairymen's Association. This followed the emergence in the 1920s of large firms that amalgamated the wholesale and retail roles, including United Dairies, Express Dairies, and Cow & Gate (Jenkins 1970). The twin phenomena of the steamship importation of dairy products, such as butter from overseas, and the expansion of the railway-based milk trade within the UK both posed a similar technical problem. Since dairy products were perishable goods, the new distances and times involved in delivering them to the final consumer posed a threat to their freshness and safety. Sourness of milk, rancidity of butter, and general risks from bacterial contamination were the most pressing problems. In the late 19th century the dairy trade experimented with a range of new technologies based on a variety of technical principles, including refrigeration, heat treatment, packaging, dehydration and chemical preservatives.

Dairy products began to be chemically preserved in the 1870s. Boric acid and other boron compounds had been proposed by M. Dumas before the French Institute in 1872 as preservatives for food, and they were patented two years later. The microbicidal action of these substances was weak, however, and the problem of applying antiseptics to foods was how to have sufficient action without being unpleasant or harmful to the consumer. In this respect, boron compounds had ideal properties ("almost tasteless in dilute solution with an absence of odour") (Rideal 1903: 139, 274). Boron compounds became widely used for the preservation of dairy products by the early 1880s. A survey undertaken by the Government Laboratory in 1900 (Maxwell Committee 1901) found 34 different kinds of boron preservative in use, mainly mixtures of borax and boracic acid. The use of chemical preservatives in dairy products had become an established technology. The use of chemical preservatives appears to have been greatest in the urban market for milk and cream (especially in London) and in the nationwide market for butter imported from overseas.

A number of physical techniques were also developed during this period. One was refrigeration, whose development had consequences far beyond the dairy industry. It had long been known that dairy products would last better in cold conditions, but machines for creating low temperatures were unavailable until the 1850s (Morris 1958). The growth of long-distance transport of milk was accompanied by the introduction of techniques to chill it before the journey. In the 1860s George Barham of Express Dairies modified a capillary cooler used by a London brewery and applied it to his business (Jenkins 1970). Refrigerated holds made it possible to export an improved and uniform butter from Australia and New Zealand after 1890 (Fussell 1958).

In 1864-65 Louis Pasteur developed a practical method of preventing abnormal fermentations in wine by heating it to 122-140°F (50-60°C). Such heat treatment, soon known universally as pasteurisation, destroyed undesirable microorganisms. Milk was first commercially pasteurised in Germany in 1880, somewhat later in Denmark (Society of Dairy Technology 1966). The first city pasteurisation plant was established by Nathan Strauss in New York in 1893. In the UK heat treatment was established somewhat later. By 1910 only a few firms in London had adopted either pasteurisation or sterilisation, by 1921 around half of London's milk was pasteurised, and by 1925 "practically the whole" of milk sold in London was pasteurised (Whetham 1976; Jenkins 1970: 40).

Dehydration was another technique. Drying as a method of food preservation had been known since ancient times. New techniques of scientifically controlled dehydration began to be applied to dairy products in the second half of the 19th century. The main problem was the dramatic impact on appearance and taste of the products, but for milk some cheap and effective technologies were developed. Condensed milk, for example, was first manufactured by Gail Borden in New York state in 1860. Powdered milk was first patented in England in 1855 by F. S. Grimwade. Large scale production did not start until the end of the century (Morris 1958).

Packaging and hygiene relied on the exclusion of bacteria after the application of techniques such as heat treatment or dehydration. Although George Barham of Express Dairies first bottled milk in 1884 with a wired-on cap, until the 1920s pasteurised milk was commonly delivered in the traditional manner, dipped from a churn into jugs provided by the buyers. Mechanical systems for bottling and sealing were introduced by the large companies in the 1920s. That decade saw an integration of the technology of pasteurisation, refrigeration, and bottling accompanied by the amalgamation of wholesalers and retailers (Whetham 1976).

All of these technological options remain active within the dairy industry today — with the exception of chemical preservatives. This technology, although well established, was eliminated during the first quarter of the twentieth century. The Public Health (Milk and Cream) Regulations of 1912 prohibited the use of any preservatives in milk (or cream containing less than 35 percent milk fat) and the Public Health (Preservatives etc. in Food) Regulations of 1925 prohibited the use of any chemical preservative in cream and butter. Most recent historical accounts wrongly suggest that the abandonment of the chemical trajectory was unproblematic and straightforward. "Very properly most of these preservatives are now excluded from foods" (Tilbury 1980).

In reality, the use of chemical preservatives in dairy products was not a marginal, short-lived phenomenon but a mainstream technology used by the UK dairy industry for 50 years. Not surprisingly its abandonment was a protracted and controversial process. It was paralleled by the emergence of pasteurisation as the dominant technology in the dairy industry. The ascendancy of this physical trajectory was neither obvious nor without conflict.

The rapid growth of chemical preservatives in the dairy trade had aroused a great deal of controversy. New scientific professionals specialising in public health, public analysts and medical officers of health especially, tended to be critical of the chemical option because it was seen as a technology that posed possible risks to the consumer. The assessment of this risk and its implications for choice between alternative technological trajectories was, however, a complex and conflict-ridden matter. One point of debate was that the chemicals themselves were harmful to health; another was that their use could mask unhygienic practices in the milk trade and even disguise bacterial contamination in the food itself. The assessment of the toxicity of boron preservatives soon ran into the difficulty of demonstrating chronic versus acute effects and the significance of low versus high doses. The first suggestion that borax when taken in small and repeated doses through the consumption of food could produce toxic effects was made by Le Bon in 1879. Eleven years later, at a meeting of the Society of Public Analysts, Hehner argued at a more general level that it was "inconceivable" that preservatives which vitally affected the protoplasm of bacteria would at the same time be utterly inert upon human protoplasm (Hehner 1890). The suggestion was that because of their very mode of action preservatives should be regarded with suspicion.

The Local Government Board of 1891 differed sharply from this approach. The quantity of preservative used was given much greater emphasis and proof of injuriousness was demanded before prohibition could be contemplated. *The Lancet* established a Special Sanitary Commission on the Use of Antiseptics in Food. After considering the opinion of a number of medical experts on the question of injuriousness of preservatives, the Commission concluded only that there was "considerable diversity of opinion" on the hazards to health posed by the use of preservatives in food (Sanitary Commission 1897).

Another issue was the indirect consequences of preservatives. As early as 1882 it was argued by a Public Analyst that "all additions to milk in the form of antiseptics should be looked upon with disfavour, for by their use cleanliness in the dairy would not be such an essential as it is now" (Wynter-Blyth 1882). Other scientists also suggested that the use of preservatives to prevent milk turning might encourage the use of "dirty rooms and dirty utensils" (Dupré 1884). This suggestion touched on the very sensitive issue of hygiene in the dairy trade, particularly regarding the provision of milk in the large towns. As with the problem of direct toxicity of preservatives it was a very difficult allegation to prove or disprove, though there was some striking evidence. During the summer months there tended to be a higher level of infant mortality and prevalence of so-called summer diarrhoea, which became very marked during hot weather. There was a particularly bad epidemic in the hot years of the late 1890s, as well as a marked peak in the drought of 1911. The total infant mortality in London during July-September of that year was more than double that of the equivalent period in the preceding and following years. Infected milk was considered to be an "obvious" factor (Whetham 1976).

The promoters of chemical preservatives in dairy products included a section of the technical community drawn principally from medicine and from industrial chemistry. In Britain the stature of Joseph Lister, who had pioneered the development of antiseptics during 1865-68, led to a generally favourable attitude to the wider application of this technology, including the preservation of food. It has been commented that "food manufacturers of that time were perhaps more impressed by the work of Lister on antiseptics than they were by the studies of Pasteur" (Tilbury 1980). Lister himself had lectured to the Pathological Society on the preservation of milk in 1878 (Jenkins 1970). The Lister Institute played an active role in the development of boron preservatives for use in dairy products in the 1880s and 1890s. From the 1880s through to the 1920s, then, there was a coalition of forces — sections of the technical community and of the dairy trade as well as the suppliers of the preservatives themselves — actively promoting the chemical trajectory in dairy food preservation.

Promoters of the alternative path of refrigeration and pasteurisation were also drawn from both the technical community and the dairy trade. As we have seen, it was initially the new public health professions of public analysts and medical officers of health who led the argument for the choice of physical over chemical technologies. In the late 19th century their emphasis was upon hygiene, rapid transport, and refrigeration as the trajectory requiring support.

The response of some medical experts and government officials attempted to confine the debate to one of risk assessment within the chemical trajectory alone. On the other hand the public health professionals like Hehner and Cassal were very conscious of the limits of such an outlook and the likelihood that debate would become bogged down in inconclusive arguments about accurately measuring such risks. As early as the 1880s they had begun to pose the issue in terms of choice between the different technological trajectories available for preserving dairy products. In 1887, they acknowledged the uncertainty of evidence concerning the safety of boric acid, but stressed that there was an alternative technology (refrigeration) available which would avoid the doubt attached to the use of chemical preservatives (Cassal 1887). Three years later, Hehner argued similarly that since there were alternative techniques of preservation such as the exclusion of germs or refrigeration, the addition of chemical preservatives should be avoided: "If preservation could not be effected without the addition of some foreign material, the benefit to mankind of preventing good food substances from decomposition would doubtless be greater than the slight physiological evil effect of the antiseptic itself." However, since there were alternatives, then chemical preservatives should be prohibited (Hehner 1890).

These two leading critics of chemical preservatives clearly perceived that the fundamental issue at stake was the selection of alternative technological trajectories rather than simply the condemnation of chemical methods. Ultimately this same recognition by the regulatory authorities would lead to the complete abandonment of the chemical option for dairy products in favour of pasteurisation and refrigeration.

The official government inquiry in 1901 did not reach a firm conclusion on the matter of risk. The 1901 Departmental Committee concluded that there was evidence pointing to an injurious effect of boron-treated milk upon the health of very young children. It also agreed that an objection to the use of preservatives in milk was that "it may remain sweet to taste and smell and yet have incorporated disease-germs of various kinds."

The industrial promoters of chemical preservatives included the manufacturers themselves. The Boron Syndicate was active in promoting the boron preservatives, for example by circulating information to all doctors throughout the country in 1900. Within the dairy industry at the turn of the century there was support for preservatives across the different sections of the industry: the producer, the wholesaler and the retailer. While there probably was a greater propensity for their use by the small retailers who have no cold storage larger wholesalers and retailers used them as well. (Maxwell Committee 1901)

Until the turn of the century the main element of the physical trajectory in the UK was refrigeration. Some of the new retailers of milk were enthusiastic about the new refrigeration technology and encouraged its adoption. George Barham, for example, founder of Express Dairies, delivered a paper to the Food Committee of the Royal Society of Arts on new refrigeration methods (Jenkins 1970). By 1900 another of the large London dairy companies, the Aylesbury Dairy Company, was keen to declare that it could rely on refrigeration and that it used "no preservatives whatever." Similarly a number of farmers, represented by the Royal Agricultural Society, were proud of their ability to send milk to London from as far away as mid-Staffordshire by ensuring that it was carefully strained and cooled by means of water (Maxwell Committee 1901). The British Dairy Farmers Association and the annual Dairy Show, both established in 1876, were two important institutions for demonstrating new methods and appliances and spreading the latest information (Morris 1958).

The 1901 Departmental Committee accepted the need for a preserving agent for perishable food due to the urbanisation of the population and the importation of food. The total prohibition of preserving methods was rejected since it would cause "serious results to the public health." There was an awareness that the preservation function could be achieved through physical technologies, i.e. the agent need "not necessarily [be] chemical." Rather than this awareness enabling a more general choice between two different technological trajectories it was instead seen as merely allowing the control of specific applications of preservatives. The choice of chemical preservatives over physical methods was seen as one essentially for the trade to make. The only issue for the Committee was whether any of the substances used were "likely to be so harmful to the consumer as to call for their prohibition, limitation, or declaration" (Maxwell Committee 1901, paragraphs 68-70). In other words, the Committee perceived the issue as simply one of risk assessment.

For butter the conclusion was that "there exists no sufficient reason for interfering to prevent the use of boron preservatives" by the trade. A level of 0.5 percent boric acid was defined as the technically necessary amount "sufficient for the purpose of preserving butter." Milk was treated differently, for it was seen as a special case for a combination of reasons: it formed a very large proportion of the daily food of the public, it was important to the nutrition of infants and young children, preventing the addition of excessive amounts through successive treatments was difficult, and chemical preservatives masked the effects of unhygienic practices. The central concern was the level of possible consumption ("the large quantity which may be taken into the system of the consumer") and therefore the higher level of possible risk.

The Committee considered the practicability of prohibiting preservatives in milk and the use of alternative preservation methods. It had "no doubt whatever" about the feasibility of supplying large towns with fresh milk without preservatives. Refrigeration was clearly seen as the practical alternative used by large numbers of dairy businesses in Britain. That chemical methods were used by only half of the London dairy trade and that many producers and wholesalers did not use preservatives at all demonstrated the current practicability of alternatives. In spite of the committee's cautious approach its recommendations were not given legal force for more than a decade. Eventually, in 1912, preservatives were banned completely from milk. It seems likely that the political catalyst was public concern over the safety of milk with the rise of infant mortality in the summer of 1911. Boron preservatives continued to be permitted in butter.

Government and its advisors in the first years of the 20th century had not fully understood nor accepted the argument for favouring the physical trajectory over the chemical. The prohibition of chemical preservatives along with the promotion of physical methods had been passed over in favour of a focus on regulation in terms of limits to levels, declarations of presence, and control of specific applications of chemical preservatives. The prohibition-and-promotion perspective had failed for two reasons. First was a division within the technical community, partly on disciplinary grounds. The medical establishment was suspicious of the new public health professions. As well, biological science had not matured sufficiently to deal with problems spanning chemistry and biology. Different disciplinary groups, including chemists, medical specialists, and physiologists, vied for dominance in this field. The second reason was the diversity of business interests shaped by a mixture of economic and structural factors, along with different views on the public desirability of the competing preservation technologies. These same factors also suggest why chemical preservation failed to become hegemonic. An uneasy coexistence of the two trajectories was the result.

From early in the new century pasteurisation began to be promoted increasingly vigorously. Having had his daughter die of bovine TB, Nathan Strauss, who established the first American pasteurisation plant, became a passionate propagandist for heat treatment, which he advocated at many meetings in Britain (Jenkins 1970: 40) By the second decade of the century two far-reaching clean milk campaigns were launched to promote the benefits of hygiene, refrigeration, and pasteurisation. The National Institute for Research in Dairying played an active role from 1912 onwards under its head, the bacteriologist, Stenhouse Williams. The National Clean Milk Society was founded in 1915 by agricultural administrator, Wilfred Buckley, who in 1923 convened a National Conference on Pasteurisation, at the Guildhall, London (Jenkins 1970). Still, there was not a complete consensus on the merits of pasteurisation. Some of the criticisms of it were similar to those directed at chemical preservatives, i.e. that it would disguise unhygienic practices. "Pure milk from healthy cows, properly cooled before transport, bottled and promptly delivered was the objective of many reformers" (Whetham 1976).

A number of the dairy companies began to adopt pasteurisation. The Cooperative Movement became increasingly involved with dairy products, and many of its local coops established pasteurisation plants. In the early 1920s the largest retailers and wholesalers, including Express Dairies and United Dairies, built major new plants that integrated pasteurisation, refrigeration, and bottling. United Dairies, then the largest combined wholesaler/retailer in the country, took a "strong lead" in pasteurising. The firm even established an engineering subsidiary, United Dairies Equipment Company, which became the leading national manufacturer of heavy duty bottling equipment and industrial refrigeration plant. These developments represented a decisive commercial recognition that the future lay with this physical trajectory.

By 1923 the assessment of risks and technologies had changed decisively. The new Departmental Committee now accepted as its starting point that physical technologies were in principle preferable to chemical preservatives. It was "self-evident" that "it is undesirable to add to food any material not of the nature of food" because of the possibility of both direct and indirect risks to the consumer. The principal question it asked was how far can preservatives be dispensed with? This approach was in sharp contrast to that of the 1901 Committee, which had asked rhetorically "are any of the substances...used for preserving foods...likely to be so harmful to the health of the consumer as to call for... [regulation]?" The 1923 Committee was now confident that for many foods "trade methods could be adjusted...so that preservatives could be eliminated." It argued that the presumed necessity for their use in Britain was largely due to a shortage of cool storage and transport. This recognised that the earlier failure to exercise a decisive choice over the trajectory of preservation technology had inhibited the adoption of refrigeration. If preservatives were controlled then this would create a pressing demand for the extension of such facilities.

The 1923 Departmental Committee considered that the vast majority of experimental work on the risks of boron preservatives was of "not much value" in deciding whether small quantities over a long period of time were harmful or not. The main experimental evidence that the

Committee regarded as significant showed that "boric acid was excreted more slowly than it was absorbed, so that it accumulated in the system." The Committee's view was a cautious, rather than a definitive, negative and it classified the boron preservatives in the second of three groups, ordered according to their relative undesirability. In 1925 regulations were passed that completely prohibited the use of chemical preservatives in all dairy products. In spite of the wide concern about boron preservatives and the extensive investigations, at no point were there authoritative conclusions as to their harmfulness. Their abandonment was not therefore a simple consequence of their toxicity.

In sum, then, the government policy developed in the mid-1920s was clear support for physical preservation and prohibition of chemical preservatives across the whole of the dairy industry. The policy arose because the idea of exercising such a choice had been accepted as a legitimate social decision, and the pattern of technological change in the industry made such a choice feasible. These two factors can each be related to underlying patterns of change. The limitations of risk assessment were more clearly recognised with the higher status of the public health professions and the maturing of biological science. Otto Hehner, the Public Analyst, was a member of the 1923 committee, as was Frederick Gowland Hopkins, the eminent biochemist. During the 1920s the structure of the dairy industry had become more integrated and concentrated, facilitating the application of refrigeration and pasteurisation in bulk.

Ironically, the very diversity of technical opinion and business organisation both facilitated the criticisms of the chemical trajectory and permitted its persistence for such a long period. Even while it was undergoing industrial concentration in the 1920s, the dairy industry remained very fragmented as to its representative organisations and opinions on the desirability of preservatives. Faced with a decisive attack on the chemical trajectory its defenders appeared as only a sectional interest rather than one with broad-based industrial or technical legitimacy. Therefore although an attempt to resist the new regulations was made through the government's Milk Advisory Committee it was not a serious challenge and was readily marginalised.

The presence of a strong external constituency articulating the concept of technology choice and promoting the physical trajectory provided the necessary context for government to use its legal powers to prohibit chemical preservatives. The culture of public health and technological choice, promoted by the public health professionals, influenced the development of feasible physical technologies by a range of industrial interests. Government was therefore able to use the regulatory instrument to force technological change.

Comparisons Between Bread and Butter

Both of the cases cover a period of over 50 years from the first introduction of a chemical process until its regulation by government. Bleachers and improvers were first introduced at the turn of the century, and the Bread and Flour Regulations were passed in 1963. Boron preservatives were introduced in the early 1870s, and the Public Health (Preservatives in Food) Regulations were passed in 1925. Both chemical technologies were criticised for possible consumer risks, and physical alternatives were promoted following assessments by expert advisory committees. The use of both flour treatment agents and dairy preservatives provoked public campaigns against their use. In both cases the evidence of human risk was uncertain, and the physical alternatives were shown to be commercially viable. Yet in one case the chemical trajectory prevailed; in the other it was replaced by the physical trajectory.

The key differences between the two cases revolve around the organisation of industrial interests and the expression of technical opinion. The ability of government to promote the development and adoption of 'consumer-safe' technologies was inhibited in the case of bread by the dominance of a highly organised and technically sophisticated industrial interest. It was facilitated in the case of butter by the prominence of a professional public health constituency with support for the physical alternatives amongst sections of the industry. These differences seem to explain the fate of the technology assessments undertaken by government's advisory committees.

References

Amos, A.J. 1952. "Flour and Bread Improvers and Flour Bleachers." *Chemistry and Industry* (10 November): 946.

Bentley, H.R. 1950. "Action of Nitrogen Trichloride on Certain Proteins." *Proceedings of the Royal Society* (B137): 403.

Cassal, C.E. 1887. *The Lancet* 2: 392.

Departmental Committee. 1924. *Final Report on the Use of Preservatives and Colouring Matters in Food.* London: HMSO.

Dodds, S.C. 1960. "Preservatives Sub-Committee Report on Flour Improvers," in: R. Groves (Chairman) *Food Standards Committee Report on Bread and Flour.* London: HMSO.

Dupré, A. 1884. *The Analyst* 9: 133.

Fussell, G.E. 1958. "Growth of Food Production," in: C. Singer (ed.). *A History of Technology: Volume V: The Late 19th Century.* Oxford: Clarendon.

Hamill, J.M. 1911. *Local Government Board: Report on the Bleaching of Flour and the Addition of So-called 'Improvers' to Flour.* London: HMSO; Hansard, 1927: 348-9, 2049-50.

Hehner, O. 1890. *The Analyst* 15: 221.

Hickinbotham, A.E. 1938 and 1939. *Bread and Flour Regulations.* Public Record Office: MH56/303.

Hughes, T.P. (1987)."The Evolution of Large Technological Systems," in: W.E. Bijker, T.P. Hughes and T.J. Pinch (eds.). *The Social Construction of Technological Systems*, Cambridge, Mass.: MIT Press.

Jago, W., and W.C. Jago. 1911. *The Technology of Breadmaking.* London.

Jenkins, A. 1970. *Drinka Pinta: The Story of Milk and the Industry that Serves it.* London: Heinemann.

Kent-Jones, D.W. 1930. *KJ Continuous Thermoprocess.* Messrs. Woodlands Labs/Henry Simon.

Langrish, J. et al. 1972. *Wealth from Knowledge: A Study of Innovation in Industry.* (Part 3(4), BBIRA: Chorleywood Bread Process). London: Macmillan.

Maunder, P. 1970. *The Bread Industry in the United Kingdom.* Loughborough: University of Loughborough.

Maxwell Committee 1901. *Report of the Departmental Committee appointed to Inquire into the Use of Preservatives and Colouring Matters in the Preservation and Colouring of Food.* (Cd. 833). London: HMSO.

Mellanby, E. 1946. "Diet and Canine Hysteria." *British Medical Journal* 2: 885.

Ministry of Food. 1953. *Aeration Patents.* Public Record Office: MAF 256/205.

Ministry of Food. 1954. *Toxic Substances: Informal Conference on Flour Improvers: Aeration Process.* Public Record Office: MAF 256/20.

Ministry of Health. 1939. *Bread and Flour Regulations, Food and Drugs Act 1938, Section 30.* Public Record Office: MH56/303.

Ministry of Health. 1952. *Flour Improvers: Particularly Agene 1946-52.* Public Record Office: MH56/296.

Monro Committee. 1927. *Ministry of Health: Report of the Departmental Committee on the Treatment of Flour With Chemical Substances.* London: HMSO.

Morris, T.N. 1958. "Management and Preservation of Food," in: C. Singer (ed.). *A History of Technology: Volume V: The Late 19th Century.* Oxford: Clarendon.

NABIM. 1939. *Memorandum on the Draft of Bread and Flour Regulations proposed to be made by the Minister of Health, 12 May 1939.* Public Record Office: MH56/303 Bread and Flour Regulations, Food and Drugs Act 1938 Section 30.

Rideal, S. 1903. *Disinfection and the Preservation of Food.* London: Sanitary Publishing Co.

Sanitary Commission. 1897. "The Use of Antiseptics in Food." *The Lancet* 1: 56.

Simon, H. 1893. *The Present Position of Roller Flour Milling.* Manchester: Simon.

Society of Dairy Technology. 1966. *Pasteurising Plant Manual.* London: Society of Dairy Technology.

Tilbury, R.H. 1980. "Introduction," in: R.H. Tilbury (ed.). *Developments in Food Preservatives - 1.* London: Applied Science.

Weedon, B.C.L. (Chairman). 1971. *Report on Additives in Bread and Flour.* London: MAFF.

Whetham, E. 1976. "The London Milk Trade, 1900-1930," in: D.J. Oddy and D.S. Miller (eds.). *The Making of the Modern British Diet.* London: Croom Helm.

Wynter-Blyth, A. 1882. *Foods: Their Composition and Analysis.* London: Griffin, 266.

Part III

Experiments with Social Learning

Experiments with Social Learning

Introduction

Social learning processes are put high on the agenda by the CTA approach. Jelsma and Herbold propose and evaluate several conditions for successful learning. First, learning processes require openness and access. Herbold recounts the planning and construction of a new waste disposal site in Germany. He shows how, after initial public resistance, it was possible to start up a learning process which resulted in viable solutions precisely because an open planning process involved active public participation. In this process technical experts can remain key figures but they must take the public's risk perceptions seriously. Solutions and management strategies should become part of a negotiation process. To achieve collaborative decision making, interaction between experts and the public must consist of dialogue between the different parties involved, and not just one-way education.

Jelsma's analysis of biotechnology developments makes clear that such a dialogue will only take place when all participants have access to the information they need. When critical information is kept away uncertainty increases about the motives and aims of the other parties. This uncertainty stimulates an antagonistic game in which positions are defended and in which information becomes a key to win the fight. If information is more freely available, however, participants could develop a cooperative game with viable solutions of benefit to all. A second precondition for actors to participate is that it has to be a meaningful dialogue connected to concrete decisions and providing access to decision-making arenas. Something must be at stake; otherwise there is no interest in participation and interaction.

Third, learning requires orchestration and coordination. Jelsma's case and others in this volume make clear that such coordination will sometimes come about as an unintended consequence of interaction.

In the first period of biotechnology development in the laboratories (1973-83), coordination was achieved through actions of experts with a high reputation. They managed interaction between researchers and between the research community and outsiders, government and environmentalists. In the second phase (from the mid 1980s on) no such leading actor was available. Government tried to do the job but failed.

Jelsma maintains that government agencies are not well suited to the task of coordinating learning processes. Who should play this role? Remmen suggests, on the basis of a case-study of the introduction of cleaner technologies in the Danish fishing industry, that interorganizational networks could carry learning processes and in this way provide coordination. Herbold discusses societal experiments that could perform the same job (see Schot et. al. 1994 as well). Experiments can be envisaged as niches in which technologies are specified, needs defined and concretized, and user representations are put to test. Experiments make it possible to establish an active search and learning process. Indeed, Herbold argues that implementation processes of new technologies are increasingly in the nature of social experiments (see also Krohn and Weyer 1994). For instance, the testing of industrial plants and new computer devices is riddled with uncertainties related to upscaling and the unpredictability of social behaviour. Such testing, Herbold suggests, is characterized by verifiable hypotheses, an organized research process, and the definition of an experimental situation.

Finally, social learning can be enhanced when technical devices make it possible to reverse decisions. In Herbold's case, the disposal site was designed to allow for changes to reduce risk in later stages. Such novel technical fixes resulting from learning processes also appear in the analyses of Jelsma and Downey. Thus, design options which combine implementation and flexibility should be actively sought. But in most cases flexibility/reversibility will only be partial and in some cases even impossible. Inflexibility is often needed to make technology robust.

Until now we have mainly discussed learning in terms of stimulating more interaction and feedback. Akrich extends this analysis by showing how innovators construct user representations – before they interact with any users. These user representations are scenarios of a new world around a proposed technology, and in this way resemble the concept of actor-world introduced by Callon (1986). The existence of such anticipation substantially changes the character of learning. When the new technologies are developed and introduced very little attention is given to these existing user representations and often diffuse scenarios. They remain unarticulated, unless something happens that obstructs progress or is otherwise seen as a barrier to be overcome. The point for CTA is that one need not aim at developing new user representations or scenarios of a future world but first and foremost at making existing user representations and scenarios explicit, monitoring them and coordinating their multiplicity.

Remmen develops a useful tool to shape social learning processes and help articulate user representations and scenarios of future worlds: the dialogue workshop. In these workshops actors discuss new technologies focussing on problems, potentials, visions, alternatives, barriers and ways to introduce changes.

References

Callon, Michel. 1986. "The Sociology of an Actor-Network: The Case of the Electric Vehicle," in: Michel Callon, John Law and Arie Rip. *Mapping the Dynamics of Science and Technology: Sociology of Science in the Real World.* Basingstoke: Macmillan, 19-34.

Krohn, Wolfgang, and Johannes Weyer. 1994. "Real-Life Experiments. Society as a Laboratory: The Social Risks of Experimental Research." *Science and Public Policy,* 21 (3): 173-183.

Schot, Johan, Remco Hoogma and Boelie Elzen. 1994. "Strategies for Shifting Technological Systems. The Exemplar of the Automobile System." *Futures,* 26(10): 1060-1076.

7 Learning About Learning in the Development of Biotechnology

Jaap Jelsma

> "There are some people who just cannot be approached in a rational fashion. They don't want to have answers, they really don't. They just want to ask questions, and they want to be afraid. They really have not been receptive to anything that's been told at them. At that point what can you do?"[1]

"Raging Diggers uproot biotechnology test plot." "Raging Potatoes ruin field experiment with genetically engineered crop." "Virulent Viruses cut down modified corn plants." Such headlines appear now and then in Dutch newspapers. During the night and on weekends, field test plots with genetically modified crops, especially potatoes, have been damaged on several occasions by activists identifying themselves by these illustrious names.

In a manifesto sent to the press, the Raging Diggers explained their campaign against biotechnology. The Diggers denounce a technical fix for problems like hunger and pollution, which they see as the result of disturbed relationships between people and the environment, between people and animals, and between the over- and underdeveloped nations. The Diggers criticize biotechnology being promoted as a cure for these problems. Because biotechnology has historically aggravated these imbalances, it can only add to such inequalities. For the poor have no money to develop their own biotechnology, nor to buy products developed by others. In fact, the Diggers argue, technology does not aim at a better life at all, but serves competition. Anyway, a few monopolies will dominate food production. Although biotechnology has drastic societal consequences, public control of its development is largely absent. Therefore, the Diggers declare, we have disrupted the field tests not only to spark debate, but also to undermine biotechnology.

Are campaigns like this a relevant phenomenon for those who are interested in CTA? Rip and Van den Belt (1988) suggest that understanding the dynamics of learning processes is essential to improve their outputs. From this perspective the question becomes: do the campaigns exemplify aspects of the learning process a society goes through when it incorporates a new technology? The answer depends on what one considers a learning process. Conceived in its widest sense, a learning process involves societal interactions that have certain modifying effects on a technology that — as a result of this interaction — is becoming embedded in society. Such effects are not always immediately apparent, and often require close study to be revealed.

Some effects are apparent from the Dutch campaigns against field tests with genetically modified plants. The development of certain modified crops has been delayed. Firms have suffered damage, but from their reactions it is clear they will not move abroad unless things get much worse. More important, the campaigns prompted the media, previously silent about field tests, to provide detailed information to the public. In addition, public interest groups with a critical attitude towards biotechnology received media attention. Finally, the campaigns have made the proponents of biotechnology even more apprehensive of the public's attitude towards biotechnology. The campaigns enhanced the public's awareness that there is "another side" of biotechnology. Consequently, industry, as well as certain parts of government, may be more willing to communicate with consumer organizations and public interest groups, and to take part in CTA-like experiments.

The effects of campaigns by groups such as the Raging Diggers will not be easy to trace, and are not expected to be radical. They probably will not stop biotechnology from further entrenchment in society; rather the opposite will be true. But the resulting technology may be different, being built on broader compromises than otherwise would be the case. For instance, a kind of biotechnology may result with more attention to its effects on the Third World, and more scrutiny of its environmental consequences. In having such effects the campaigns, together with many other interactions accompanying the embedding of biotechnology and having similar effects, may teach us a lesson. Studying such interactions carefully will increase our understanding of the social entrenchment of new technologies as the outcome of learning processes. It is to the latter aim that I want to contribute in this article, by reviewing the first two decades of biotechnology from the viewpoint of a learning process. Therefore, I will not strive for a complete description of events, but focus on the relation between social interactions in the establishment of biotechnology and the transforming effects of such interactions on the technology itself.

The First Stage of the Learning Process: 1973-1983[2]

From 1973 to 1975 the impacts of recombinant DNA were mainly dealt with by the community of molecular biologists who had developed the technique. For these experts, the novel possibility of placing DNA fragments from one genome into another evoked an ambivalent reaction, one of fascination mixed with concern. That this technique promised striking advances in knowledge was immediately clear. But at the same time, the molecular biologists realized that the hybrid DNA molecules might engender new and unpredictable hazards. A new biohazard was, the biologists guessed, not only a problem for themselves. Since the matter of risk was expected to stir concern outside science too, scientists felt compelled to come to terms with the potential risks of the new technique because they did not want others to do it.[3] This meant that they had to tackle the uncertainty surrounding the risks of the new tool. In other words, they had to reduce the equivocality of the situation to a workable level, by finding some kind of problematization which opened possibilities to act on the problem in a responsible and credible way.[4] Otherwise, they would not able to break the deadlock of "we cannot do anything because it might be hazardous, but if we do not do anything we will never know whether our concerns are warranted." That is, they had to formulate a strategy for learning about the unknown without precipitating disaster. This strategy for learning was constructed in a number of steps, in which the risk problem was bounded so that it became compatible with existing knowledge, and controllable by a technical fix. Along this path, however, important elements of the problem got lost, became invisible, and therefore untreatable.

The first step in making the risk problem manageable was done by cutting off the moral dimension, and by associating part of the novel molecules with known hazards while leaving the rest alone. The experiments that transferred genes from known pathogens or those coding for toxins had to be considered as potentially hazardous. The moratorium proposed in 1974 in the famous letter of Berg et al. was founded on this problematization, and persuaded scientists not to carry out such potentially hazardous experiments. Thus, for the Berg committee, the imaginable hazard was the production of DNA constructs in which the involved genes coded for disease or hampered the treatment of disease. In consequence, all other recombinant DNA experiments were left unproblematized and open to experimentation. Berg himself was quite clear about the moral and ethical side of the problem. About a possible stop in his own experiments he said in 1974: "I'd stop if there were a sound practical reason, but not if it were an ethical judgment" (quoted in Lewin 1974). Thus in the Berg letter the carrying out of risky experiments was problematized as posing a technical problem of the control of hazards.

The moratorium on broad classes of recombinant DNA experiments was only a short term solution to create room for a more profound attempt to handle the risk and to secure the future of DNA research. This attempt was successfully carried out at the Asilomar conference in 1975. At this conference, the outline for a safety regime was created by a preselected group of reputable scientists from different countries. At the start of this conference, social and ethical issues, like genetic engineering of humans and the possible military use of recombinant DNA, were explicitly banned from the agenda. With respect to the risk problem, the conference struggled with two rival definitions of the problem (and their respective constituencies). The more radical definition was supported by the younger molecular biologists who stressed the novelty of man-made recombinant molecules. According to this view, a safety regime was proposed in which recombinant DNA experiments were rank-ordered in classes according to the level of ignorance about the consequences of such experiments; these classes had to be matched with increasing levels of physical containment. A working group of virologists proposed a rival definition built around four risk-categories already existing in cancer virology. Underlying their proposal was the tacit assumption that the hazards under discussion were similar to the hazards of cancer viruses. Eventually, the virologists' proposal was taken over by the conference. A breakthrough in the conference was the concept of "biological containment," i.e. constructing combinations of specific host-bacteria and DNA vectors with sharply restricted capacity to spread outside the laboratory. Together with existing means of physical containment, biological containment was a technical fix that, combined with the safety levels proposed by the virologists, stood out as an attractive solution for the risk problem.[5] These measures together created a safety regime, in which learning about the true hazards of recombinant DNA could occur without evoking disaster.

The task of elaborating the rough safety concept of Asilomar into a detailed regulatory regime fell to a committee established by the U.S. National Institutes of Health (NIH), the body responsible for implementing the resulting guidelines in academic molecular biological research. The Recombinant DNA Advisory Committee (RAC) consisted mainly of molecular biologists, i.e. of technical experts. However, to work out the Asilomar agreement was not a straightforward technical task at all. According to the chairman of the RAC, the conclusions of the committee were reached "in part intuitively, in part by analogy, in part by hunch, and in part by consensus [...]."[6] (DeWitt Stetten, quoted in Krimsky 1982, 157). These negotiations took place within a small group of technical experts selected from a community just beginning to consider a very complicated matter. The matter was new and confusing to them, but the strategic meaning was quite clear. Thus the outcome of the deliberations within RAC, although perceived by the scientists as a technical achievement (a set of guidelines based on intricate details of molecular biology) was and could only be political in nature, and open to challenge in the wider arena of molecular biologists. This was precisely what happened.

The first set of guidelines drafted by a subcommittee was criticized inside and outside RAC as watering down the recommendations of Asilomar and eventually dismissed by the full RAC. A revised set of guidelines was accepted by a majority vote of molecular biologists during a turbulent meeting, and NIH issued the so-called Hill guidelines in June 1976. Thus the drafting of regulation is best conceived as politics expressed in technical terms; in the case of recombinant DNA, in the terms of molecular biology. The struggling of factions motivated by different identities, ideals and histories, is a characteristic feature of such politics. In the community of molecular biologists, three factions were involved: (i) young radicals inspired by the ideals of the 1960s and the emerging environmental movement; (ii) conservative elders wanting to protect the freedom of science, and who saw regulation as an unnecessary and politically motivated constraint; and (iii) pragmatic centralists who recognized the potential hazards, but wanted practical solutions to save the promising future of recombinant DNA research.[7]

This perception of the process in political terms was not the dominant view in the community of molecular biology. Though some actors had an eye for the political character of acting on the risk issue (see Wade 1975), others stressed the traditional distinction between the normative and the empirical (Davis 1976). The effect of denying the political character of risk regulation is an important question for CTA, explored by Wynne's contribution to this volume, and it is an issue to which I will return.

After the NIH guidelines were issued in 1976 the community of molecular biologists threatened to split along the faction lines sketched above. However, a new consensus emerged during the next two years around a redefined and curtailed problematization; consequently, broader issues and their constituencies became drop-outs to the debate on DNA. Several factors contributed to this development.

In the first place, the Asilomar agreement and the resulting NIH guidelines were not stringent enough for some molecular biologists. Some of the more radical of them, not being invited to Asilomar or to the RAC, started criticizing the guidelines in public. Focusing on public health, they questioned whether the safety regime was sufficient in the light of potential ecological and evolutionary effects, and whether biological containment was effective enough. Serious doubts were cast on *E. coli* K12 being the suitable host, even in its enfeebled form. Others questioned the rationale behind the experiment's rank-ordering.[8] Hazardous social effects of potential applications of DNA technology were also put on the discussion agenda, and it was even questioned whether a group of technical experts was entitled to decide on such political matters. Along these lines, recombinant DNA became connected to the issue of democratization of science, pushed vigorously by Science for the People, whose spokesmen included several prominent molecular biologists. Thus scientists voiced serious doubts in public whether the new learning regime was sufficient to avert disaster. Consequently, the image of a unanimous community carefully and responsibly coping with the external effects of its professional activity, fell to pieces.

The breakdown of consensus in the community of molecular biologists opened up the debate to outsiders. Outsiders gained official recognition when the director of NIH organized a broadly conceived hearing, to which not only scientists but also environmental and consumer organizations were invited. In consequence, the network of actors involved in the recombinant DNA issue extended beyond the core set of molecular biologists. Gradually, a full-scale public debate developed. Recombinant DNA became an important issue for the media, which created a forum for the quarrelling scientists and their constituencies. Views on the risks of recombinant DNA were no longer expressed only at scientific meetings, but also in city halls, in newspapers and on television. To the displeasure of the leaders of Asilomar: "What began as an act of responsibility by scientists [....] has become the breeding ground for a horde of publicists" (Stanley Cohen, quoted in Krimsky 1982, 197).

The experience of working with the new regulatory regime added to the growing equivocality of the situation. The "silent majority" of scientists at the bench had rushed into the field of recombinant DNA research as soon as the guidelines were issued, and followed the debate at a distance. The RAC office was overwhelmed with requests for approval of planned experiments, and it often took months before such approvals were awarded. Leading investigators resented being cross-examined by a media-covered Senate hearing for starting an experiment in anticipation of a permit from RAC. Such incidents changed the attitude of the investigators towards the guidelines. Instead of being praised for their self-imposed precautions, they felt themselves shackled with nasty and superfluous rules.

In the summer of 1977 Congress considered several bills to regulate recombinant DNA research. The guidelines, already cursed by many investigators, might acquire the force of law and become implemented by the hated "bureaucrats." The scientists faced an unpleasant turn of events: the threat of intervention in science despite the fact they were resigned to constraints on research in the light of intangible risks. Thus the battlelines began to shift: "set safe standards" became "protect science from direct regulation" (Krimsky 1982).

Control of the situation was regained by the building of a new consensus in the scientific community, based on a restricted problematization and some tentative results of learning about effects of DNA recombination. This cognitive restructuring helped technical experts to re-establish their core position in the network and, once again, silenced discussion of the broader issues. The recombinant DNA debate narrowed down into a highly technical discussion on risk, from which non-experts soon dropped out. In addition, leading molecular biologists began to shift their opinion on the risks. Their former attitude of "proceed with caution" shifted to a position of "our former concerns were probably unwarranted."[9] To counter the threat of legislation, a lobbying campaign was started by a coalition of scientific societies led by the American Society of Microbiology and the prestigious National Academy of Sciences. In effect, the community of molecular biologists closed ranks. A resolution sent by the NAS to Congress attacking the proposed legislation as overly restricting recombinant DNA research, was even signed by Sinsheimer, one of the most ardent critics of the guidelines (Baltimore 1977).

The shift in opinion on risk was legitimized on several grounds. One argument gaining force was that no adverse effects had appeared even with large numbers of recombinant experiments. A second consideration was that unwarranted restrictions could force "scientists to leave the country." Furthermore, the molecular biologists strengthened their case by referring to evidence purporting to show the harmless character of most of the recombinant DNA experiments. It is important to emphasize that most of this evidence was not *new* evidence. The original problematization was flexible enough to replace, by shifting the elements and accents, an impression of uncertainty by one of near certainty. An entirely different picture of risk emerged by shifting attention from possibilities to probabilities, from "poor understanding" of phenomena to "a basic understanding," and from "survival" of *E. coli* in the environment to "colonization" in the human intestines.

These shifts were meaningful and credible within an implicit narrowing of the risk problem. Comparing the original (1974) and revised (1977) problem definition makes this clear. The original definition emphasized a wide range of potential irreversible, long-term ecological effects associated with all kinds of recombinant molecules. By comparison, the revised definition stressed the possible pathogenicity for humans of a specific microorganism.[10] In effect, the complicated risk problem was rephrased into one key question: could *E. coli* K12 be converted by recombinant DNA into an epidemic pathogen? Through this re-definition, the risk question could be connected to the paradigm of microbial pathogenesis, i.e. to irrefutable and firmly established concepts and principles. On this basis, the risk question could credibly be answered with a firm "no."[11] Experiments not covered by the new risk definition — including large scale experiments and deliberate releases of recombinant organisms into the environment — remained forbidden, and their potential consequences unanalysed.

With the intangible risks of a new technology now linked to familiar, well-investigated phenomena, the restructured problem definition quickly gained support in the community of molecular biologists. Looking back, their earlier concerns were readily perceived as naive, though understandable considering the field's ignorance about the mechanisms of evolution and pathogenesis (about which the molecular biologists could claim "we now know the facts"). This re-appraisal changed the meaning of those actors who refused to align themselves with the majority position. Once, those who defended the broad problematization were accepted as important critics. Now, they were seen as dissidents stubbornly clinging to vague, unworkable and obsolete views (Berg and Singer 1976).

The revised view on risk became increasingly stabilized in the late 1970s and early 1980s. The NIH guidelines were relaxed several times, since no hazards had manifested themselves. Further, the safety regime designed around the host *E. coli* K12 was extrapolated to other host-vector systems, whose safety was assumed on the basis of analogy.

In Congress, the successive bills introduced to control the safety of recombinant DNA research became ever more modest. The first successful commercial applications of recombinant DNA, announced at press conferences and Congressional hearings, built momentum for the new field of "biotechnology." Industry voluntarily complied with the safety regime governed by RAC. In the early 1980s, the ban on large-scale experiments and deliberate releases was lifted. The first field tests were announced in what seemed to be a peaceful and effective regulatory climate.

In terms of learning, the outcomes of the first stage are ambivalent. By the late 1970s, the initial confusion and fear of disaster had been replaced by an articulation of the risk problem by molecular biologists. This articulation helped formulate a strategy for action, based on a technical fix (forms of containment) coupled to a classification of risk. The strategy established an opportunity to work with the novel molecules within a contained space. In this contained space, learning about the hazards of these molecules could occur without causing disaster. The paralysing deadlock about the risk problem was broken.

This success had its price, however. The scientists were too eager to create a bounded area of certainty, a safe territory in which they could hunt for success without being troubled by outsiders. The broader issues of ecological and social effects were effectively ignored, and thus remained unanalysed and inaccessible for learning. In other words, a reduction of breadth in the learning process was the price of restoring the experts' control of the situation. The result divided the actor field into winners and losers, with the "broader" issues and their constituencies on the losing side. The latter dropped out of the network, taking with them any pressure to learn on these themes. These issues became invisible, but not (as we will see) irrelevant.

Although recombinant DNA is often cited as a success story of coping with uncertainty (Morone and Woodhouse 1986), we must have doubts whether much learning about hazards took place under the regime established. The regime mainly functioned to facilitate the rush for exciting discoveries. And potential hazards were not at all exciting to study for the actors who now dominated the network. Rather, there emerged a resistance against learning about hazard.[12] Indeed, within the revised problematization, all uncertainty had been removed. In addition, revisions of the guidelines were not based on systematic risk research but rather on mere passive and confirmational learning ("we see no dangerous thing happen"). There were no systematic attempts to falsify the reassuring assumptions. To re-start learning, access for new actors and issues had to be forced on the network in power.

The Second Stage of the Learning Process: Mid-1980s to the Present[13]

In the early 1980s, applications of biotechnology with relevance for agriculture — especially plant biotechnology and biopesticides — reached the development stage. It was clear that recombinant organisms, microorganisms as well as plants, would leave the protected laboratory environment to be tested in the field. These organisms would be deliberately released into the environment, a new domain which goes beyond the jurisdiction of the regulatory regime controlled by the molecular biologists. The environment is covered by federal laws enacted in the 1970s to control pollution, especially by toxic chemicals and pesticides. Thus the ongoing development of biotechnology crossed the boundary between two regulatory regimes. Under the federal environmental regime, the molecular biologists became exposed to outside intervention. For the environmental laws allowed citizens to force state regulation on any activities with environmental consequences, i.e. on the activities of the molecular biologists which had so far developed under the protection of the scientists' self-regulatory regime.[14] As we will see, the losers of the first round seized this opportunity successfully to bring biotechnology within governmental control.

Another contextual change in the early 1980s was the Reagan administration. In rhetoric and in policy, technology gained a strategic meaning for sustaining American leadership in the world (as Downey also indicates in his essay). Biotechnology was seen as one of the spearheads of America's military might as well as commercial competitiveness. For that reason, state-of-the-art biotechnology should not "be sold out cheaply," nor should "industry in pursuit of biotechnology be shackled," a presidential advisory committee told the White House. The committee stressed the importance of promoting biotechnology, urging the government to protect America's leading edge in this field by an export control policy (U.S. Government Interagency Working Group on Competitive and Transfer Aspects of Biotechnology 1983). In this logic, updating and streamlining of regulation relevant for biotechnology gained strategic importance for U.S. "competitiveness."

The first requests for field tests were put on the RAC's table in 1982, and the first field test was approved the following year. Carried out by scientists of the University of California, the test consisted of a series of field experiments with a bacteria called "ice minus" genetically modified to diminish frost damage to certain crops. The RAC's quick approval reflected the consideration that such bacteria *already* existed, produced by conventional means (chemical mutagenesis).

In the fall of 1983, while preparations for the field tests were underway, a number of organizations headed by Jeremy Rifkin's Foundation on Economic Trends brought the NIH into court. The lawsuit charged the NIH with not submitting a full Environmental Impact Statement concerning experiments with recombinant organisms outside the laboratory. In addition, the lawsuit challenged the RAC's competence to deal with the risk of such experiments; for instance, no ecologists were on the committee. The court determined that field experiments with recombinant organisms were not allowed while Rifkin's charge was pending. The decision was explicitly confined, however, to NIH-sponsored researchers. The private sector remained free to do field tests; formally, they had nothing to do with NIH.

The consequences of this court challenge were profound. It reintroduced uncertainty in several ways. In the first place, the RAC's monopolistic role as the regulator of biotechnology was jeopardized. The court's decision made clear that permission from RAC was no guarantee that a field test could be carried out in practice. Second, the court inadvertently created two regimes for the field experiments, dependent on who carried them out. For NIH-sponsored scientists, such experiments were forbidden for the time being. On the other hand, scientists in the private sector did not face any regulatory hurdle to do the same experiments. This duality was made clear when Advanced Genetic Sciences (AGS), a biotech firm which sponsored the University of California project, asked the RAC permission to field test "Frostban," a prototype of the ice minus bacteria. After some hesitation, the RAC granted the permission. Two weeks later, Rifkin went into court again to get the earlier decision extended to the private sector.

Thus by the legal actions of outsiders, stabilized roles in the regulatory regime were broken up. Indeed, confusion arose about where to go for a reliable permit to do a field test. In other words, a well-defined situation with respect to regulation suddenly changed into an undetermined one. This change created room for new actors bringing new problematizations, values and interests to make themselves seen and heard, and forced the actors already present to re-evaluate their views, and to make them explicit to the newcomers. A new arena opened up for a struggle about regulating the risks of biotechnology. The second stage of the learning process had begun.

The renewed struggle about regulating the risks of biotechnology soon became coupled to other complex policy issues, including America's competitiveness, free science vs. bureaucracy, environmental protection, and ethics. This coupling expanded the number of actors who perceived biotechnology regulation as touching their interests, and who started to take action on the risk matter. Consequently, the regulation of biotechnology itself became a multi-actor strategic game.

In this game, a key point of dispute was the ecological effects of introducing genetically modified organisms (GMOs). Unlike the situation in the 1970s, the GMO problem could not be narrowed down to the chance of one type of a crippled microbial host escaping from a contained laboratory. Field tests entailed a whole range of modified organisms and the environment. To have the intended effect, GMOs should be robust and must survive for some time.

In subsequent debates on field tests, ecologists emerged as the antipodes of the molecular biologists. The ecologists stressed the incapability of existing ecological knowledge to make general predictions about the effects of deliberate releases of GMOs; such effects could only be investigated on a case-by-case basis. The molecular biologists replied that this was exactly what the RAC was doing. Some of them re-introduced their reassuring interpretation of evolution theory as formulated during the first stage of the learning process.[15]

As in the 1970s, Congress became a player too. The Senate and the House organized hearings, while the House Committee on Science and Technology published an influential report about regulation. In this report, the risk of releasing GMOs was qualified as a "low probability/ high consequence risk." To deal with this novel form of risk, the House Committee recommended that the Environmental Protection Agency (EPA) regulate commercial biotechnology, and that a new interagency committee should coordinate the development of regulation. Until this regime was put into effect, no field tests should be carried out.

Industry and science resisted these ideas in the beginning. By broadening the RAC, they hoped to save the existing regulatory approach based on expert review. A great advantage of this approach was its low bureaucratic load: for all field tests only one agency had to be contacted. Because of the practical impotence of the RAC, however, firms as well as scientists gradually began to turn to the EPA for permission for field tests. EPA was totally unprepared to take over this task, and acted reluctantly at first. Then, in the fall of 1984, the EPA issued an interim regulation for genetically modified microbial biopesticides. A new actor had been forced into the regulatory arena. The EPA was destined to play a central role in the discussions which were rapidly heating up.

An important question in the expanding debate was whether these novel risks should be regulated on the basis of a new biotechnology law, or on the basis of existing statues. Most jurists who spoke out on the matter favoured the latter option, which also received wide support in Congress. The EPA revised its view expressed in the 1970s that a new law was necessary to regulate biotechnology adequately. Accordingly, the Reagan administration decided there would be no new law for biotechnology.

Now the problem shifted into a new shape. In which ways should existing statues be implemented to cover the risks of a technology for which they had not been designed? This question had a socio-cognitive character indeed. On the cognitive level (the formulation of the statutes, i.e. their scope, concepts and terms) all kinds of translations had to be made. For instance, the Toxic Substances Control Act was enacted to regulate chemicals, i.e. dead substances. Now it had to cover living biologicals too. Could "chemical substance" be expanded to living things? A host of such translation and definition problems almost drained the life out of EPA, which had the most difficult of these jobs. Terms from the contexts of pathology and taxonomy, such as "pathogen," "species," and "genus" (the latter needed to define "intergeneric substances"), had now to be unequivocally defined in a juridical context. "We realized two years ago that the term 'pathogen' just did not work well", said an EPA administrator. "Defining 'pathogen' is like defining pornography: we all know what we're talking about, but it is hard to define it" (J. Moore, quoted in Fox 1987). Clearly, defining "deliberate release into the environment" was also difficult.

On the social side, all these definitions interfered directly with various investments by the different actors. Industry felt confronted with "a bewildering matrix and perplexing set of definitions," and did not find the promised road map "clarifying the regulatory path that a company with a new product would follow." A director of the U.S. Industrial Biotechnology Association (IBA), a lobby of the biotech industry, wrote: "The current definition of deliberate release is so elusive that design of manufacturing processes and facilities remain problematic. If facilities must be designed for *no* release (rather than for very low levels of release) it could be very expensive" (A. Goldhammer, quoted in Fox 1986). For the biotech industry, the crucial question was whether (novel) *products* of biotechnology would be regulated — or the *processes* by which they were produced. The latter regime would cast a much wider regulatory net than the first, and industry passionately lobbied against it.[16] In the end EPA shifted its position and chose the former.[17]

The mobilizing effect of definitions also touched the government agencies' mutual relations. In overlapping domains of regulation, e.g. between chemicals and pesticides, definitions determined the borderlines of authority between — and sometimes within — agencies. To mitigate coordination problems, in 1984 the White House Office for Science and Technology Policy established an interagency committee, which subsequently evolved into the Biotechnology Science Coordinating Committee (BSCC). Although the committee was intended to clarify scientific questions relating to the design of regulation, it was unable to function because of internal discord. At the end of 1988, EPA point-blank refused to attend any further BSCC meetings. While its charter assumed that BSCC would be able to distinguish scientific issues from policy issues, most regulatory questions are in fact mixed questions of science and policy which cannot easily be separated.[18] The agencies quarrelled further in public, and the learning process became deadlocked.

As in the 1970s order was restored by redefinitions (on the level of problematizations) and by creating a technical fix. On the level of definitions the novelty of biotechnology was downplayed, suggesting that existing juridical frameworks were adequate to regulate it. For example, in the preamble of the 1984 regulation proposed by EPA "an explosion in our understanding [....] which has introduced a new and profound dimension into the field of classical genetics" emerged two years later as "an extension of traditional manipulations that can produce similar or identical products." More importantly, EPA, under the severe criticism from science and industry, reduced the scope of its draft regulation.

The appearance of technical fixes was a response of the biotechnology community to the safety demands of public interest groups such as Rifkin's. Technical fixes are being developed for detection and for eco-safeguards, i.e. biological containment to be applied in field situations. In the first line, EPA has started to support academic projects aimed at environmental detection of genetically modified organisms (GMOs), while the food multinational Monsanto has developed monitoring technology for GMOs as part of its biological pesticide programme.[19] For eco-safeguards a promising approach is designing microorganisms with built-in mechanisms for self-destruction after the organism has completed its intended task in the environment. These result in so-called suicide systems (Molin et al. 1987). Safeguards for biotechnology are themselves based on intricate genetic engineering. As safeguards for biotechnology become available, their uptake in biotechnology trajectories will probably become a social condition for further development of these trajectories.

Lessons from the Case on Learning

In the foregoing I have described relevant experiences from the first two decades of the development of biotechnology. In this section I will examine what we can learn from this experience, focusing on learning processes, then conclude with recommendations for enhancing learning.

Learning is constrained by uncertainty and strategic behaviour of actors
The scientific breakthrough of DNA recombination created new and unpredictable opportunities for human activities in science and technology. Such novel activities generate uncertainty in three different respects. One source of uncertainty was ignorance about potential effects. In the beginning, knowledge for the assessment of different kinds of potential effects (on human health and the environment) was completely lacking, which resulted in rampant speculations. It took quite some time to reach a consensus among core actors about how to curtail the problem of effects to manageable dimensions (first a curtailment of risk, and then a narrowing down further to health risks), so that existing knowledge could be mobilized to solve it. However, this translation of existing knowledge — from the domain of evolution and infectious diseases to that of risk questions in biotechnology — created new controversy, initially adding to the uncertainty. Moreover, though the banning of ecological effects from the risk agenda had advantages in the short term — it made the risk problem manageable — uncertainty related to such effects returned forcefully as soon as biotechnology came out of the labs and into the fields.

When concerns about the risks of field tests increased, the realization that inadequate knowledge existed for the assessments of such risks increased too. The establishment of risk assessment programmes can be traced to this realization. Researchers participating in such programmes play on these uncertainties and ignorance, not least to legitimate their funding. The uncertainties reported are of both paradigmatic and empirical nature. "Predictive ecology" is still beyond the horizon, and most of the basic methodology for obtaining data to fill up the lacunae is still being developed.

A second source of uncertainty was the question of how to regulate. To set clear standards and to develop transparent procedures, one needs a clear definition of the hazards and of the concepts on which such a definition is based. But in the beginning of biotechnology, clear definitions and concepts had yet to be developed. More vexing, existing concepts like "species" and "pathogen" were put into question by DNA recombination.

Another pressing question for the U.S. — with respect to the field tests — was whether new regulatory frames should be developed, or existing laws should be adapted to cover biotechnology. How regulation for a new technology should relate to existing regulatory regimes is often a difficult puzzle. For implementation immediately touches the competence and authority of the current institutions. There are inevitably coordination problems, too. As technology becomes more and more internationalized, regulation should be as unitary as possible. This means that federal rules should be attuned to local regulations and to emerging international regulations. Within the European Community, with its many member states, the coordination problem is at least as difficult as in the U.S. Unsolved coordination problems create uncertainty for all actors who want to invest in the new technology. A distinctive problem with gene technology is that safeguards and monitoring devices use the same technology (i.e., gene splicing) as that whose safety is under discussion. In this manner, the development of safeguard technology adds fuel to the controversy, and requires itself to be modified in the light of these concerns.

A third source of uncertainty relates to strategic behaviour in the complex actor field in which biotechnology takes shape. Different kinds of strategic games are going on in this field, which can (for analytical purposes) be described as a triangle between regulators, firms and public interest groups each with different agendas. Each party in the triangle depends on the others for reaching its goals (or at least part of its goals), and there are indications enough that most of them are aware of that.[20] Thus parties are, to different extents, considerate with the others.

On the other hand, parties have high stakes in the regulatory process. If they cannot realize all their goals, they want to realize as many as possible. This leads to forms of behaviour like threatening the others, "teaching them a lesson," etc. Even the cooperation in risk-assessment programmes might be interpreted as a strategy aimed at rolling back regulation. With such an array of stated and unstated goals the shaping of biotechnology is a "strategic game" *in optima forma*. This strategic character is an underlying source of uncertainty in the entire learning process. It makes it difficult to be sure about what actions of others mean precisely, and about how to react to them.

Learning processes need intentional management
Though problems relating to uncertainty and strategic behaviour are not easy to handle, in biotechnology so far these problems have been overcome. In most industrialized countries, regulatory regimes have

gradually emerged and started to function, while risk-assessment programmes are ongoing. In turn, disruptive incidents diminish, campaigns and lawsuits from activists become less frequent, and trust between actors rises. Products have begun to trickle down the novel regulatory machineries. Scientists from industry and academe, though at some places trying to making regulatory procedures more convenient, have at least accepted them. That is, learning has taken place.

These learning processes did not proceed by themselves; they required orchestration and coordination. In the first stage (1973-83) this role was played in the U.S. by reputed molecular biology experts, i.e. the group around Paul Berg. These scientists were quite aware that if they did not deal with the risks of recombinant DNA research, outsiders (i.e. "bureaucrats") would intervene. Accordingly, the "deans of Asilomar" had to coordinate different interaction processes within the scientific community (which was initially divided on the risk issue, and on how it should be regulated) and with outsiders to whom the scientists needed to remain credible in handling the issue, i.e. with specific parts of bureaucracy (NIH), environmentalists and Congress.

In the second stage (from the mid-1980s on) an effective orchestrating body was absent in the U.S. The Office for Science and Technology Policy tried to play this role, but failed to create cooperation between the various governmental agencies designing regulation for biotechnology. Social interest groups were neglected. After some time, EPA even dropped out. The whole process of regulation ground to a halt. Eventually the regulation of the first field test took five years. In Germany, delays in designing regulations caused some German biotech firms to locate facilities overseas. Such delays strained the learning process. It meant that uncertainty was prolonged, that scientists lost hope that their field test could be carried out in the next growing season, and that design of manufacturing processes could not take place. Understandably, actors lost confidence in the process.

Seen from a CTA perspective, controversies are not problematic *per se*, but they are a risky way of learning. Unless controversies are managed in such a way that they yield useful outcomes, and that frustration and alienation of important actors are avoided, the learning may be counterproductive. If actors emerge from a controversy with adverse attitudes and negative experiences, such a frustration-laden history can block cooperation for years. For a number of the participants, the controversy in the 1970s set the stage for the difficulties encountered in regulating the field tests in the 1980s.

Government is not well suited to do the management job
In most countries, government is a conglomerate of agencies competing for competence, power, and sometimes survival. Such governmental agencies often have quite different and conflicting missions. There are contesting factions within such bureaucratic bodies, too. Regulating the second stage of biotechnology development has been strongly retarded by fights within the U.S. bureaucratic apparatus. Since biotechnology is seen as a strategic technology, and since it covers a very broad range of products and processes, almost the full width of the bureaucracy became involved in the regulatory process.

The mission of governmental agencies can be ambiguous, creating confusion about how to interpret the agency's behavior. A mission to protect the environment (for example) requires both promoting and restricting biotechnology. The policy of an agency with such a mission often is perceived as zig-zagging or even hypocritical; there was much brain-racking over EPA's "true agenda," for instance. The U.S. National Institutes of Health (NIH), another important player in the biotech game, has an ambivalent mission, too. On the one hand it is supposed to play an important role in boosting the competitiveness of the U.S.; on the other hand it regulates hazards from biotechnology. The fact that separate parts of government are themselves important players makes it difficult to appoint one of these bodies as a coordinator. It also means that outside parties — often rightly — do not see a governmental body as impartial, which undermines its credibility.

All these factors together put severe constraints on the role of government as an orchestrator of fruitful learning processes. Other institutions may be better suited to this task or could be created. Recently the Dutch Ministry of Agriculture drastically changed its biotechnology policy. A much broader input into policy making is now being realized by organizing round tables and workshops in which public-interest groups and consumer organizations participate alongside different kinds of experts.

General Conditions for the Enhancement of Learning

Create openness and access
If it is true that public reservation towards biotechnology is based on fear, then handling public fear is a point for CTA. Which transformations in the process of technology development are needed to overcome this fear? The general political answer to this question is that regulation should reassure the public that "biotechnology is politi-

cally, ethically and practically controlled" (Ken Collins, European Parliament[21]). But curing public fear about technology by regulation may be counterproductive. The public rather suspects that in fields where much has to be regulated, there is much to be feared. For CTA regulation *per se* is not enough. It is at least as important that technology takes shape in a process which is as open and transparent as possible. This guideline should apply not only to the construction of regulation, but also to the way it is practised and communicated. Key factors with respect to openness are the production and handling of information, access to negotiation and decision-making arenas, and communication structures.

Openness is often at odds with the strategic character of technology development. Firms — and sometimes nations — want to protect their competitive positions by keeping information classified. At first sight, this seems to be a roadblock for transparency. But many corporations have found that their credibility is damaged by being secretive, and that only small parts of risk dossiers are really sensitive. Indeed in an attempt to dampen local resistance at the test spot, Monsanto disclosed (with the exception of a few pages) its entire classified 800-page risk assessment of the modified bacterial pesticide (Sun 1986). Transparency should not, however, be left to the strategic whims of applicants. Regulation should specify — as the EC regulation on field tests does — who should supply which information, which information should be public, and which may be protected. Openness can further be enhanced by facilitating access to information through hearings and round tables for relevant segments of the actor field. In such activities, the public should also be encouraged to be transparent, i.e. to come up with motivated comments. Patent law should foster the early disclosure of sensitive information.

Access to the decision-making arena is also an important feature of openness. In the U.S., the process in which regulation was developed has been rather closed. Industry and public interest groups have regularly complained of being "not at the table." In other countries, including the UK, the regulatory committees for biotechnology have a much broader membership. This may be an important reason why the process of regulation of biotechnology in Britain runs rather smoothly.

When communication among parties involved in technology development is lacking, uncertainty increases about the true meaning of rival parties' behaviours. This stimulates suspicion as well as strategic reactions and counter-reactions. Thus CTA should promote communication throughout the process of technology development.

Create and maintain rich networks on different levels in emergent fields
A number of the foregoing conditions for learning can be fulfilled by the creation and management of networks. Coordination and communication can be improved by deliberately creating networks between relevant actors and organizations to reduce uncertainty and therefore diminish strategic behaviour. In terms of the techno-economic network model (Callon 1992), the different poles of such networks should be developed, as well as connecting networks between the poles. For CTA purposes, the model should be extended with at least a "regulatory" or "public interest" pole. Such networks should function across the different levels of policy making and implementation, i.e. not only on the federal level, but also on local levels where field tests or facilities are planned. Thus CTA management could be conceptualized as network management. Network management was what the "deans of Asilomar" did effectively, and which was absent in the second phase in the U.S.

Network management from a CTA viewpoint should focus not only on convergence in networks, but also on variety and flexibility. Thus networks should maintain overlaps with other networks, and should be prevented from becoming isolated or disconnected. As soon as all resources and interests relating to a technology are satisfied through one and the same network, an inflexible situation emerges. In that case, a lobby monopolizes an irreversible technology, blocking access for new actors. The presence of competing paradigms is a favourable situation for learning, as long as there is a open dialogue and a good balance of cooperation and competition, and of discord and consensus. It follows that an array of constituencies must be created or strengthened, asymmetries in positions and resources must be countered, balances of power controlled. For instance, the realization of extended peer review (Ravetz and Funtowicz 1989) to check experts and avoid the production of one-sided information, is a matter of network management. For the same reason of preserving variety, the process of technology development should not be too selective; losers must not too quickly be forced out of the network. They add to the richness of the network: the options available, the information produced, the assumptions made explicit, the values to choose from. In short, network management is essential for CTA, but not every network is good for CTA. Thus CTA-relevant networks should protect *and* manage cultural pluralism (Schwarz and Thompson 1990).[22]

The networks in which biotechnology has been developed up to now have not always met these criteria. In the strategic context of biotechnology development, it has been obviously difficult for promoters and critics of technology to remain reflexive, let alone to entrench reflexivity in the shape and management of the networks. Such structural reflexivity is necessary for learning to continue and for its outcomes to be evaluated. Especially destructive to this goal has been, in the early stages of biotechnology development, the demarcation of scientific issues from political issues. Conceiving the problem in "separate spheres" stimulates the premature rise of separate networks. Now, in the later stages, there is a struggle going on again to unite promoters and sceptics of biotechnology.

Technical fixes are important too
Not only do social means like network management create conditions for learning; technical means can also do this job. In the biotech case, technical means supported the social ones in generating a situation in which learning became possible. In the early period the discussion focused on the nature of the effects of recombinant DNA experiments: could they lead to disaster? This very question could not be answered because of ignorance. Thus the debate ran into a deadlock, generating only a clash of paradigms and increasing politicization. As soon as the risk problem was (re)structured — requiring network management — on a *limited* class of effects, technical means (physical and biological barriers) could be committed to the problem to create a contained situation in which trial-and-error learning could occur. Thus the focus shifted from an "analysis without strategy" (Morone and Woodhouse 1986) to developing a strategy for learning, made possible by averting disaster through application of technical means (containment). In the second period, a similar role was played by the development of detection technology (like application of harmless markers for monitoring of GMOs) in the environment.

Notes

1. Stephen Lindow (University of California, Berkeley), originator of research on Pseudomonas syringae which led to the "Ice Minus" strains (Frost-ban) used by Advanced Genetic Sciences (Oakland, CA), quoted in Van Brunt (1987).

2. For an extended analysis of this period and for references, see Jelsma and Smit 1986 and Wright 1986. Historic data are derived from numerous conference proceedings and many journals, the most important of which are: *Science, Nature, New Scientist, Trends in Biochemical Sciences,* and *Chemical and Engineering News.* The consulted books are Wade 1977, Lear 1978 and Krimsky 1982.

3. It is quite clear that the genuine concerns among scientists about potential risks were mixed with concern about their freedom to do research. In the concluding section of a memorandum to the signatories of the Berg letter, Roy Curtiss, a microbiologist from the University of Alabama, warned about the establishment of central regulatory agencies if scientists were too

slowly addressing the problems and dangers of their research. He reminded his fellow scientists of historic cases like radiation hazards and experiments with humans and animals (see Curtiss 1974, 15-16). Bernard Davis, a Harvard bacteriologist, commented on the Berg letter by stating that "the initiative to police ourselves [...] may help to avoid external imposition of excessively severe restrictions" (Davis, 1976).

Thus at the Asilomar conference (see below), decisions were reached "in the explicit awareness that science no longer enjoys the automatic favor of governments and society, and that if the scientists present failed to regulate themselves in an evidently disinterested manner, others would do so for them." One of the prominent scientists present, Sydney Brenner, illustrated this feeling by saying: "we live at a time where I think there is a great anti-science attitude developing in society [...] we must be seen to be acting" (see Wade 1975).

4. The notion of equivocality reduction as a condition for social interaction is borrowed from Karl Weick; see Jelsma and Smit 1986.

5. It can be shown that biological containment indeed functioned as a technical fix. As long as this fix was not credibly shown to be realized, the moratorium could not be lifted, and regulation was difficult to design. Since attempts to develop a crippled strain of *E. coli* ran into numerous snags, the committee which had to write acceptable regulations from the Asilomar guidelines found itself in a difficult situation. The final production of a crippled strain which seemed to meet all the requirements strongly facilitated the adoption of a set of strict regulations (see Norman 1975).

6. For scientific advisory committees, this practice is rather rule than exception. Questions to be decided in regulatory science generally involve subjective judgement, and even policy, to compensate for the absence of hard knowledge (Jasanoff 1990: 94).

7. The three categories are based on a qualification of actor views expressed and reported in the vast literature about the subject of regulation of genetic engineering. For example, I see Jonathan King and Richard Novick as important actors in the radical group, and Jim Watson, Joshua Lederberg and Bernard Davis as leading representatives of the establishment. The group around Berg can be considered as the pragmatic centralists, while Roy Curtiss can be seen as moving from an initial radical position to a centralist one in 1977.

8. A vocal critic of the NIH guidelines and of genetic engineering in general was Robert Sinsheimer, a biophysicist from the California Institute of Technology (Caltech). His main opponent was Bernard Davis, a bacteriologist from Harvard Medical School. For a comparative description of their opposing arguments, and references, see Jelsma and Smit 1986, 726-727.

9. Certainly the shift in opinion of Curtiss, who had been one of the most outspoken of the "proceed with caution" people, made an impact. He informed the director of NIH about his change of mind in an open letter that received world-wide publicity (Curtiss 1977).

10. The shifts in the problematization of risk in this period are demonstrated in Jelsma and Smit 1986, 729-731. This demonstration is based mainly on a detailed analysis of Curtiss's argument, comparing his revised view on risk with his original stand on the issue before 1977.

11. This "no" was the reassuring outcome of the risk workshop at Falmouth, Mass., in June 1977. The workshop's outcome would later be cited repeatedly as a justification for subsequent relaxations of the NIH guidelines. For an in-depth analysis of the argument underlying this "no," see Krimsky 1982, 217-232.

12. In 1978 and 1979 Curtiss, who had constructed the "safe host" as a basic element of the NIH recombinant DNA guidelines, wrote several letters to the director of NIH. In these letters Curtiss reported new evidence showing that the safety of the biological containment fell short of his earlier expectations. In contrast to his earlier letters, these critical letters fell on deaf ears.

13. Unless reference is made to other sources, the data for this section have been derived from the journals *Bio/Technology*, NIH's *Recombinant DNA Technical Bulletin*, *Science* and *Nature*.

14. In terms of social rule systems theory, one could say that the molecular biologists lost power by the shift forced upon the regulation of biotechnology towards the "rule systems" of federal laws. According to this theory, a particular rule regime entails an inherent distribution of social power (Burns and Flam, 1987).

15. This interpretation comes down to the claim that recombination of DNA by humans cannot contribute new viable combinations to the gene pool, since in the long process of evolution all feasible combinations of DNA fragments have been generated and selected already. In the 1970s this claim, vigorously forwarded by Davis, met strong criticism by Sinsheimer.

16. According to the process approach, any product from processes changing the "natural gene pool" because of "deliberate genetic intervention" is covered by regulation. In the product approach, only those of such products which have potential for hazard — i.e., intergeneric microorganisms and intrageneric pathogens — are subject to regulation (see Office of Science and Technology Policy 1984, 1986).

17. EPA justifies its switch to the product-regulation approach in: Office of Science and Technology Policy, 1986, 23315.

18. BSCC's demise must be ascribed to its attempt to mix two types of interagency coordination which probably cannot be combined: (i) identification of issues, exchange of information, fostering discussion between agencies, i.e. *process* management, and (ii) imposition of certain policies on agencies intended to move the Executive Branch in a unified and consistent direction, i.e. management of *policy content*. Moreover, BSCC became embroiled in policy matters because it was lobbied to do so by industry and the Food and Drug Administration (Shapiro, 1990). To be successful, scientific advisory committees should keep their distance from policymakers as well as from social pressure groups (Jasanoff 1990).

19. In the field of detection technology, the main strategies are to develop marker systems, selective media and gene probes, and the application of methods based on immunofluorescence and immunoradiography (McCormick 1986)

20. To give an example, the secretary of the Dutch licensing committee on biotechnology (VCOGEM) said in an interview: "Of course one cannot afford, as a committee, to say to investigators: please come with your proposals, and subsequently reject them. Only one proposal was rejected [so far...]" (*De Volkskrant*, 16 March 1991, p. 19). EPA too depends on firms as well as the public for approval of its regulation; disapproval would undermine its position in the interagency struggle about regulation. At the same time, the agency needs biotech researchers for scientific input into its design of regulation; see *Bio/Technology* (1989), Vol. 7, p. 555.

21. Ken Collins, member of the European Parliament, Chairman's Address at the Opening Plenary of "Products, Regulators and Politics," First Biotechnology Europe Conference, The Hague, 22-24 June 1992.

22. A useful treatment of the tension between conflict and compromise is offered by Weick (1979: 220). Weick argues that the presence of conflict signifies the retention of heterogeneous responses and preferences, all of which may be adaptive under some circumstances. Therefore, to retain the capacity for generating adaptive responses in future situations, it is important to preserve polarity as long as it is serving this purpose. Weick makes this **argument** with respect to group dynamics, but its value should be considered on the level of (segments of) society as well.

References

Baltimore, D., et al. 1977. Resolution of the National Academy of Sciences, April 26. Washington DC: National Academy of Sciences.

Berg, P., and M. Singer. 1976. "Seeking Wisdom in Recombinant DNA Research." *Federation Proceedings* 35 (14): 2542-2543.

Burns, T.R., and H. Flam. 1987. *The Shaping of Social Organizations/Social Rule Systems Theory*. New York: Sage.

Callon, M. 1992. "The Dynamics of Techno-Economic Networks," in: R. Coombs, P. Saviotti and V. Walsh (eds.). *Technological Change and Company Strategies*. London: Academic Press, 72-102.

Curtiss, R. 1974. Memorandum. The University of Alabama, Department of Microbiology.

Curtiss, R. 1977. Letter to the director of NIH. Birmingham, Alabama, April 12, 1-14.

Davis, B. 1976. "Novel Pressures on the Advance of Science." *Annals of the New York Academy of Sciences* 265: 193-205.

Fox, J. L. 1986. "IBA: A Few Reservations of U.S. Rules." *Bio/Technology* 4: 932.

Fox, J. L. 1987. "The U.S. Regulatory Patchwork." *Bio/Technology* 5: 1273-1277.

Jasanoff, S. 1990. *The Fifth Branch: Science Advisers as Policymakers*. Cambridge, Mass.: Harvard University Press.

Jelsma, J., and W.A. Smit. 1986. "Risks of Recombinant DNA Research: From Uncertainty to Certainty," in: H.A. Becker, and A.L. Porter (eds.). *Impact Assessment Today*. Utrecht: Jan van Arkel, 715-741.

Krimsky, S. 1982. *Genetic Alchemy; The Social History of the Recombinant DNA Controversy*. Cambridge, Mass.: MIT Press.

Lear, J. 1978. *Recombinant DNA, The Untold Story*. New York: Crown Publishers.

Lewin, R. 1974. "Ethics and Genetic Engineering." *New Scientist* (October 17): 163.

McCormick, D. 1986. "Detection Technology: The Key to Environmental Biotechnology." *Bio/Technology* 4: 419-423.

Molin, S., P. Klemm, L.K. Poulsen, H. Bielh, K. Gerdes and P. Andersson. 1987. "Conditional Suicide System for Containment of Bacteria and Plasmids." *Bio/Technology* 5: 1315-1318.

Morone, J.G., and E.J. Woodhouse. 1986. *Averting Catastrophe: Strategies for Regulating Risky Technologies*. Berkeley: University of California Press.

Norman, C. 1975. "Genetic Manipulation: Recommendations Drafted." *Nature* 258: 561-564.

Office of Science and Technology Policy. 1984. "Proposal for a Coordinated Framework for Regulation of Biotechnology." *Federal Register* 49 no. 252: 50856-50907.

Office of Science and Technology Policy. 1986. "Coordinated Framework for Regulation of Biotechnology Part II." *Federal Register* 51 no. 123: 23302-23393.

Ravetz, J.R., and S.O. Funtowicz. 1989. "Usable Knowledge, Usable Ignorance; A Discourse on Two Sorts of Science." Paper presented to Colloque Internationale "Les Experts Sont Formels," Arc et Senans, 11-13 September.

Rip, Arie, and Henk van den Belt. 1988. *Constructive Technology Assessment: Toward a Theory*. Enschede: University of Twente.

Rodgers, M. 1979. *Biohazard*. New York: Avon Books.

Shapiro, S. A. 1990. "Biotechnology and the Design of Regulation." *Ecology Law Quarterly* 17 no. 1: 1-70.

Schwarz, M., and M. Thompson. 1990. *Divided We Stand: Redefining Politics, Technology and Social Choice*. New York: Harvester Wheatsheaf.

Sun, M. 1986. "Monsanto Opens Files on Genetic Release Test." *Science* (7 March): 1065.

U.S. Government Interagency Working Group on Competitive and Transfer Aspects of Biotechnology. 1983. *Biobusiness World Data Base*. Washington DC: McGraw-Hill Publications.

Van Brunt, J. 1987. "Environmental Release: A Portrait of Opinion and Opposition." *Bio/Technology* 5: 558-563.

Wade, N. 1975. "Genetics: Conference Sets Strict Controls to Replace Moratorium," *Science* 187: 931-935.

Wade, N. 1977. *The Ultimate Experiment*. New York: Walker and Company.

Weick, K.E. 1979. *The Social Psychology of Organizing*. Reading, Mass.: Addison-Wesley Publishing Company.

Wright, S. 1986. "Recombinant DNA Technology and its Social Transformation, 1972-1982." *Osiris* Second Series Vol. 2: 303-361.

8 User Representations: Practices, Methods and Sociology

Madeleine Akrich

Technological developments impinge on a society composed of complex and many-sided users moving within a heterogeneous set of relationships. Users are simultaneously consumers, citizens, members of a family, of a religious faith, of a social club, of a political party, of a profession, or whatever. In many cases, technical design debates turn on the need to arbitrate between the varying positions of the user which are more or less incompatible within a single technological system. Accordingly creating a market invariably means tuning in to several wavelengths simultaneously. The way an innovation takes hold cannot be reduced to any one model, e.g. the rational economic consumer, the status-seeking purchaser or the inveterate trend-follower. Instead, the success or failure of innovations frequently depends on their ability to cope with dissimilar users possessing widely differing skills and aspirations. The micro computer market is a very instructive example here. The importance of users in technological development has been recognized by a number of economists. In particular, they have shown the following:

1) The process of innovation does not end when a product is launched. The final destiny of an innovation depends largely on the recruitment of first users (David 1986a,b) and on their ability to generate a whole series of improvements affecting not only the proposed technology but also elements of its environment and/or complementary technologies (Rosenberg 1976 and Teece 1986).
2) In certain technological areas users play a predominant role in the conception and diffusion of an innovation while entrepreneurs mainly take care of producing the device (Von Hippel 1976 and 1988).
3) The development and commercialization of innovations are based on interactions between the innovators and their potential users, a process which enables both sides to improve their knowledge and which allows for small adjustments to the products (Andersen and Lundvall 1988; Meyers and Althaide 1991).

In this chapter I will extend this economic analysis of users in technological development. Economists mainly focus on how the interaction between innovators and certain users allows for adjustment between an offer that already exists and a demand which will only define itself during the interaction. I go one step further and argue that innovators are from the very start constantly interested in their future users. They construct many different representations of these users, and objectify these representations in technical choices.

The different forms of user representation generated during innovation processes are of critical importance for constructive technology assessment (CTA) which aims at modulating technical change. CTA must seek to discover how it is possible, in the processes of developing and marketing a technological system, to reconcile what may be barely compatible and even conflicting user representations. For example, many opinion polls show that a majority is in favour of reducing the country's energy consumption (whether on the grounds of economy, of environmental conservation, of national self-sufficiency, or whatever). A mere appeal, however, to people's sense of responsibility is not enough to produce a new type of energy-saving user. What is needed are application systems superimposing the user representations of personal comfort (entailing high energy consumption) with careful consumption of energy. Realizing this superposition requires the possession of representations (in a cognitive and political sense) of what the supposed users are and what they want (Compare Downey, this volume).

In this chapter, I shall describe the variety of techniques employed by system designers to construct and then to appropriate user representations. This analysis will make clear that the challenge for CTA is not so much devising new user representations, for there are many of them already. CTA must discover methods for coping with the many existing representations, so that they ultimately converge in a way combining different representations and enabling the initial development programme to go forward. Going back to our energy-saving example, questions need to be answered. First, how to bring users who may be insensitive to national policy demands and beset by a variety of more immediate problems and constraints, to take an interest in energy-saving systems? Second, once energy-saving systems have been adopted, how to ensure their use will in fact save energy? This is important, because users may always deform or defeat the initial purpose of a technological system (see Akrich 1992 as well).

Throughout the chapter three case studies will be discussed.[1] All three are standard technological systems for the mass market:

- an interface terminal linking a TV cable network to its subscribers, known as the CA unit ("coffret d'abonné") and basically designed for TV programme selecting;
- a range of telephones known as the "Contact Ambiance" range and designed for household use;
- a domestic computerized system known as "Securiscan," which has a number of applications ranging from house surveillance to programmed heating and remote-control of automatic devices, adaptable according to user specifications.

The CA unit was supported by the French Telecom administration and developed in close collaboration with the manufacturers, whereas the other two systems are entirely private developments. These three systems illustrate that the creation of successful artifacts depends on the ability of innovators to generate user representations and integrate them into their designs, and that alternative strategies are available for aligning the various user positions. When analysing these strategies, I shall extend the scope of the innovation process beyond its generally accepted boundaries, to encompass marketing and policy decisions that illuminate lines of thought and action relevant to a constructive technology assessment.

User Representation Techniques

In this section I describe all the techniques for producing user representations observed in our research. Our approach diverges fundamentally from that of the market researcher who deals with only a fraction of the full range of methods deployed in connection with designing and developing technological systems. We deliberately refrain from making a prima facie distinction between techniques "legitimized" by a formal scientific and conceptual basis, and those of a more empirical kind lacking any such basis. In our approach, we consider that the mere fact that an argument introduced into the project discussions was put forward in the name of the user makes it relevant to our purposes. This brings us to classify the techniques observed in two categories: *explicit techniques*, based on special skills or qualifications in the area of defining or interpreting consumer representations; and *implicit techniques*, which rely on statements made on behalf of the users.

Explicit Techniques

Market Surveys
Market surveys enjoy a very special status. When questioned about users, the authors of an innovation immediately quote the market

survey as a pillar supporting their perception of the future users of their systems. When pressed for specifics, however, they become more evasive and produce nothing more than a few broad generalizations hinting at the potential strong demand for products of a certain kind. Researchers who persist by trying to obtain a copy of the survey report are likely to be disappointed: in only one of the three cases investigated were we able to find any written trace of such a study. Our experience is that this strange state of affairs reflects that, in many cases, these studies are used to persuade the higher decision-making authorities to support the project, but that these studies are rarely consulted during the subsequent development phase. In other words, actual practice is based on a division of responsibilities between the technical and the marketing sides, the former handing over a complete product package that the latter are expected to sell "as found." In these conditions, the market survey may be used to identify buyers, or to put the final touches on the sales argument, but it certainly has nothing to do with the design and development of the actual product.

The one market survey report we obtained, for the "Contact Ambiance" telephone, confirmed this interpretation. It was issued when the product design phase was nearly completed. Its salient indicators were of the socio-economic type ("originality" of outward appearance and price). The report did not include technical recommendations, thus leaving every discretion to the designers. The purpose of the market survey was not to reveal opportunities for creating a new demand, but (assuming an existing demand) to identify anything which might restrain the buying impulse.

Consumer Testing

When a product emerges from its design cocoon, its designers often organize tests directed to a sample group deemed to be representative of its future users. The nature and coverage of such tests can vary a great deal and, in the case of the CA terminal, two different types were used. The first type of test was conducted with personnel of the sponsoring administrations and the manufacturer. The aim was to find out which of several different presentations being considered for the unit would meet with the widest approval. Other questions related to the place reserved for the unit within the "domestic layout arrangement." The survey was of the "democratic" type assuming the existence of an average standard of taste, minimizing the number of dissatisfied users, and weighting heavily the user motivation factor.

The consumer test of the "Contact Ambiance" telephone consisted of introducing strongly contrasting outer presentations of technically identical products. This was in direct contradiction with the prior assumptions underlying the test described above. Thus, it is obscure why the test was done in the first place. The markets targeted by the other two products were entirely dissimilar. For instance, it was not the interface terminal (CA unit) itself which was marketed, but the whole system carried by the video communications network. This case makes clear that the nature of a test depends on the kind of market that is expected. In other words, the implementation of a particular consumer testing technique amounts to making one or more assumptions about *which* market is expected to expand.

The second type of test was contracted out to specialized institutes independent of the project sponsors. The purpose was to confirm the CA unit's ergonomic validity. At the same time, the interviewees were asked to give opinions concerning the CA unit, its place within the household, and the services to which it gave access. Here, the user was a "system operative" approached through a "school exam" test. The way the questions were formulated, presented and contextualized, situated the examinee as a user of the technological system. Tests of this type cannot be expected to cover, once and for all, the whole range of problems likely to face the user in contact with the system. Generally speaking, the "classroom" situation places special emphasis on some problems, whereas some others arising only when the network is in full-scale operation go unnoticed.

One set of unnoticed problems evolved around security procedures. In the proposed CA unit, access to services subject to surcharges was protected by a secret code word which had to be keyed in whenever these channels were called up. This enabled the subscriber to control access to these surcharged services, e.g., by preventing children in the house from running up unwanted bills. Once the user selects a restricted-access channel, if he presses the "delete" key before entering his code word in full, he is returned to the general-access channel and has to recommence the entire procedure including selection of the restricted channel. Here, the system assumes that the user has abandoned his intention to call up the service in question, either because he does not wish to incur the charge or because he finds himself ineligible (not possessing the code word). This situation is not foreseen in the test scenarios designed to verify understanding of operations instructions and procedures. The test assumes that, in all cases, the user knows what he wants and is duly entitled to obtain it. The test does stimulate a rather different situation, however, namely that the user has committed an error when entering his access code, but still wishes to call up the restricted channel by re-entering the code correctly. In this situation, being returned to the general access channel is found either a nuisance or an inexplicable aberration.

Thus, although such tests may often have the outward appearance of an encounter between humans and machines, we must not forget that they do not cover the full extent of the relationships between the two. They rely on "typical" and simplified situations for the purposes of demonstrating particular user responses. This makes them efficient, but at the same time limits their true relevance.

Feedback On Experience

Through contacts with after-sales services, equipment suppliers, and other customer services, the "real" users feed back information to technological system designers. But this information is twice filtered: by the users themselves, who pass on only those remarks they consider relevant to their relationship with the system itself or with its agents; and by the latter, for similar reasons. In the cases we studied, this feedback was one of the factors causing modifications. For example, it was only in real-life situations that a problem with the aiming of the CA unit remote-control pad was revealed. The pad was effective only if pointed directly at the unit itself. The test conditions, although designed to challenge the competence of prospective users, totally failed to detect problems of this kind. Subjects of testing are much more concentrated on this particular task than users, sitting comfortably at home in front of their TV screens. The "laboratory" situation imposed by the test protocol was unrealistically strict. The real-life user refuses to be represented as a dedicated keyboard operator. While using the remote-control pad he wants to be able to maintain normal relationships with his environment: to take part in a conversation, to glance at the programme schedule, to help himself to a drink or a bit of food, and so on.

In the case of the "Contact Ambiance" telephone, the after-sales department insisted on adding a bracket enabling the user to put down the handset, without hanging up, which is needed, for instance, when he has to go and look for information requested by his correspondent. The designers had represented and met consumer-purchaser wishes such as easy storing, by providing a wall-mounted telephone (not a usual option in the business world to which the manufacturer has hitherto been accustomed). But they had not simulated a comprehensive user universe, which might have revealed the unfortunate spectacle offered by a telephone dangling forlornly in space. Although, as our case studies show, feedback of after-sales agencies can suggest design changes, it is not always possible to modify the system to meet every complaint or suggestion. In some cases this could create serious difficulties on the production and organizational sides.

Implicit Techniques

Market surveys and testing have the common aim of seeking to simulate the prospective user's behaviour and reaction. They present "typical" scenarios acted out by people deemed to be representative of real-life users. However, they are not the only producers of user representations present in the design-development processes. Feedback via after-sales services provides much valuable additional input. Other less-formal techniques can be identified as well. These techniques have in common that they actually address the "real" users, but rely on spokespersons of three general types: designers, expert consultants, and other products.

"I..."
Reliance on personal experience, whereby the designer replaces his professional hat by that of the layman, is a much more common device than might be thought at first sight. When there is no other available means of bringing in the end-user, or when organized test procedures seem too complicated or too expensive, arguments uttered from the founts of inspiration can carry a certain amount of conviction. In any event, reliance on personal experience was used throughout the design and development phase of the CA unit. This was no doubt due to some degree of isolation of the design team. It was heavily dominated by engineering specialists, with nobody to represent the marketing side, no true ergonomics expert (even if some engineers thought they could handle this aspect), no media representative, and no operator of a comparable service.

The Experts
The user is brought into the equation without ceremony but with great impact depending on the way that a company or project team organizes the development of new products. The process is initiated by working procedures which enable effective collaboration between the technical and the marketing departments although, as we found in the "Contact Ambiance" study, this does not always occur. In that case, the marketing side won authoritative status only after a hard struggle. It is not so much that marketing people are more authorized to represent the users than their technical counterparts. In fact, giving too much

authority to marketing people who by definition are more concerned with the user's market role (purchaser, consumer etc.) may result in neglecting the user in his technical role (machine operator and "fixer" etc.). In our cases introducing this particularly technical facet of the user occurred only in a late stage of product development, i.e. at a point where it was not easy to revise the technological options. These findings show that "the user" is not a single entity taken on board when the project is launched, but a set of disparate characteristics which will not necessarily merge into a tight configuration ready to accommodate the definitive end-user. The presence in the project organization of a wider range of disciplines, when they all contribute to formulating the technical options, may expedite and facilitate the amalgamation of these disparate characteristics.

The production of the user's manual for the CA unit was an illustration of how user representation of experts can be brought into the project organization. A special group composed of new kinds of specialists, separate from the design team proper, was set up to write the instruction manual. In particular, the new group included a specialist who had been connected with the first French experiment with optical-fibre cable networks (Biarritz pilot experiment) and another who had contributed to the user's manual for the Minitel system. Thus, the project organization "co-opted" experts to produce user representations on the basis of their experience of user relations in similar technological systems.

Other products
Yet another option to elicit user representations is to adopt or reject representations present in products considered to have something in common with the innovation project in hand. This assumes the existence of relationships between the technical options under consideration and the type of users of the equivalent product. The adoption or rejection of an existing formula is in itself a more or less explicit decision concerning the type of user being targeted and his/her tastes, expectations, etc. Making such a decision requires calling up the particular representation of the user incorporated in the comparable product. We found examples of this in all three of our case studies.

From the outset the Securiscan project, a system for controlling all kinds of domestic applications, was defined in *opposition* to existing systems. The aim was to develop a "wire-free" installation, whereas all the alternatives involved wiring networks.[2] This approach was based on the marketing concept that additional wiring constituted, on both financial and aesthetic grounds, a major buying constraint. It was argued that the market would expand significantly once this constraint had been removed.

The manufacturer of the "Contact Ambiance" telephones referred back to options for products he had already supplied to business users, sometimes incorporating the options and sometimes discarding them. For example, amplified reception had become a standard feature of business installations and it was retained for the top-of-the-line domestic model. This decision sharply contrasted with the general trend to omit amplification at that end of the domestic market. However, for reasons not made entirely clear, the sound amplification needed to be of higher quality for domestic than for business purposes. Finally, the "confidential" function, also a very common feature of business installations (it cuts off the microphone) was omitted for the domestic models on the grounds, assumed specific to this market, that it fulfilled no useful purpose since users did not trust it and continued to cup their hands over the microphone.

Finally, the CA unit keyboard as well as the user-CA dialogue was designed on the same lines as that of the Minitel terminal, as opposed to that of TV remote-control pads. For instance, it features a key labelled "envoi" (used to validate the instruction just keyed in) which has no counterpart on the TV controls. The fact that these options were taken almost without discussion is not surprising when we know that CA unit designers had previously worked on developing the Minitel. At each stage of design, they set the ground rules governing interactions between the user and his terminal and thereby incorporated in the system a particular concept of that user's aptitudes, frames of reference, and behaviour patterns.[3]

Summary

Without aiming at an exhaustive account, we have drawn up an inventory of methods employed by system designers to develop, promote, and impose a particular representation of the user. We have been concerned to show that despite apparent differences each method aims at articulating and developing user representations. Not all methods reviewed are of equal force. Contrary to what is expected and put forward by designers, implicit methods seem to be more powerful than explicit ones. As well, different methods correspond to different problems and circumstances. We particularly wish to stress the point that, even if our use of the term "user" tends to suggest the contrary, none of the constructed representations is merely the representation of a particular facet which the "complete user" may display in a given situation. From this standpoint, the non-discriminating quality of the word "user," which is also an obstacle to its scientific aptness, well illustrates the incomplete synthesis between all the different postures offered by the representation techniques deployed and, sometimes, materialized in technological options. The problem for successful design as well as for CTA is how to deal with this proliferation of "users," each of whom corresponds to a specific situation. At some point, if they do not merge into a whole, they must at least become coherent with each other.

The Articulation of User Representations

From Local to Global Rationality
Arbitration between different representations of the user which may lead to conflicting pressures, or to the maintenance of certain outmoded user representations throughout the development process, is a problem that innovators have to deal with all the time. It would be unwise to overrate the rationality of the choices made. Often they are rationalized afterwards, when the corporate management (or for that matter the sociologist) asks the project leaders to describe how the project evolved. Situations of severe disagreement, for instance, taking the issue outside the laboratory or company bounds, also impose greater coherency of explanation.

We should not, however, regard these *ad hoc* decisions as purely arbitrary. Each of them can usually be justified at the local level by the configuration at a given point in time, i.e. the positions of the actors involved, the material factors at issue, and the nature of the problem that brings the actors together. Still, local rationality cannot explain in every case why some demonstrably relevant system characteristics were ignored. We have to look at the accumulation of separate decisions which builds up to an outcome that takes its final shape only when the whole process is complete. There is no absolute technological determinism at work in design processes. Each new option will rely on decisions taken at an earlier stage, but in most design processes there is enough latitude to neglect (some) elements of earlier decisions.

For instance in the CA unit case, the cable TV users in Montpellier were surprised and disappointed because they could not use their videorecorders as previously. They encountered three kinds of constraints. First, connecting the CA unit, the TV receiver, and the videorecorder was by no means a straightforward matter. It sometimes involved switching the connections depending on whether a programme was recorded or was viewed after it had been recorded. Second, it was not possible to view a programme other than the recorded one at a given time. Third, preprogrammed recording (i.e. in the user's absence) entailed severe constraints. When we try to explain these apparent design flaws, we come up against a paradoxical situation. On the one hand, a multitude of decisions affecting separate features of the CA unit can explain the failures. These *ad hoc* decisions were fully justifiable in the specific situation. Only the reconstruction by the analyst (prompted perhaps by end-user complaints) imposes a coherent sequence which led to a disqualification of the videorecorder. On the other hand, it cannot be said that the design flaws and subsequent disqualification were at any time a deliberate intention of the designers, or could be easily foreseen.

This raises the problem of how to simulate a confrontation between different user representations in the design-development phase and, from there, how to devise compromise solutions which ensure their superposition. In the following we will show that several different strategies can be applied to combine and superimpose separate representations of the user. These stategies involve not only a reallocation of product-definition responsibilities but also a rigidification of the technical objects conducted on a very different time schedule.

The needed compromise of user representations entails the alignment of networks. When we designate the user as a "customer", "telephone subscriber", "town-dweller", "parent", or whatever, this immediately calls up a picture of a whole set of actors, objects, and networks of relationships binding them together. The term "telephone subscriber" brings with it objects called "telephones", whose keyboards are laid out in a particular way, "actors" like the Telecom administration personnel, bills made out in a certain way (fixed component and pro rata charges, etc.) and a definition of the contractual relationships and responsibilities involved. In the case of the CA unit, the "telephone subscriber" reference invokes all these features either directly or by inference. As is already put forward in actor-network theory and shown for many cases, a new technological system will succeed only when it is able to attract a whole universe: a network of socio-technical relationships has to be put together, persuaded, and enlisted. In the final analysis, verifying the viability of the proposed combination of user representations entails determining if a system is able to relate harmoniously to appropriate networks, and ensuring that the various implications of the proposition are not conflicting and do not introduce intolerable stresses or constraints. This brings us to the various strategies available for achieving network alignment.

Aligning the Socio-Technical Networks

Three different strategies were observed in the course of the case studies performed: 1) relying on the technological system itself, 2) delegating reconciliation of representations to intermediaries belonging to established socio-technical networks, and 3) "bridging" networks put in place concurrently with the system's development. They differed according to the basic means adopted for the purposes of reconciling the varied facets of the user.

Delegating the alignment task to the system means endowing it with a number of physical features that enable it to assume all the functions this involves. This was the strategy adopted, almost to the point of paroxysm, in the case of the CA unit. It could be said that the system's designers produced a marvellous machine capable of coping with extreme market segmentation by narrow differentiation of both products and subscriber characteristics. They relied on the Minitel example which seemed to sanction a form of technological determinism. They failed, however, to construct a supply structure capable of giving proper shape to this new type of audiovisual market. From the very outset, the project's stated objective was to set up a multi-media system (hi-fi sound, TV, data-transmission, telephone etc.). This was technically feasible only through the use of optical fibre. The system was interactive enabling the user to communicate with the network operator. It could accommodate a range of scales of charges (basic subscription plus a large number of options, as well as restricted-access viewing for services open only to selected user categories (e.g. programmes produced by pharmaceutical companies and reserved for medical practitioners). This complex configuration meant that the CA unit had to handle transactions on two levels, the technical (selection and reception of programmes by users as desired) and the contractual (effective protection of restricted access services, clear notification of financial commitment undertaken by subscribing to particular services). The designers thus had to make provision for a number of access keys (numerical code words) for general use, for admission to restricted-access or surcharged programmes, and for preventing unauthorised use of the unit.

The access key for surcharged programmes was not included in the original plan, but gradually emerged as the solution to a problem that the designers had to face: How to ensure that a user acknowledges his debt when making a financial commitment to the network operator? As a solution, the designers incorporated on the front of the CA unit an additional feature, a "surcharge display" which informs the user about the selected programme's tariff status. The user signifies the acceptance of the conditions displayed by pressing twice on a key labelled "envoi." This is the formal equivalent of a signature and ensures that the "viewer" and "customer" facets of the user coincide. The second facet covers the contract relation with the network operator.

Once this point had been dealt with, yet another problem arose: the customer is already defined by another procedure. The signing of the subscription order and the additional procedure brings no guarantee that this signatory is the *same* user as the one who pressed twice on the "envoi" key. The possible non-coincidence of these two actors may give rise to difficulties, for example, when subscribers contest the amounts charged to them. The designers thus added a "paying access key" that was supplied to users on request, and whose purpose was to guarantee the existence of a legally responsible person backing up the user who presses twice on the "envoi" key. This grants power of decision to duly authorized members of the user household, freely conferred by the contract signatory, with the latter remaining solely responsible in the eyes of the network and its operator. This is a particularly clear illustration of the way in which the strategy adopted by the unit's designers tends to achieve the coincidence of different representations of the user entity by incorporating appropriate features in the technological system itself.

The "range of products" approach is another way of delegating the task of reconciliation to the artefact. In the case just discussed, a single artefact capable of coping with very different situations, expectations and requirements had to be developed. An alternative strategy consists in determining coherent sets of user characteristics. For example, if the business user and the domestic user are not likely to converge, a successful approach might be to develop two similar but not identical products, each tailored specifically for the identified user type.

On the whole, strategies that delegate reconciliation functions to the system hardware are based on a "technological determinist" concept of the innovation process. They elevate the innovator to the status of a demi-god, capable of foreseeing and handling user reactions. The process of identifying user representations is fully concentrated in the design phase. This involves the risk (which, in France at least, has often materialized) of ending up with a kind of technological monster, extremely sophisticated but finally quite ineffectual because it is unable to attract the users for whom it was intended. This excessive sophistication is not due to a pure love of technology, blind to the mundane conditions of the real world, but to an undiscriminating determination that the system itself should cope with the whole infinite variety of possible situations it might have to address.

Reconciliation of divergent user representations can also be delegated to intermediaries, the second general strategy. Such intermediaries might be "outsiders" regarded as able to provide entry to established socio-technical networks. These networks are themselves identified with the circulation of specified items (or systems) between actors whose own identities are defined in terms of that process of circulation. If such networks are to be able to channel the new system towards its intended users, a mutual rapprochement must be effected between the users and the system. This is the task the intermediaries have to perform.

The Securiscan system used a strategy of this kind. The designers started with the idea of a high-tech product adaptable to individual user specifications and distributed through supermarkets. The designers soon realised that such a configuration assumed a variety of user representations which were unlikely to converge in a single individual. These representations were those of the gadget-lover, the do-it-yourself addict, the person preoccupied with security, the computer specialist, the supermarket customer, the middle-to-high income earner, and so on. At a fairly early stage the decision was taken to abandon the distribution through supermarkets and instead rely on conventional dealers who were better able to sound out their clientele. The dealers were to work with any interested parties to devise a system configuration corresponding to their individual requirements, keeping within the limits of the technical possibilities. In addition they installed the system, programmed its implementation, provided after-sale services, and encouraged users to extend their installations over time.

A strategy of this kind assumes a special design approach. It must allow and enable the intermediary, after appropriate coaching, to make the necessary adjustments between user and system. The Securiscan designers had to satisfy just as much the user wanting to run his heating system with optimum efficiency, for example, as the one who is afraid of burglars, or wants his lawn to be watered when he is away, or his shutters to be closed automatically as well as to meet any *combination* of these requirements. In response the final version of Securiscan is basically an open-ended system comprising three types of components:

- a central programming unit governing the overall installation set in place by the intermediary;
- end-applications equipment (e.g. switching mechanisms with control doors or taps);
- interface elements which transmit the central unit's instructions to the applications.

It was an interactive system. The applications equipment could feed back information (such as thermostat readings) to the central unit, which then made decisions. The above example shows that the strategy consists in appointing intermediaries to organize the demand side, ensuring the convergence of disparate user representations. This strategy requires that product design must not be the sole prerogative of the people engaged in the design-development phase proper, and that the distributors must be qualified to make any finishing touches required.

The last of the three possible strategies identified was creating a new network. When the Contact Ambiance telephone was being brought out, telephone sets as such were not recognized as retail items in the French market. They were supplied by the Telecom administration to all new subscribers and thus considered as a natural extension of the telephone network itself. The problem thus arose of how to bring about the existence of a "telephone buyer." Initially, the manufacturer thought that the mere fact that people bought products of all kinds was sufficient to ensure that they would also buy telephones. Accordingly the intention was to sell the Contact Ambiance telephones through the mass retail outlets. However, although the telephone was a familiar object, a set of problems emerged which could not be solved through this style of marketing. In supermarkets, for instance, the telephones were displayed "unescorted." How could potential customers be persuaded that there was any point in buying telephones when they were already supplied by the Telecom administration at a nominal charge and replaced at no cost whenever necessary? And how could they be convinced that the products on display were of high standard and fully compatible with on-going plans for expanding the telephone service, then in full swing? The strategy adopted by the manufacturer focused on two main points: securing approved status for the "Contact Ambiance" range from the Telecom administration; and creating a specialised sales network.

In this case we do not have a pure example of creating a specific network because features of the "intermediaries" strategy were also present. The manufacturer sought to superimpose the "telephone subscriber" and "telephone buyer" with the aid of an existing network: the customer relations agencies of France Télécom whose function was to manage subscriber accounts at the local level. The manufacturer formed these into a distribution network, training their personnel in sales techniques, and providing them with sales literature and display cabinets. Since France Télécom was not prepared to act as the exclusive sales agents of one manufacturer, competitors could use the agencies as well. This had the advantage of presenting a wide range of choice and reinforcing the credibility of all, as well as of discouraging the creation of rival networks.

In this particular instance, the product was not significantly redefined by the actors downstream of the original designer. Some changes were made. For example, the instruction leaflets initially intended to help the user left alone with his new instrument, were rewritten to make room for the interaction between the untutored buyer and a sales agent with specialist knowledge in telecommunications. The convergence of the two facets of "subscriber" and "buyer" was achieved partly through combined action of the administrative and the commercial networks, and partly due to the increasing diversification of the products comprising the range, in terms both of their aesthetic qualities and their functions.

Although considered separately for convenience of description, the three strategies reviewed above are rarely encountered in their pure form as is indeed demonstrated by our third example. In general, we find a mix of strategies with varying emphasis. The idea that a novel artefact can by itself achieve convergence among the universe of user representations is no more than an engineer's fantasy. Equally, it would be vain to pretend that a salesman can save an artefact that is unable to sustain *any* of the relationships that it seeks to develop. Developing and incorporating user representations during the design stage is, therefore, not merely a matter of forming a realistic image of the potential user. It is above all a matter of forming a binding association between technological objectives and marketing strategy. In other words, achieving the convergence of all the different user representations involves gaining control over both the technical and the commercial sides, and acting on the interface between them.

Conclusion

The methods employed for bringing the end-user into the design process are varied. They all define ways of representing users, and each produces a stylized picture of the situation. The main problem for design, and CTA, is not so much multiplying the number of approaches and resulting user representations for these already abound. To strengthen the design process by incorporating a multiplicity of user representations, the main challenge is to coordinate the application of the various methods and reconcile their results.

It is therefore productive to devise assessment tools that can be used to draw up a chart for each phase of a project. The chart could show all the states of the user representation, how they have been constructed, and whether they are made into "given" assumptions by technological options already applied. The purpose of this exercise is threefold: First, it could verify the coherency (and potential for consolidation) of all the varied user representations as the project goes along. Second, the strategy should continually monitor potential ways to achieve convergence among disparate states of the user representation. Finally, it could help build a strategy for further project development. To develop this part of reconciling different user representations a bit further, I want to emphasize the binding complementarity of the decisions regarding the technical content of an innovative product and those shaping its marketing strategies.

For CTA it is of special importance to find ways of ensuring that certain user representations — which would otherwise not be considered by the innovators and the entrepreneurs — are taken into account. Conceivably, the whole range of methods and resources already in use, including financial incentives and standards, could superimpose neglected user representations. In addition, public authorities should identify, create and/or use a number of mediators between innovators and end-users to redefine the demand and thus allow new user representations to be considered. To implement this it will be necessary to develop a sociology of marketing and distribution able to become an integral part of the design, development, and innovation process.

In a way the redefinition of demand has already been extensively dealt with by economists. They pose the problem by asking, what mechanisms would correct shortcomings in the market? But the way we view these mechanisms, and the place given to different actors in the debate, can be considerably transformed if we are conscious of the omnipresence of the users in the process of conception, and of the main resulting problem: how to make the different representations coherent, or, in other words, how to construct "interessement devices" (Callon, 1986, 1987) which permit articulation between the largely heterogeneous techno-economic networks.

Notes

1. Joint research project culminating in a report edited by Dominique Boullier (1989).
2. Meaning that the new system should be able to work off the wiring already installed on the premises.
3. Industrial standards also impose certain assumptions concerning user characteristics. Once they have been adopted, they leave system designers with limited scope for discussion.

References

Akrich, M. 1992. "The De-Scription of Technical Objects," in: W.E. Bijker and J. Law (eds.). *Shaping Technology/Building Society: Studies in Sociotechnical Change.* Cambridge, Mass.: MIT Press, 205-240.

Andersen, E.S., and B.A. Lundvall. 1988. "Small National Systems of Innovation Facing Technological Revolutions: An Analytical Framework," in: C. Freeman and B.A. Lundvall (eds.). *Small Countries Facing the Technological Revolution.* London and New York: Pinter Publishers, 9-36.

Boullier, D. (ed.). 1989. *Genèse des modes d'emploi: la mise en scène de l'utilisateur final.* Rennes: CCETT-LARES.

Callon, M. 1986. "Some Elements of a Sociology of Translation: Domestication of the Scallops and the Fishermen of St. Brieuc Bay," in: J. Law (ed.). *Power, Action and Belief: A New Sociology of Knowledge?* Sociological Review Monograph. London: Routledge and Kegan Paul, 196-233.

Callon, M. 1987. "Society in the Making: The Study of Technology as a Tool for Sociological Analysis," in: W.E. Bijker, T.P. Hughes, T. Pinch (eds.). *The Social Construction of Technological Systems: New Directions in the Sociology and History of Technology.* Cambridge, Mass.: MIT Press, 83-106.

David, P.A. 1986a. "Understanding the Economics of QWERTY: The Necessity of History," in: W. N. Parker (ed.). *Economic History and the Modern Economist.* Oxford: Basil Blackwell.

David, P.A. 1986b. "La moissonneuse et le robot. La diffusion des innovations fondées sur la micro-électronique," in: J.-J. Salomon and G. Schmeder (ed.). *Les enjeux du changement technologique.* Paris: CPE-Economica.

Meyers, P.W., and G.A. Althaide. 1991. "Strategic Mutual Learning Between Producing and Buying Firms During Product Innovation." *Journal of Production and Innovation Management* 8: 155-169.

Rosenberg, N. 1976. "On Technological Expectations." *The Economic Journal* 86 (September): 523-532.

Teece, D.J. 1986. "Profiting from Technological Innovation: Implications for Integration, Collaboration, Licensing and Public Policy." *Research Policy* 15: 298-305.

Von Hippel, E. 1976. "The Dominant Role of Users in the Scientific Instrument Innovation Process." *Research Policy* 5: 212-239.

Von Hippel, E. 1988. *The Sources of Innovation.* Oxford: Oxford University Press.

9 Technologies as Social Experiments. The Construction and Implementation of a High-Tech Waste Disposal Site[1]

Ralf Herbold

For several decades, the introduction of new technologies, such as biotechnology and nuclear technology, has been accompanied by wide-ranging discussions about the risks involved. This has recently led to a new self-perception of society as an experimental society. In this chapter, a case study from the field of waste management, I will argue that the implementation of new technologies in fact increasingly takes on the character of a social experiment.[2] In this field the characteristics of the classic laboratory experiment have been integrated into the planning process. Like the classic laboratory experiment, the social experiment is characterized by verifiable hypotheses, an organized research process, and the definition of an experimental situation.

Within an industrial context some authors have identified similar developments. These include non-linear learning (Meyers 1990) and innovation as a recursive process (Asdonk et al. 1991). Here, the implementation of new technologies is increasingly guided by the insight that the performance of new technological systems is only partly predictable. Consequently, feedback loops become a fundamental element of exchange between suppliers and customers. New technologies are thus products of a social learning process.

Some forms of experimental implementation may be opposed by society because of perceived risks, and such opposition cannot be neglected. Indeed, issues of safety relating to the social impacts of new technological systems cannot be mediated solely by assessing objective risks and isolating technical arrangements from technical systems (see Wynne, this volume). Instead, social negotiations about the perceptions of risk and the conceptions of society may be necessary to make such controversies productive (in the sense of social learning) and to avoid ritualized and unproductive confrontations. This chapter explores the idea of social experimentation through a case study of the planning of a new waste disposal site. Beginning with a situation of political conflict and technical uncertainty, we can observe the social conditions needed to turn social experimentation into a productive social learning process.

Society as a Laboratory

A technological invention is not restricted to the construction of an artifact, but also comprises the design of social action. The technology producer always constructs a socio-technological connection that contains a number of social components besides the strictly instrumental ones. A small part of these connections can be found in the operating instructions; the greater part consists of often implicit assumptions about the artifact's utilization as well as the range of interactions with other artifacts and/or participants (see Akrich, this volume). It is hardly possible to examine these assumptions or hypotheses under controlled laboratory conditions. Therefore, technological inventions are increasingly tested via experimentation in a social context.

The testing of socio-technological systems through experiments on pilot projects is burdened with uncertainties resulting from 'scaling-up' and the unpredictability of social actions. Even laboratory testing of pilot projects cannot completely replace large-scale technological implementation. Whereas the ideal laboratory contains dangers apart from the outside world, conducting real-life experiments inevitably loads some dangers on the society and the environment. Nevertheless, renouncing such experimentation results in a loss of options, foreclosing choice of ecologically attractive alternatives.[3] The construction of a new disposal site discussed in this chapter was a large-scale, real-life technological experiment.

The experimental character came about because the construction process embodied the following three characteristics: a) deficit of knowledge, b) innovative hypotheses, and c) ignorance about the experimental results. First, the deficit of knowledge included, for instance, the interaction of contained substances, the time necessary to neutralize wastes ("Abbaugeschwindigkeit"), the emergence of percolation water and its composition, as well as the barrier effects of artificial and natural sealings. Especially for disposal sites whose contents are not sorted out, accumulation and long-term effects can hardly be extrapolated from laboratory experiments. The effects of these deficits in knowledge are aggravated by unique local conditions. These include the specific mixture as well as the long-term changes of the waste, and so experiences from one disposal site are only partly applicable to another.[4] To overcome these deficits in knowledge and to avoid the problem of the impossibility of modelling reality within the laboratory, monitoring research is conducted in existing plants. Despite these endeavours most waste scientists would agree that "reality is already beyond research" (Franzius 1980: 8).

Second, the underlying hypothesis in discussions on new disposal sites is that new ideas on technical equipment and organization can reduce risks and increase safety. Such new concepts include basic or surface sealings, dry depositing, and the separated deposition of waste products. It is expected that with these provisions past mistakes can be avoided. Nevertheless, it can hardly be predicted whether the interactions of the system's technical and social components will take place as planned. For example, a well-organized depository ("Geordnete Deponie") may stop functioning not only due to a defect in the barriers, but also because waste products are incorrectly declared or regulations are disregarded by the operating crew. The additional installation of safety provisions, designed to prevent known or anticipated disruptions, merely reinforces the point that a new large-scale technical plant is always its own experiment.

Third, the operation of a disposal site cannot be completely described since all eventualities cannot be anticipated. Only putting the disposal site into action can generate the experience necessary to test the hypotheses. Complete anticipation is impossible because, among other factors, the future composition of the waste is unknown. It can be influenced by new waste products as well as by new strategies for preventing wastes.[5] Such uncertainties (and their interactions) make clear that there is "ignorance" not only about the social and technical *processes* constituting a waste disposal site, but also about the *results* of the experiment.

The Planning of a Modern Disposal Site

Only recently has the dumping of garbage become a subject matter for scientists and technicians.[6] Not until the end of the 1950s did the developing waste sciences initiate a professional discussion that put aesthetic and hygienic problems in the centre of attention and at the same time conceptualized 'waste as a risk.' Studies in the 1960s concerning the contamination of surface and ground water around garbage dumps presented strong evidence of their noxious environmental effects. Finally in the 1970s, the emerging ecological movement made waste into a matter of public discussion. Further debate and research rendered obsolete the prevailing practice and theory, that a disposal site is harmless after it had been filled in and recultivated or could be left to the "self-healing natural energy" ("den selbstheilenden Naturkräften überlassen," Salomo 1985:66).

The conviction grew that waste disposal sites are a long-lasting danger,[7] and that the most modern techniques should be used to ensure their safety. This conviction was expressed in the German Federal Waste Disposal Act that came into force in 1972 and emphasized the public good and the harmlessness of waste disposal. These two aspects became prerequisites for official approval of new plants.[8]

Two further aspects of the "Abfallbeseitigungsgesetz" are of special interest for my argument: the scientification of disposal site techniques, as well as the formalizing and rationalizing of the decision procedure. The law marks a caesura. Evolutionary learning-by-doing is gradually replaced by a systematic and methodically controlled approach to develop new and safer techniques. The growth of waste heaps and the public's increasing risk awareness creates a pressing dilemma for the planners. On the one hand planners are responsible for clearing away the garbage, unless they want to run the risk of a waste disaster. On the other hand their ability to act is increasingly limited by legal requirements, public attention, and the legal rights of affected persons. The search for a disposal site location that fulfils the formal and informal criteria is already an extremely complicated and time-consuming process. Facing these circumstances, planners needed new planning processes and the ability to realize them in political concepts. In the case of Bielefeld, a drastic impetus was needed to start this process of rethinking.

Mistrusting the Planners

In 1983 an abandoned waste site was found underneath a new housing site in Brake, a suburb of Bielefeld.[9] After the failed attempt by politicians and the city administration to downplay the danger and thus avert the obligation to act, they reluctantly ordered an examination of the area, which revealed a large health risk due to high concentrations of polychlorinated biphenyls.[10] A screening of the 185 residents found 44 cases of chronic liver damage, which was attributed to PCB-contaminated soil. A series of public actions by the residents and political discussions forced the city to buy the housing site in 1985.

(The original owners received compensation to buy new estates and houses.) The problem began when an industrial disposal site, in use until the 1970s, was declared a residential development area by Bielefeld's administration. (The city's development plan was annulled by a court in 1985.) As a result nearly the entire population mistrusted the administration, which is understandable considering their hesitation in handling the problem. The next local elections took place during the scandal and ended the Social Democratic Party's traditional control over the city council. The Green Party had gained so many votes by supporting the citizens' initiative politically, organizationally and financially, that they formed a coalition with the Social Democratic Party. The coalition agreement created an environmental department that the Green Party was to run. In 1986 Dr. Uwe Lahl, an independent chemist with a strong reputation in the environmental movement, was selected by the city council to head the new department.

The city administration had started planning for a new disposal site in 1979 under the impetus of estimates that the existing disposal sites would be filled up by 1992. At first, the administration used a conventional planning model: the administration evaluated different locations, made an advance selection, and ordered an expert's report to compare the sites. Participation of the population was not intended. The planning process took place entirely within the administration. The final choice would be legitimized by a political vote by the parliament.

A voting system was used to compare different locations and two particularly appropriate locations emerged.[11] In further discussions only these two were considered. For the highest-rated location, a geological report was ordered to investigate the water conditions. According to this report, published at the end of 1985, the location was especially suited for the construction of a disposal site, because of the absence of any water deposits. The planners hoped these reports would constitute the basis for a political decision. But the reports were published by the local newspapers and criticized by the affected citizens. The geological report was criticized especially since the examination was made when there was very little precipitation. These problems, in conjunction with the public mistrust as a result of the abandoned waste site scandal, made the situation extremely difficult for the disposal site planners.

Restoring the Ability to Act
The search for an acceptable location was threatened by the public criticism of the experts' reports. Up to this point, the administration had been guided by a conventional solution, in which the landscape is

levelled out with garbage. The ground- and surface-water was to be protected by sealing the ground and collecting the percolation water. At the beginning of the second phase in 1986, a novel initiative came from the new head of the environmental department. Lahl, the independent chemist, formulated prerequisites for a new technical solution that resulted in a totally new disposal site conception. The prerequisites were *repairability*, *controllability* of the disposal site to guarantee the possibility of shutting off the plant in case of emergency, and *retrievability* of the contents to take into consideration the discovery of new utilizing options and risks. The formulation of these far-reaching demands, combined with a credible promise to make the planning steps transparent and to facilitate participation in further planning, diminished the danger of a politicized planning process and secured the administration's ability to act.

In the 1980s a discussion on the feasibility of these novel conceptual demands took place in the community of waste science. These demands suggested a temporary depository and a planning process that would allow for feedback of the results of disposal site operation and subsequent changes. Parts of these concepts and ideas adopted from the nuclear waste field were dismissed, however, as impracticable or undesirable for economic reasons. The discussion alternated between the poles of a temporary depository with new constructing measures and a conventional final depository. A combination of both ideas also developed.[12]

Constructing the Site
In the further course of the planning process, the administration de-emphasized the narrow decision on a technical concept and instead pursued new ways of fostering public participation. The new technical prerequisites introduced by the administration were well received by the public, and they became the basis for a contest. This kind of competition has been practised for a long time in fields such as architecture and art, but for the waste field such a procedure was a novelty.[13] Two renowned external experts assessed the submissions. Among these were conventional final depository solutions, which could not meet the prerequisites. One temporary depository solution became the top choice, but did not seem practicable for legal reasons. The design finally chosen consisted of a combination of these two concepts and included a number of new technical features. To fulfil the prerequisite for 'retrievability' waste was deposited into isolated cubes, with extensive documentation of the deposited substances and their localization. A free-hanging roof construction, the size of 25

soccer fields, was designed to prevent percolation into the disposal site. The construction of a repairable and walkable basic sealing was made possible by the 'multi-barrier-concept.'[14]

The design embodied a far-reaching consensus of all participants, and thus formed a solid basis for further negotiations and constructive criticism. For the now-enlarged group of planners it was the starting point for the technical planning of the plant, while it was a minimum but acceptable solution for the affected persons. The design was widely accepted by the general population as well. The conflict-subject 'disposal site' disappeared almost completely from public discourse. Bielefeld's citizens and enterprises were now guaranteed a waste management solution valid for a long period of time.

Typical for the chosen procedure identified by the planners themselves as 'open planning' was that it had an informational as well as justifying character. The legitimacy of decisions that followed planning was increased by the enlarged democratic participation.[15] The new course of action included, for instance, informing the local press and making the material about the technical planning available to make decisions transparent. Further, a participating process was institutionalized by public hearings and discussions between planners and affected persons.

The technical planning was split into several phases. The commissioned city engineering office made interim outlines, which were checked by risk analyses that included technical and ecological risks. They were carried out by external experts, who helped formulate new suggestions as well. This additional input gave the administration flexibility with the citizen's initiative as well as in subsequent decisions. The planning process was brought to a conclusion in the summer of 1990 by initiating the legally necessary approval procedure. The submitted documents detailed the technical side of the artifact as well as its operation and after-care. The proposed plant had a size of 1,000 x 600 metres with a volume of 5 million cubic metres. Based on today's figures, it is big enough to take the wastes of a region with 500,000 inhabitants over a period of 28 years, with a cost of between 600 and 1,000 million DM.

The Bielefeld model[16] of planning a large-scale technological plant can be summarized in the following aspects:

> a) It makes knowledge about the risks of new technologies the starting point of the planning process, and encourages the affected persons to participate in the search for possible solutions.
> b) It allows for feedback of interim planning results. This deals with the problem that new knowledge will be gained during the operation and,

thus, will become available during the planning process but that it cannot be anticipated.
c) It limits the effects of a wrong decision by the use of technical concepts like 'retrievability' or 'repairability.'

This model requires high investments. This is, however, nowadays broadly accepted. At the same time it includes an imperative obligation to act according to the same standards for the next decades.

Conclusion

The success of the open planning began with stimulating a public discourse about the circumstances of waste production and management. In the new style of decision making the experts remained the key figures but took the risk perception of affected persons seriously. As a direct result, different problem-solving strategies were taken into account. Specific models of waste management and technical solutions (which usually are the *starting* point for planning) became a matter for social negotiation and collective value learning (see Wynne, this volume). NIMBY (not in my backyard) ceased to be a promising tactic; instead, the decision to build a waste disposal site at a certain location was considered legitimate by the public. Trust in the authorities was reestablished at the same time. The subsequent discussion focused on the reduction of risks and involved active participation of the public in the technical planning. This involvement placed a lasting credibility pressure on the city administrators. For example, they had to take into account that the affected persons simply demanded extended security standards far beyond the technological state of the art.

The public acceptance of the location procedure and its results can partly be explained by its perceived voluntary character. Psychometric studies have shown that voluntariness is an important attribute of public risk acceptance.[17] Another important dimension of risk acceptance is reversibility. In general, trial-and-error procedures and experimentation with technologies in society are not acceptable to the public. In this case, however, it was possible, as in a classic laboratory experiment, to combine experimenting with reversibility.

Decision making on risky technologies is often typified by one-time decisions made variously on the basis of political expediency and on the basis of methods that purport to anticipate and calculate risks, such as technology assessment and risk analysis. In the Bielefeld disposal-site planning, the risk analyses were instead used in the technical constructing process as well as in the operation of the plant. In this way, risk analysis served as a constructive input to the decision process, and the start of a learning process. Society has to accept risks as part, even as a precondition, of this learning process. In an experimental society, science cannot contain hazards in laboratories.[18]

Another result was that the parameters of risk analysis could be checked in a real life implementation. The documentation of the site's construction, depository techniques, deposited substances and their localization, as well as continuous measuring programs made the site utilizable as an object for research. Observation and experience were coupled. The site is of special interest to the waste sciences which gain their material from the work on large-scale technical systems because so far only qualitative statements could be derived from existing sites.[19] The experimental design of Bielefeld's waste disposal site can thus be seen as a contribution to the progress of science.[20]

The classic planning dilemma is that both the renunciation of options as well as the introduction of new technologies increases risk. This dilemma cannot be solved by calculating risks and trying to anticipate them. The only way out is to take seriously the experimental character of the introduction of technology and to make the process socially transparent. By integrating reflexive mechanisms (self-observation, feedback of operation experiences, reactions after changes of substances) and by providing 'intervention-reserves,'[21] technologies in operation can be modified to reduce if not eliminate risk. The experimental character of implementation can be described by three features:

1. The construction of technology takes place within an open social process. Decision making solely on the basis of prognostic methods (TA and risk analysis) is renounced.
2. Technology becomes an object of continuous scientific reflexion.
3. Intervention-reserves and reflexive mechanisms embody the on-going effort to keep technological options open.

Constructive Technology Assessment aims at influencing technology while it is under construction. This case has shown that CTA could be achieved through introducing technologies as social experiments. This will make it possible to test underlying hypotheses, and for that matter user representations (see Akrich, this volume) and to bring this knowledge back into the process of technology development.

Notes

1. This article is a product of a close collaboration with Wolfgang Krohn, Johannes Weyer, and Ralf Wienken.
2. For this argument see Krohn and Weyer (1989). They even argue that the results of technical-scientific activities in the past may retrospectively become experiments, as by the examples of DDT, dioxin, and the ozone layer.
3. This type of non-option is discussed in Herbold, Krohn, and Weyer (1991) for the development of safety in passenger airplanes.
4. This problem is not solved yet; Stief (1991: 33) summarizes: "Every municipal disposal site is unique."
5. An actual example may illustrate how fast premises for waste-management concepts have to be reformulated: new laws concerning the extensive usage of packaging materials may endanger the existing paper-recycling system.
6. See for details Herbold and Wienken (1993).
7. Unlike nuclear wastes many conventional wastes do not even have half-lives. Because of this, Schenkel from the Umweltbundesamt (German Federal Environment Agency) speaks of 'modern pyramids': "Landfills are secular events. To build them is equivalent to the construction of pyramids." (Schenkel 1986a: 46). See also Schenkel 1986b.
8. The 'state of the art' as the embodiment of this demand initiated in this case a self-dynamic process of developing stricter technical standards. The technical planners sought official approval by exceeding these standards. The approving authorities, however, demanded an increase of security standards.
9. An account about this case is found in Baumheier (1988).
10. The account in Levine (1982) can be seen as paradigmatic for politicians' and administrations' management of abandoned waste sites.
11. For this planning the criteria were availability of transportation infrastructure, distance to housing areas, obstacles for acquiring the area, existence of biotopes, aesthetic and geological aspects, and size of the disposal site.
12. "Of course, the planned disposal site is technically as safe as it is possible today. But it is not a final depository since there are simply not sufficient concepts for a final depository available. These security concepts would have to be intact for centuries. The *Mietdeponie* (a special form of temporary depository) takes this seriously. There is only room for now not usable materials until there will be a chance to make them usable again... But we do have an excuse for not realizing this. It is quite sure that we would have had done too much by trying to build such a plant five years ago: technically, financially and first of all politically i.e. consciously. And such a concept means a lot. For instance the industry would have to accept that they no longer can simply deposit their wastes but would have to pay a 'rent' for the use of the plant. You have to make that understood first ..." (Interview with Lahl, 14.02.1991, in: Herbold, Krohn, and Weyer 1992: 193)
13. Denzin (1989: 60) reports that a waste disposal concept contest was held in 1987: "Contests...are especially conducted for innovative projects. The application to the disposal site is logical, although as far as we know it has not been done before."
14. This technique is hardly tested but can be taken as an outcome of a closure process that will lead to stabilization by implementation.
15. Lahl, the head of Bielefeld's environmental department, himself gives a hint in this direction: "The open planning process has the advantage that the planners produce a public picture quite different to that of a conventional administration. We haven't been the administration that followed a 'closed shop'-policy without informing the citizens and tried to push through

something with secret diplomacy and a privy councillor mentality as in an authoritarian state. In my opinion a procedure like the one we used in our planning process, concerning a plant that is of public interest, is also a fight over the public opinion in a city." (Interview with Lahl, in: Herbold, Krohn, and Weyer 1992: 203)

16. See Nieling, Peters, and Wiebe (1989) who promote this model.

17. For a brief overview see Wynne in this volume.

18. This is confirmed by Perrow's (1984) analysis that complex systems inevitably produce unexpected interactions. However, for him this is a negative feature of these systems because accidents become unavoidable. Our analysis shows that this is only one side. The positive side is that such unexpected interactions are needed and can help create safer systems.

19. The attempt to gain knowledge from abandoned waste sites was not successful because of the ignorance about contents, construction, and filling techniques of disposal sites.

20. Parts of this conception have been constantly demanded by the Umweltbundesamt. The renunciation of complete documentation and measurement of disposal sites is responsible for problems in development that research has to face.

21. Lange, the head of the technical planners, summarized this for the actual planning: "We have separated the disposal site into eight squares. Each will be filled within four years. Knowledge gained from the first, will be the input for the next. For instance there could be a new conception of the basic sealing or modifications of the roof. And we are able to react if there are changes in the composition of wastes." (Interview with Lange, 17.12.1990, in: Herbold, Krohn, and Weyer 1992: 212 f.)

References

Asdonk, Jupp, Udo Bredeweg and Uli Kowohl. 1991. "Innovation als rekursiver Prozeß: Zur Theorie und Empirie der Technikgenese am Beispiel der Produktionstechnik." *Zeitschrift für Soziologie* 20: 290-304.

Baccini, Peter, and Walter Ryser. 1988. *Reaktordeponie und Endlager*. Münsingen: Administration "Swiss Workshop on Land Disposal of Solid Wastes."

Baumheier, Ralph. 1988. *Altlasten als aktuelle Herausforderung der Kommunalpolitik*. München: Minerva Publikation.

Denzin, Manfred. 1989. "Planung einer Deponie auf der Grundlage eines Ideenwettbewerbs," in: Forschungs- und Entwicklungsinstitut für Industrie- und Siedlungswasserwirtschaft sowie Abfallwirtschaft e.V. *Zeitgemäße Deponietechnik III*. Berlin/Bielefeld: Schmidt-Verlag, 53-89.

Franzius, Volker. 1980. "Schadlose Ablagerung," in: Umweltbundesamt (ed.). *Materialien zu Stand und Entwicklungstendenzen in der Abfallwirtschaft und -beseitigung in der Bundesrepublik Deutschland*. Teil 8. Berlin: Umweltbundesamt.

Herbold, Ralf. 1991. "Technikentwicklung und Risikodiskurs im Abfallbereich." Bielefeld: Universitätsschwerpunkt Wissenschaftsforschung (unpublished).

Herbold, Ralf, Wolfgang Krohn, and Johannes Weyer. 1991. "Technikentwicklung als sozialer Prozeß," in: B. Joerges. *Wissenschaft, Technik, Modernisierung. Verhandlungen der Sektion Wissenschaftsforschung der Deutschen Gesellschaft für Soziologie beim 25. Deutschen Soziologentag in Frankfurt, Oktober 1990*. Berlin: Wissenschaftszentrum, 76-95.

Herbold, Ralf, Wolfgang Krohn and Johannes Weyer (eds.). 1992. *Technisches Risiko und offene Planung: Strategien zur sozialen Bewältigung von Unsicherheit am Beispiel der Abfallbeseitigung*. Bielefeld: Universitätsschwerpunkt Wissenschaftsforschung/Fakultät für Soziologie, Wissenschaftliche Einheit 'Wissenschaft und Technik.'

Herbold, Ralf, and Ralf Wienken. 1993. *Experimentelle Technikgestaltung und offene Planung: Strategien zur Sozialen Bewältigung von Unsicherheit am Beispiel der Abfallbeseitigung*. Bielefeld.

Krohn, Wolfgang, Johannes Weyer. 1989. "Gesellschaft als Labor: Die Erzeugung sozialer Risiken durch experimentelle Forschung." *Soziale Welt* 40: 349-373.

Levine, Adeline G. 1982. *Love Canal: Science, Politics, and People*. New York: Free Press.

Meyers, Patricia W. 1990. "Non-linear Learning in Large Technological Firms: Period Four Implies Chaos." *Research Policy* 19: 97-115.

Nieling, T., J. Peters and Andreas Wiebe. 1989. "Integration der Umweltverträglichkeitsprüfung in den Planungsablauf einer Deponie." *Abfallwirtschaft-Journal* 1: 54-58.

Perrow, Charles. 1984. *Normal Accidents: Living with High-Risk Technologies.* New York: Basic Books.

Salomo, Klaus-Peter. 1985. "Technische Möglichkeiten zur Sanierung gefährlicher Altlasten." *Müll und Abfall* 17: 61-70.

Schenkel, Werner. 1986a. "Ziele künftiger Abfallwirtschaft." *Müll und Abfall* 18: 41-47.

Schenkel, Werner. 1986b. "Entwickelt sich die Deponie zur Pyramide des Konsumzeitalters?" in: Trägerverein des Instituts für wassergefährdene Stoffe e.V. an der Technischen Universität Berlin (ed.). *Symposium: Die Deponie - Ein Bauwerk.* Berlin: Trägerverein des Instituts für wassergefährdene Stoffe e.V. an der Technischen Universität Berlin, 95-102.

Stief, Klaus. 1980. "Gedanken zur Ablagerung von Abfällen." *Müll und Abfall* 12: 240-249.

Stief, Klaus. 1989. "Deponietechnik im Umbruch - Nachbesserung bestehender Deponien," in: Forschungs- und Entwicklungsinstitut für Industrie- und Siedlungswasserwirtschaft sowie Abfallwirtschaft e.V. (ed.). *Zeitgemäße Deponietechnik III.* Berlin/Bielefeld: Schmidt Verlag, 7-31.

Stief, Klaus. 1991. "Haben Deponien für unbehandelte Sonderfälle eine Zukunft?" in: Klaus Stief and Klaus-Peter Fehlau (eds.). *Fortschritte der Deponietechnik 1990.* Berlin: Schmidt Verlag, 27-52.

Wynne, Brian. 1988. "Unruly Technology: Practical Rules, Impractical Discourses and Public Understanding." *Social Studies of Science* 18: 147-167.

10 Pollution Prevention, Cleaner Technologies and Industry

Arne Remmen

The report of the Brundtland Commission *Our Common Future* has called for a sustainable development "that meets the needs of the present without compromising the ability of future generations to meet their own needs" (World Commission on Environment and Development 1987:43). At the same time Danish environmental authorities have given higher priority to the development and introduction of so-called cleaner technologies which reduce pollution at the source. Cleaner technologies are an important means to meet the growing demand for prevention of environmental problems. The new orientation toward cleaner technologies is an attempt to address the shortcomings of environmental efforts which until now have focused on thinning, purifying and disposing of waste. This strategy often only transformed one pollution problem into another. For example, at municipal treatment plants polluted waste water was transformed into sludge and then into a disposal problem.

The shift toward pollution prevention has a broad political basis in the Parliament of Denmark. The former Minister for the Environment, Lone Dybkjær, has formulated the environmental challenge in the following way:

> Our environmental policy is in a transition in which we partly have to complete the efforts in progress in the form of cleaning and purifying, and partly have to implement a readjustment where an increasing part of the environmental efforts is directed toward introduction of cleaner technologies and resource-preservative technologies. It is a question of integrating environmental protection into everyday life, into the households and into production and social life. Only in this way can we secure a sustainable development and a responsible utilization of resources. (Ministry of the Environment 1989: 3)

Implementing this transition is difficult because the environmental problems of the past require cleaning commitments in addition to current environmental investments. A similar dilemma between the actual political priority and the spending of funds has been pointed out in relation to the environmental policy of the Netherlands (Dieleman et al. 1991). In 1986 the Danish Ministry of the Environment began a development programme for cleaner technologies, which includes subsidies for industry and for institutions of research and development to encourage innovation and implementation of cleaner technologies. No support is given to end-of-pipe technologies. According to this programme "cleaner technology means that pollution and waste related to the production, use and disposal of products are minimized to the extent possible as close as possible to the source. This implies that the product or production process is changed so the overall environmental pressure from the circulation of materials and substances in the society is minimized" (Ministry of the Environment 1990: 3). In this definition working conditions in industry are left out. Experiences with cleaner technologies have shown, however, that working conditions influence pollution profoundly.

The shift required by cleaner technologies to an anticipatory and preventative approach makes new demands on companies (both employees and employers), relationships between companies, environmental authorities, and research institutes. Based on experiences with projects on cleaner technologies in the Danish fish-processing industry I will argue in this chapter that successfully implementing a prevention strategy requires an analysis of the different parties involved, their relationships and the search and learning processes which occur during the experiments. As will become clear, pollution prevention requires specific methods. Research on policy aspects must interact with the innovation and implementation of technologies. The shift toward prevention also requires a shift from traditional technology assessment to constructive technology assessment (CTA).

Technology Assessment in Denmark

In the 1970s research on technology assessment in Europe and the U.S. primarily aimed at illuminating the direct and derived *consequences* of technology. Methodologically such research focused on developing general checklists for consequences at different aspects and levels (Hetman 1973). The basic idea was to establish a better basis for governmental decision making on the implementation of new technologies, and hence to alleviate the most detrimental effects of technology. "Decision making processes related to technology were considered to be rational, sequential, and containing separate decisions; during the process, the results of technology assessment were often ascribed a decisive role" (Smits and Leyten 1987:18). However, it proved difficult to determine the consequences precisely and to weigh different consequences against each other. In addition, introducing changes is difficult since the technology in question is already developed (Collingridge 1980). This *reactive* approach has furthermore focused on technical solutions to the consequences of the technology applied.

As a result of dissatisfaction with this approach, a more *comprehensive* and *proactive* technology assessment tradition emerged in Denmark in the 1980s. This approach examines negative effects as well as traces positive potentials and alternative actions. Comprehensive assessment aims at formulating the needs and demands for technologies that are yet to be developed and implemented. The interests of various parties are recognized, and consequently the process of choosing solutions is perceived as a political struggle. Methods from "future research," such as scenarios and Delphi techniques, are applied primarily within the frames set by technological development. One of the aims has been to promote technologies whose effects are assessed as beneficial to society.

Based on practical experiences a *constructive* approach to technology assessment has gradually developed during the late 1980s in Denmark. Such a constructive approach aims at integrating social and environmental criteria right from the beginning, i.e. to include these criteria in research, construction, and planning. It emphasizes potential options and constructively tries to "anticipate hindsight" by ensuring that, for example, environmental considerations are taken into account in the decision-making process. It involves an *interactive* approach with participation of all relevant groups. Thus, such a constructive assessment approach must alter the traditional focus of both consequence and comprehensive assessment, where technology is perceived as both the problem and the solution (Remmen 1990).

Integrating constructive technology assessment into the process of innovation and implementation is not easy. Traditional comprehensions of these processes have been linear and static, and the choices made are considered irreversible. This has to be replaced by a dynamic and iterative comprehension of technological change. Such an understanding can point out various possibilities of choice and action.

The constructive approach evolving in the Danish context resembles the idea of constructive technology assessment (CTA) advocated by

Rip and Van den Belt (1988) and Schot (1991 and 1992). Taking a governmental perspective Rip and Van den Belt discuss three CTA strategies:

* To support the development of desirable technologies;
* To regulate or tax ("punish") undesirable technologies;
* To influence transformations of technology that occur anyway.

In my opinion the most interesting, new and productive strategy is the third one, where the goal becomes the indirect influencing of ongoing technological changes, leading to a multi-actor decentralized form of control (as argued by Rip et. al. in the concluding chapter of this volume). This form of control requires a democratization of technological change with an active participation of relevant groups (see chapters of Callon and Coombs). Constructive technology assessment, however, cannot cover all cases of technological change. Some technological decisions are irreversible, because they involve a choice between completely different alternatives. In these cases a democratic debate has to be initiated prior to a political decision. CTA, which aims to integrate social and environmental criteria during innovation and implementation, has limited value in such cases.

Traditionally, technology fixes the agenda for development projects and consequently technological projects are perceived as programming tasks. Goals and means are well-defined, and uncertainty, learning processes and conflicts are not associated with the change. In this view development projects are a step toward implementation. In many cases, however, it turns out that implementation processes are rather slow and chaotic. Disagreement about goals and means will arise, and the planned solution turns out to be insufficient (Christensen 1985).

Conflicts concerning goals and uncertainty about means are in fact basic conditions in technological projects. To democratize technological change, research methods must be generated that are capable of establishing a basis for exchange of knowledge between technicians, users, researchers and other relevant groups. A promising method to shape CTA is as a social experiment (see chapter of Herbold). Any approach must appreciate that negotiations and experiments are key issues in influencing technological change. Experiments and negotiations among the parties concerned can bring social and environmental considerations into the innovation and implementation processes.

Experiments offer opportunities for all interested groups to assess the consequences and possibilities of certain technologies before a final decision is made. Experiments increase the possibility of defining the problem more precisely and assessing alternative solutions. By gaining practical experience it becomes possible to facilitate productive negotiations. This has been the case both in the Scandinavian approach to system development (Ehn 1988 and Andersen et al. 1990) and in the Danish experiments with information technology.

Social experiments with information technology have spotlighted the interaction of social needs and technology. Experiments can result in a specification and disclosure of needs of involved parties, and assess whether the technology applied can meet those needs or if a further technological (or social development) is required (Qvortrup 1984 and Cronberg et al. 1991). Social experiments make it possible to establish an active search and learning process which brings new actors on the scene and stimulates the integration of new considerations in the innovation process. In this way experiments are an attractive and powerful tool for CTA.

In social experiments researchers play an active role. The researcher not only observes the experiment but actively takes part as supplier of data and ideas, as well as partner in dialogue (Duelund 1991 and Storgaard 1991). CTA research could in fact be labelled as dialogue research, in which the researcher has to balance the roles of participant and observer.

Summarizing, the current three approaches to technology assessment are (Remmen 1991):

1. Consequence assessment	Reactive - oriented towards solutions (= to limit detrimental effects)
2. Comprehensive assessment	Proactive - a given room for solutions (= to promote desirable technologies)
3. Constructive assessment	Interactive - future room for solutions (= to influence technological change)

The three approaches are not necessarily mutually exclusive. In other words, constructive technology assessment needs to evaluate the consequences and potentials as well, but CTA must go one step further and become an integrated part of the processes of innovation and implementation.[1]

Thus, on the one hand the constructive approach forms a break with the previous research traditions of technology assessment (see appendix 1). But since the three approaches focus on different aspects they can be used complementarily (see appendix 2). The insights for example from a consequence or a comprehensive technology assessment can become an important part of the constructive technology assessment. In the TOR project we have applied such a combination.

Technology and Resources

Technology and Resources (TOR) is a research project funded by the county council of Northern Jutland to analyse the possibilities of innovation and implementation of cleaner technologies in its fish-processing industry. The long-term goal of the TOR project is to strengthen the capacity and enhance the resources of enterprises to integrate environmental (and working conditions) criteria in the planning, production, and innovation process of cleaner technologies. This chapter will describe the results of the consequence and comprehensive assessments in the TOR project. The *consequence* assessment included study of relevant literature, archives and documentation as well as interviews with key persons within the branch. The consequence assessment resulted in insight into existing technology applications in the fish-processing industry as well as consequences for environment and working conditions.

The *comprehensive* assessment used the same methods for data collection. In addition, methods from future research such as scenario techniques were applied for developing insight into future regulation as well as future technological development in the fish-processing industry. The comprehensive assessment analysed possibilities for cleaner technology projects in the fish-processing industry, and of present and future public regulation of working conditions, emissions, and waste streams. The investigation also clarified how problem definitions and the strategies of existing development projects have limited the diffusion of cleaner technologies.

In the fish-processing industry a lot of water is used as a result of veterinary demands and the requirements of product quality as well as production flow. It has been assumed that when abundant water is used, the product will be nicer and the process proceeds more easily. The resulting waste water has never been seen as a problem. It is simply perceived as delivering back to the sea some of the organic material that the fishermen have caught. The environment is not polluted with any sort of poisonous waste as in other industries. Moreover, at a national level the fish-processing industry's discharge of nutrients and organic materials is small when compared with agriculture and wastewater treatment.

Traditionally, environmental regulation for fish processing has been based on "recipient quality planning" and an estimation whether the sea where the waste is discharged can thin and decompose the organic material. Along the North Sea (Skagen, Hirtshals and Hanstholm), which is considered sufficiently "robust," the fish-processing industry had permission to discharge waste directly without purifying until 1995. But the industry along the Limfjorden and the Kattegat, characterized as "vulnerable," has been forbidden to discharge its waste directly into the sea. These plants are to be connected to municipal wastewater treatment plants. With the passing of the "Aquatic Environment Plan" in 1987 the discharge demands were tightened allowing government to command individual fish processors to use best available technology. The Environmental Protection Agency estimated that the best available technology may result in a 30-50 percent reduction of discharge of nitrogen and phosphorus as well as 90 percent reduction of organic waste (Environmental Protection Agency 1988b).

Working conditions in the fish industry have often been in full blaze publicity because of frequently reported accidents. Other important problems are the monotonously repetitive work at a rapid pace and bad working conditions caused by heavy lifts, noise and moisture, draught and cold. Since the mid-1980s the authorities have succeeded in reducing especially draught and cold via information campaigns and demands on the companies. The problems related to the monotonously repeated work at high pace are unchanged, however, or have even increased.[2]

In the consequence assessment we found that pollution and health safety problems vary considerably depending on the following factors:

* fish type, quality of the raw material and the degree of processing;
* the attitude and competence of the employers;
* the age and size of the company;
* working procedures, technologies and processes;
* competitive strategy (price or quality) and production capacity.

These factors show that only a part of the pollution and working conditions can be traced directly to the production technology. The consequences also depend on the choices and decisions made in relation to the factors mentioned above.

The fish-processing technology comes mainly from foreign suppliers (Baarder in Germany and VMK in Sweden) whose priorities have been efficiency rather than, say, water saving. For example, a filleting machine demonstrated in 1991 will result in the same environmental problems and working conditions as one from twenty years ago. One reason why foreign suppliers are not urged by Danish customers to produce less polluting machines is that the fish-processing industry itself does not possess the competence to make the necessary technological specifications and demands. In Denmark only a few companies and perhaps two or three consultant firms possess this competence. Another reason is that Baarder and VMK are the only machine suppliers on the world market, and in this connection Danish customers are a very small market.

To reduce the pollution problems in the fish-processing industry the Ministry of the Environment has financed several demonstration projects on cleaner technologies. Most of these projects focus on minor technical adjustments and have been executed by consultants (experts) and a single company, which has limited the diffusion of the cleaner technologies.

A pioneering effort in the field was "Cleaner technologies in the fish-processing industry" (Environmental Protection Agency 1988a). This project was a descriptive "trade mapping" consisting of an overview of places where environmental problems arise in the production flow of herring, flatfish and roundfish and a short exposition of waste treatment methods and possible cleaner technologies. The project found that in the herring industry a separation of organic waste from the process water must take place as early as possible to reduce the dissolution of waste in the discharge water. For the round- and flatfish industry the project suggested changes in cleaning routines, technical means for water saving, separation of waste from the process water and dry transport of waste.

Subsequently, technical demonstration projects have been carried out in most of these areas. For example, a project on reducing pollution from cleaning has been implemented. Moreover, projects have been conducted on exsuction and scraping of herring guts which have cut emissions by up to 35 percent. Consultants to the fish-processing industry have estimated that the application of existing cleaner technologies could reduce discharge of nitrogen, phosphorus and organic materials by 55 to 60 percent. Such a reduction exceeds regulatory requirements. The demand for 90 percent reduction of organic materials, however, cannot be met by introducing cleaner technologies (Green Waste Water Plan 1991). None of the cleaner technologies implemented thus far have considered the working conditions, and their effects (positive or negative) have not been studied. Accordingly, Industrial Health and Safety Units (common to small companies in a local area) and the employees have not been involved in the analysis and the solution of the problems.

In many respects this trade mapping was successful. It had the character of an informal sector agreement in which the environmental problems were briefly described and different technologies for their reduction were identified. The project also led to several successful demonstration projects. But our comprehensive assessment showed that two obvious and important routes for solutions were still neglected: the immediate possibilities of changing working procedures ("better housekeeping") and long-term possibilities related to new technology opportunities. This is partly a result of the technology-expert-centred approach. I will discuss both neglected dimensions in turn.

Experiences from other parts of the food-processing industry indicate that large water savings can result from simple changes in working procedures. For example, at Grindsted Products ideas on water savings from employees permanently reduced water consumption by 25 percent. The experience of Danfoss, a Danish metal industry enterprise, showed that changing the piece work system could also achieve substantial effects. The present piece work pay system is not conducive to reducing water consumption and thus pollution. No incentives are present for employees to be resource minded. Better housekeeping could include simple technical measures such as adjusting the water nozzles as well as changing old working routines and payment systems. These important steps toward reducing pollution could create a positive atmosphere for continuous improvement and prevention among employees and managers. Until 1993 employees had not participated in implementing cleaner technologies, and the possibilities for improving working conditions were not investigated.

The Ministry of the Environment's approach to cleaner technologies has a time horizon of a few years. Thus, it only includes the production processes known in the fish-processing industry today. It is clear that environmental problems have not been reduced. The extent of automation has been increased with automatic feeders and deskinning machines, which actually increase water consumption and pollution. This clearly demonstrates the problems arising when the suppliers are not actively involved in developing cleaner technologies.

For a long-term perspective, the development of cleaner technologies must include broader questions. What will happen if an increased part of processing takes place on board the fishing fleet? Or if fully automatic filleting and cutting plants are developed? Or if biotechnology is used to a greater extent? Or the refinement of the product is increased? If these questions are not considered by development projects, they may risk offering solutions to yesterday's environmental problems.

Summing up the results of our comprehensive assessment with focus on the potentials of cleaner technologies:

* The technical and expert-centred approach to cleaner technologies has led to a 40-50 percent reduction of pollution (without considering the just mentioned technological changes);

* Cleaner working procedures can contribute toward reducing consumption even more substantially. At the same time, such changes can result in better working conditions;
* A preventive approach has to incorporate the creation of incentives to continuous environmental improvements among employers and employees;
* A strategic and preventive approach demands both immediate changes in working procedures and long-term development of process technologies.

Barriers to the Diffusion of Cleaner Technologies

The comprehensive assessment surveyed different approaches to innovating and implementing cleaner technologies in the fish-processing industry. Depending on the chosen approach there will be different barriers to the diffusion of cleaner technologies. Traditionally barriers to diffusion are conceived as obstacles, e.g. technological, economic, legislative, institutional, or those relating to knowledge and attitudes (see PREPARE 1991). Such a conception focuses on the appearance of the barriers and is often rather static. CTA must go one step further and explain the reasons why barriers occur. Furthermore, barriers refer to unutilized potentials. An analysis of unutilized cleaner technology potentials will show the kind of new means that can overcome the barriers described in the following (Remmen 1992).

Few of the fish-processing companies had until 1992 implemented the simple technical measures and available cleaner technologies. More far-reaching options were even not considered. Why is diffusion so limited? Three barriers can be identified: narrow problem definition; lack of interaction between companies, suppliers and others; and the regulatory philosophy of standard setting.

Narrow *problem definition* is found among companies, in the network (e.g. consultants), and at the government level, and it can be traced to three (already discussed) characteristics. First, cleaner technologies have been defined too narrowly — the relation to working conditions has not been assessed and obvious possibilities of improvements have been neglected. Second, the problem definition has been too expert-centred with only a few consultants and senior executives involved. Thus, the implementation of a preventive strategy is not firmly anchored among employees, management, suppliers and customers. Finally, the strategy has been too technology-centred. Neither the immediate possibilities of pollution prevention through cleaner working procedures nor the long-term possibilities through cleaner processes have been examined. As a consequence, cleaner technologies have influenced only the few companies in which the projects were carried out, but not the industry as a whole.

The narrow approach to cleaner technologies has resulted in a negligence of internal barriers at the company level. Consequently, management has not defined its environmental responsibility, and the company has not prepared a policy nor established procedures for pollution prevention. The employees have not been considered as a resource, and incentives for a lasting prevention via changes in attitudes and behaviour have not been created. Our comprehensive assessment makes clear that the attitude of employers and employees, the wage system, the corporate culture, daily routines, etc. are all potential barriers to prevention. Unfortunately, at this moment the relative magnitude of each internal barrier cannot be adequately assessed, due to the lack of experiments with focus on the organization of the companies.

A second barrier can be found at the *interorganizational network* on the meso level between companies and the authorities. So far cleaner technologies have primarily been based on the ideas and contacts of consultants. Trade unions and trade associations have participated in consequential groups and have commented on the implementation of the projects only as a check on negative effects. They have not contributed to the initiating of projects. Another important missing party is the suppliers of machines and equipment. They have not been involved in innovating and subsequently diffusing cleaner technologies to the industry. One reason is political: non-Danish producers cannot receive financial support from the Cleaner Technology Programme.

The third and most important barrier to overcome is the impact of the traditional *regulatory philosophy*, which has resulted in a deadlock. Standards set have led to establishment of wastewater treatment plants, and once companies are connected to such plants their incentive decreases to save water and to reduce waste water. "We no longer have environmental problems" becomes their typical attitude. In addition, the very existence of wastewater treatment plants reduces the return on cleaner technology investments. Water savings will only lead to an increase of wastewater-treatment charges, because these plants have to be paid for and earn their money. Yet for some years the plants are necessary to achieve the mandated reductions.

This dilemma between wastewater treatment and cleaner technologies represents a confrontation between two traditions of regulation — cleaning versus prevention. This is made visible through a conflict between the fish-processing industry and municipalities along the

North Sea against the Danish Ministry of the Environment. The subject is whether the companies must be connected to these wastewater treatment plants, when greater reductions can be achieved through the use of cleaner technologies.

These three barriers indicate that a shift toward pollution prevention and cleaner technologies cannot be accomplished through traditional top-down regulation alone. Instead the effort requires new methods and strategies to integrate environmental and social considerations in production and thus create new types of interactions between authorities, companies and other groups such as suppliers. A prevention strategy demands that government rely less on standards and more on informative and organizing means, agreements between trade and government, environmental labelling, etc.

Analysis of the use of means by state authorities typically distinguishes between physical, economic, informative and normative means (Eckhoff 1983). In this view it is possible to select from a broad array of means, make the best combination and achieve the desired effects. But the selection of means has to correspond to the environmental problems as well as to the barriers to pollution prevention. The three barriers indicate that the means must establish new types of interactions between the parties concerned, and not only focus on means applied by the authorities. New opportunities for a preventive strategy have to be found at three levels: companies, interorganizational network, and authorities.

A preventive strategy at the *company* level should use a methodology that provides an overview of the character and range of problems as well as ideas for solutions.[3] The use of environmental audits has in many cases led to substantial reductions of waste, use of water, raw material, etc. (Lauritzen 1992). That companies save money on treating waste and emissions as well as on purchasing raw materials has resulted in the mobilizing slogan "pollution prevention pays." Other advantages are rising productivity resulting from more insight into the production process, increased motivation among employees, and anticipation of future environmental regulation (Fischer and Schot 1993). To realize these advantages companies have to clarify:

* Objectives (environmental policy of the company including reduction targets);
* Strategy (specifying steps to be taken);
* Organization (defining responsibilities).

Since 1990 the Cleaner Technology Programme has given financial assistance to the development of environmental auditing and to trade-consultancy services. This broader approach to prevention reflects the

Environmental Protection Agency's recognition that the diffusion of cleaner technologies must be given assistance. The method for environmental audits has been simplified as an adaption to the Danish enterprise structure with many small and medium-sized companies. In a Danish survey the same positive aspects of environmental audits and trade consultants as above are stressed. At the same time it is pointed out that preventive activities would probably not have been initiated in the companies without financial support and that employees and working conditions have still not been involved in environmental management (Christensen and Holm Nielsen 1992).

One further reason why the pollution problems of companies vary considerably is their differing environmental strategies and attitudes. Some companies only react under pressure from regulatory demands, while others try to build up credibility with customers, environmental authorities, local communities, etc. Furthermore, it makes a big difference whether the company's strategy is based mainly on productivity, innovation or credibility. These differences can be illustrated in the following way:

Table 1: Environmental Strategies of Companies

Interactive	"We prevent"	
		Credibility
Proactive	"We are ahead"	
		Innovation
Reactive	"We obey the law"	
		Productivity
Inactive	"We have no problems"	

Actors in the *interorganizational network* have to be mobilized. As the fish-processing industry gains more insight into the production process and waste and emission streams it can formulate demands on suppliers. To enhance the effectiveness of these demands the industry must formulate them. This could eventually result in cooperation between the fish-processing industry and its suppliers to address pollution. If the suppliers are not part of the preventive solutions, they will become part of the problems. Not only suppliers need to be involved. Consultants, universities, trade unions, and trade associations all have competences to be elicited as well. Since pollution prevention must be anchored among all the parties concerned, it is necessary to activate an interorganizational network (see table 2 below). Furthermore, the companies have to cooperate more on environmental issues (whereas in the fish-processing industry there has been a coordination to resist the regulatory demands on cleaning). On preventive issues and cleaner technologies the information exchange has as elsewhere been limited and unstructured (Cramer et al. 1992).

Finally *environmental regulation* itself needs to stimulate pollution prevention. Public instruments that can support prevention at the source are insufficiently developed. Until now for financial support demonstration projects and trade consultants have been the most important means in Denmark. From 1992 cleaner technologies and environmental management have become a more important part of the Environmental Protection Act (Ministry of the Environment, 1992). The intention of the government is to give higher priority to preventive measures in the environmental approval of companies. In my opinion preventive regulation has to be based on superior objectives and a list of possible means that is confirmed by a trade agreement between government and industry. A long-term, dynamic and adapted regulation is required:

* long-term: demanding a prevention plan of the companies related to the objectives and means in the trade agreement;
* dynamic: gradual intensification of standards (resource and emission demands) year by year to the pollution level required;
* adapted: to current possibilities of prevention, local conditions as well as the environmental strategy of the company in question.

Moreover, the implementation of preventive regulation has to reflect local variations in types of emissions and in their consequences. The companies need to establish an environmental policy and prevention plan based on recurring environmental audits, internal control systems, and a yearly reporting of results to the municipalities. Regulation of the municipalities must reflect the differences of the companies. Therefore, it is necessary to use different means depending on the specific problems and the environmental strategy of the company.

Until now different voluntary programmes have been established, e.g. ECO-Management, ECO-Labelling of products, Cleaner Technology Actions Plans. In this way there are positive incentives for the companies to establish a prevention strategy. Towards the companies who have an inactive or reactive environmental strategy (see table 1 above), the authorities have to use "the stick": green taxes, standards, approvals of existing and new production processes, etc. This demands a new type of interplay between companies and local authorities. The authorities must be able to be both inspectors and consultants.

Constructive Technology Assessment as Pollution Prevention

Prevention and cleaner technologies represent a shift in environmental policy. Therefore, objectives and means are not given in advance. None of the parties — the companies, the interorganizational network nor the authorities — have a precise idea of how to establish sustainable production. This uncertainty about means will require new methods to integrate environmental and social considerations into the innovation and implementation of technologies. Such integration can be initiated by dialogue research between policy researchers and the parties involved. Moreover these processes have to be organized openly as social experiments, which facilitate a gradual and progressive specification of objectives and selection of means.

Opportunities for a preventive strategy include creating and (re)shaping interactions within companies, in the interorganizational network and between companies and authorities. Such actions could be labelled as constructive technology assessment (CTA). These actions to influence interactions go far beyond the traditional role of technology assessment. Traditional technology assessment produces knowledge and information for political decisions in governments. But influencing interactions implies that CTA has to be based on a dynamic relation between knowledge and action. This relation could be organized through social experiments and dialogue workshops (see appendix 2). This will provide a *social locus* for search and learning processes and a way to cope with uncertainty connected with implementation. In addition social experiments and dialogue research will allow and stimulate bottom-up initiatives, thus eliciting interest from a broad range of relevant social groups.

As already stressed, dialogue has an important role to play in CTA. This includes mapping which actors are involved, to get insight into their problem definition and action orientation as well as their different interests and power positions. It is important to make clear who is involved in the dialogue, and what the dialogue is about. In table 2 the possible sparring partners (scenes and activities) are identified. From our experience in the TOR project with applied consequence and comprehensive assessments, we have added some examples of aspects and problems to be put on the agenda.

Consequence and comprehensive assessments can inform a constructive technology assessment, partly as knowledge on the industrial sector and the relations of power and interests among the actors, and partly as a basis for getting into dialogue with the parties concerned. While appreciating the consequences of existing technologies and the potentials for future technologies (see appendix 2), CTA has to advance one step further. CTA must emphasize the importance of search and learning processes which, via experiments and negotiations, can handle uncertainty. Bottom-up initiatives are a necessary supplement to the top-down means traditionally applied by public authorities. This involves the possibility of activating all relevant groups and in this way to influence decision making on different scenes.

Table 2: Potential Actors and Aspects of Constructive Technology Assessment

	Scene	CTA shall contribute to
1.	**The company**	**Problem analysis and solution**
	* employers	- responsibility and internal control
		- environmental management system
	* employees	- prevention teams and organization
		- change in work routines & attitudes
		- participation and influence
2.	**The interorganizational network**	**Reduction of "aspect-blindness"**
	* consultants	- cleaner working procedures
		- cleaner process technologies
	* suppliers	- changes in design & construction
		- clean technologies
	* educational institutions	- learn prevention strategies
		- new courses and further training
	* trade unions	- working conditions versus environment
		- "cleaner" wage-bargaining system
	* trade organizations	- diffusion of knowledge on prevention
		- international standard-setting
3.	**The authorities**	**Dynamic and adaptive regulation**
	* municipality	- environmental action plans
		- health and safety
	* county	- environmental certification
		- external prevention campaigns
	* state	- trade agreements/standard-setting
		- promote development projects
		- initiate "market pull"
4.	**The public**	**Enlightenment and democratic debate**
	* citizen	- knowledge on problems and solutions
		- motivation to prevention
		- change of consumer behaviour
	* media	- information and debate

The TOR project has initiated two activities involving new dialogue. First, external prevention teams conduct environmental audits and use them to draw up a prevention plan for introducing cleaner technologies. These prevention teams have started a new dialogue in the interorganizational network (see table 2). Second, a study group for shop and safety stewards in the fish-processing industry in Skagen has been established. This study group aims at exchanging experiences with prevention efforts and identifying new opportunities. It is clear that employees possess an experience-based understanding of the production process that makes them essential to prevention efforts inside companies.

In cooperation with four companies from different industrial sectors, a subsequent project aims to investigate the internal barriers and different methods for prevention by organizing dialogue workshops among employers, employees and technicians. Every group of actors has its own approach to the problem analysis (what is wrong?) and to the solution (what can be done?). The dialogue workshop can confront these different approaches, and through experiments and negotiations find the best way to prevent the problems. In addition we expect that new interactions inside the firm will result in new interactions between suppliers, authorities and companies.

Appendix 1

Traditional versus Constructive Technology Assessment

Traditional Technology Assessment	Constructive Technology Assessment
* Science is ascribed a dominant role (researchers)	* Researchers and users acting as opponents (dialogue)
* Determines direct consequences and side-effects of the technology	* Is currently concretizing objectives and means as well as consequences and problems
* Limited problem analysis	* Emphasis on problem analysis
* Focus on technical solutions	* Combine a range of possible solutions
* Result = report	* Result = design criteria + report + mediation
* Tool for decision-making	* "The catalyst effect"
* Automatic connection to parliamentary decision-making processes	* Coupling to a multitude of decision-making processes on different scenes
To find the right answers	**To ask the right questions**

Appendix 2

Dialogue Workshops

A dialogue-based research and development methodology must both promote debate on problems and visions as well as contribute to specific strategies for action. One possibility is a further development of "future workshops" which are widely used in Denmark. The three phases of future workshops are criticism, visions and realization (Jungk and Müller 1981).

When planning future workshops, or what we prefer to call dialogue workshops, themes must be created that include the experiences of each individual participant. At the same time the various issues from the three approaches to technology assessment have to be considered (see the themes below). The suggested themes must always be related to the unique experiment (problems, possible solutions, and objectives). Dialogue workshops must continuously focus on technology itself and be oriented towards actions related to problems, potentials, barriers and means. The aspect of realization may in different ways be included in each phase according to the increasing radicalness, from reactive problem orientation to interactive creation of future possibilities for action.

The challenge is to create possible realizations at different levels. The parties involved have an opinion on consequences and problems from existing technologies — but the consequences must be visible to everybody and the different parties must agree on the problems. Experiences from social experiments show the importance of beginning with "small visible successes" which can provide the specific basis for forming an opinion on the technology in question. A practical attitude and experiment by way of "trial-and-error" are an important part of dialogue workshops. Prototypes, scenarios, practical trial with various technologies, meetings where advocates of different solutions confront each other, etc. are some of the "techniques" at hand, depending on the project character and the phase of the dialogue workshop. Furthermore, it is important to emphasize the continual interaction of problem analysis and resolution. CTA can be seen as a synthesis of problems and potentials with special attention to realization.

Modes of Technology Assessment

Consequence Assessment (Critics - reactive)	**Comprehensive Assessment** (Visions - proactive)
Consequences rising from	Potentials rising from
* application of existing raw materials, chemicals and products	* research and development of new technologies
* production modes, competencies and organization of process and equipment	* new production processes, cleaner technologies, etc.
* construction methods, standards and design of existing technologies	* new materials and products,
Problems concerning:	Objectives concerning:
* pollution, harmful working conditions, poor product quality, etc.	* sustainable development, socially useful products, etc.

Constructive Technology Assessment
(Realization - interactive)

Barriers and means to
* integration of social and environmental considerations in development, construction and application of new technologies

* democratization and decentralized control

* participation of the parties involved via experiments and negotiations

Notes

1. In evaluation research a parallel development can be discerned. Here, the development goes from summative evaluation, which summarizes the effects of a programme, to formative evaluation which tries to induce improvements into the process (Herman 1987). The core of this "fourth generation" evaluation is a change toward a constructivist paradigm (Guba and Lincoln 1989).

2. The existing environmental problems and working conditions are described and analysed more precisely in Handberg et al. (1991).

3. Several methodologies are available. For example, the prevention manual published by the American Environmental Protection Agency entitled *Waste Minimization Opportunity Assessment Manual* and its European Counterpart *PREPARE Waste and Emission* manual based on a Dutch version.

References

Andersen, Niels Erik, et al. 1990. *Professional Systems Development: Experience, Ideas and Action.* London: Prentice Hall.

Christensen, Karen S. 1985. "Coping with Uncertainty in Planning." *Journal of the American Planning Association* 51 (1): 63-73.

Christensen, Per, and Eskild Holm Nielsen. 1992. *Danish Experiences on Environmental Audits.* Aalborg: Aalborg University.

Christensen, Per, et al. 1991. *Industrial Environmental Regulation and Innovation: A Case Study of the Fish Processing Industry in Denmark.* Aalborg: Aalborg University, The TOR project.

Collingridge, David. 1980. *The Social Control of Technology.* London: Pinter.

Cronberg, Tarja, et al. 1991. *Danish Experiments - Social Constructions of Technology.* Copenhagen: New Social Science Monographs.

Dieleman, Hans, et al. 1991. "Choosing for Prevention is Winning," in: *PREPARE. Manual experiences.* Ministry of Economic Affairs, The Netherlands.

Duelund, Peter. 1991. "Dialogue Research - Theories and Methods," in: Cronberg, Tarja, et al. *Danish Experiments - Social Constructions of Technology.* Copenhagen: New Social Science Monographs, 167-183.

Eckhoff, Torstein. 1983. *Statens styringsmuligheter, særlig i ressurs- og miljøspørsmål.* Oslo: Tanum-Norli.

Ehn, Pelle. 1988. *Work-Oriented Design of Computer Artifacts.* Stockholm: Arbetslivscentrum.

Environmental Protection Agency. 1988a. *Renere teknologi i fiskeindustrien* (Cleaner technologies in the fish-processing industry).

Environmental Protection Agency. 1988b. *Statusnotat - Fiskeindustriens spildevandsudledning i relation til vandmiljøplnen.* (The discharge of wastewater of the fish-processing industry related to the Aquatic Environment Plan).

Fischer, K., and J. Schot. 1993. *Environmental Strategies for Industry. International Perspectives on Research Needs and Policy Implications.* Washington, D.C.: Island Press.

Green Waste Water Plan. 1991. *(Grøn spildevandsplan).* Hirtshals Municipality.

Guba, Egon G. and Yvonna S. Lincoln. 1989. *Fourth Generation Evaluation.* Newbury Park, CA: Sage Publications.

Handberg, Sussi, Lisbeth Kromann and Søren Svendsen. 1991. *Fiskeindustrien i Nordjylland. En branchebeskrivelse.* (The Northern Jutland Fish-Processing Industry: A Branch Description). Aalborg: Aalborg University, The TOR project.

Herman, Joan, et al. 1987. *Evaluators Handbook.* Newbury Park, CA: Sage Publications.

Hetman, François. 1973. *Society and the Assessment of Technology: Premises, Concepts, Methodology, Experiment, Areas of Application.* Paris: OECD.

Jungk, Robert, and Norbert Müller. 1981. *Fremtidsværksteder* (Zukunftswerkstätten, Wege zur Wiederbelebung der Demokratie). København, Politiske Revy.

Lauritzen, Kristian, 1992. *It's Worth It: Introduction to Environmental Action in the Company.* Copenhagen: DIF - Danish Society of Chemical, Civil, Electrical and Mechanical Engineers.

Ministry of the Environment. 1989. *Rapport om Mål og Midler i miljøpolitikken.* (Report on Objectives and Means in the Environmental Policy). Preface by Lone Dybkjær.

Ministry of the Environment. 1990. *Clean Technology Action Plan 1990-1992.* National Agency of Environmental Protection.

Ministry of the Environment. 1992. *Environmental Protection Act.* National Agency of Environmental Protection.

PREPARE. 1991. *Manual for the Prevention of Waste and Emissions.* Ministry of Economic Affairs, The Netherlands.

Qvortrup, Lars. 1984. *Telematikkens betydning. Et essay om informationssamfundets politiske filosofi.* Gyldendal.

Remmen, Arne. 1990. *Konstruktiv teknologivurdering - om at komme bagklogskaben i forkøbet.* (Constructive technology assessment - how to anticipate hindsight). Aalborg: Aalborg University, dissertation.

Remmen, Arne. 1991. "Constructive Technology Assessment," in: Tarja Cronberg et al.*Danish Experiments - Social Constructions of Technology.* Copenhagen: New Social Science Monographs, 185-200.

Remmen, Arne. 1992. "Barriers and Means to the Implementation of Cleaner Technologies: Experiences from the Danish Fish Processing Industry," in: *Technology and Democracy: The Use and Impact of Technology Assessment in Europe.* Copenhagen: The 3rd European Conference on Technology Assessment, 173-182.

Rip, Arie, and Henk van den Belt. 1988. *Constructive Technology Assessment: Toward a Theory*. Enschede: University of Twente.

Schot, Johan. 1991. *Technology Dynamics: An Inventory of Policy Implications for Constructive Technology Assessment*. The Hague: NOTA, Working document no. 45.

Schot, Johan. 1992. "Constructive Technology Assessment and Technology Dynamics: Opportunities for the Control of Technology - The Case of Clean Technologies." *Science, Technology and Human Values*, 17: 36-56.

Smits, Ruud, and Jos Leyten. 1987. "Key Issues in the Institutionalization of TA," in: *Technology Assessment: An Opportunity for Europe*. The Hague: NOTA volume 2.

Storgaard, Kresten, 1991. "Dialogue Research - An Approach," in: Tarja Cronberg et al. *Danish Experiments - Social Constructions of Technology*. Copenhagen: New Social Science Monographs, 157-166.

World Commission on Environment and Development. 1987. *Our Common Future*. Oxford University Press.

Part IV

Constructive Technology Assessment: The Case of Medical Technologies

Constructive Technology Assessment: The Case of Medical Technologies

Introduction

The chapters in this section bring together many elements discussed in earlier chapters. The importance of positional identities discussed by Downey, limitations of existing TA methods discussed by Wynne and Remmen, and barriers and opportunities for learning discussed by Jelsma and Herbold are all specified for the case of medical technologies. One important lesson will be that effective CTA policies must be sector-specific.

The importance of positional identities is clear in each of the three essays in this section. New technologies induce boundary disputes between medical specialists and for that matter between them and their patients. Koch shows how the development of diagnostic ultrasound became caught up in a power struggle between physicists, biologists and clinical investigators over the control of the emerging field of biophysics. Walsh argues that doctors were reluctant to introduce the birth-control pill because it crossed the established and proper boundaries between them and their patients. A freely available pill presented a threat to their learned authority and professional status. They relaxed their opposition after adopting the view that the pill should be available only to women whose life or health were in danger — and they would make the judgement. Compared with Downey's case, this is clearly a case of actors moving away from an earlier position because it was possible to make such a change without a threat to their own identity.

Weijers argues as well that intensive insulin therapy altered doctor-patient relationships. Patients gained control over monitoring blood glucose and making day-by-day decisions on insulin dosage. In this case technology helped to create a new boundary with the development of a new internal medicine specialist, the diabetologists, and a new professional group of diabetes nurses who support and educate their patients. Other striking examples of technology-induced boundary disputes are easy to elicit. Specialities like radiology and cardiology owe their existence, in large measure, to the mastery of advanced technology by that speciality to the exclusion of others. Even specialities like renalogy, hepatology, pulmonology, haematology, and dermatology that focus detailed expertise and knowledge on components of the human body (respectively, kidneys, liver, lungs, blood, and skin), are technology-dependent insofar as special knowledge is visualized by technological means and conceptualized in technological terms. (Blume 1992: 5-37).

New technologies make it possible to develop new identities and change the geography of existing identities. Indeed a salient and structuring feature of the medical sector is the pervasiveness of high-tech artifacts and procedures and their effect on the identities of medical specialists. Such technologies are often accepted without any discussion of underlying assumptions and sometimes, as with ultrasound, without rigorous clinical trial. This unreflective technological enthusiasm is not only true for medical specialists but also for institutions ranging from the largest teaching hospitals to the smallest regional clinics. These institutions compete for patients (and doctors) often on the basis of advanced technologies. The rapid diffusion of once-exotic diagnostic-imaging technologies like CAT and MRI results in large measure from such institutional competition. Liability considerations, especially in the United States, reinforce this tendency. A related point is that 'good health care' is typically identified with the technologically most sophisticated care. (Blume 1992: 157-224)

Another distinctive and institutional structuring characteristic of medical technologies is that the relations between producers (instrument makers, drug companies) and users (patients) are mediated by a powerful third party: the medical profession. Doctors of course are not a homogeneous group; medical specialities differ widely in their dependence on continual technological innovation. But in most cases medical doctors play a classic intermediary role between developers and users of new technologies including organized patients' groups. Essays in this section make clear that this situation has several major side effects on technical development. Most strikingly, producers of technology do not meet cost-sensitive consumers in classical competitive markets. Instead they meet a health care system that is very receptive to new technology, is engaged in an institutional and professional competition based on new technology, and shows therefore a widespread preference for new technology.

On the other hand, users are relegated to a passive position in at least four respects. First, assumptions about diagnosis or therapy as well as user representations (compare Akrich) built into machines are invisible to them. Walsh argues that most manufacturers and doctors think that they know what patients want. Second, users are not in the position to ask questions about technologies to make these assumptions visible since they rarely interact with manufacturers. The key interaction for shaping technologies is between medical specialists and manufacturers. Third, specialists are not trained to share their expert knowledge with patients and few doctors ever act as a nexus for transferring user preferences to manufacturers. Finally, users do not pay (or do so in an indirect way via insurance hospital charges, and doctor fees). There is no adequate mechanism to link health-care consumers' cost sensitivity to decisions about technology development.

These side-effects of medical sector structuring have led to a system without appropriate incentives for cost reduction. Perversely, the actors interested in cost reduction — insurance companies, governments and increasingly in the U.S. large employers — do not have instruments to foster cost sensitivity in technology development. To remedy this gap they have stimulated the development of medical technology assessment (MTA). Koch opens her chapter with a critical overview of the practice of this field. She observes that the medical research and policy communities have become fixated on quantitative studies of the safety and efficacy of medical technologies. Randomized controlled clinical trials (RCT) have become the 'gold standard' for drug assessment. Yet RCTs not only are time consuming and expensive but also in the end give conclusive information only about the use of a particular technology (not any modifications), on a particular patient population (not the general population), in the hands of experienced clinical researchers (not general practitioners).

In addition current assessments of medical technology, as both Koch and Weijers observe, scrutinize a technology only after its development to a marketable stage, and usually after clinical application. Such a policy serves as a 'gatekeeper,' preventing the adoption of new technologies rather than encouraging the development of alternatives. Deployment of MTA does not change the underlying dynamics of technical change in the medical field. In this respect MTA resembles conventional TA and has similar shortcomings. For changing the dynamics of a sector other types of measures are necessary.

Koch, Weijers and Walsh present various useful ideas for such measures. First of all, decision-making structures should not be set up to make only a one-time, go/no-go decision on the critical question of whether the new technology or practice will accepted by a research council or be covered by third party payment. This is an inappropriate structure because the decision must be made at one point in time, while the technology is (still) under development; such a structure cannot encompass or influence choices made earlier in the design phase, the technology's subsequent development, or its broader effects on the health care system. Rather decision-making structures should be arranged in a dynamic way, to monitor medical technologies over their entire life-cycle and thus influence these technologies while they are invented, developed, and diffused into practice. More specifically, Weijers recommends that decision makers organize a sequence of decision-making steps to recognize the different phases and natures of introduction, experimental application, adoption and modification processes.

Second, data other than those on mere cost-effectiveness are needed in the decision-making process on a continual basis (rather than as again one-time input). A short list might include data on user-representations in certain technological options, the existence of various options, social changes implicated in new technologies (such as developments of new specialists and new skill requirements), new treatments that may evolve from use of a certain technology, and ethical dilemmas. Of course cost data are important as well and so MTA will still have an important role. (Compare Remmen who gives traditional TA methods an analogous complementary role). Finally, users and citizens – who are ultimately paying the bills through insurance premiums – should have access to the decision-making arena. They will likely ask for data beyond costs as well. This will enable the development of "dialogic" communities, a term introduced by Koch, in which users will be in the position to question underlying assumptions of new technologies. Such communities will help users (and others) to articulate their concerns, needs and values, and in this way be focal points for the collective value learning process Wynne has asked for in his contribution.

It does not follow from these recommendations that building an entirely new decision-making structure in the medical area is necessary. On the contrary, existing structures can be shaped toward this end. Koch locates the United States National Institutes of Health as a major player that could serve as a dialogic community, by better organizing interactions between experts and others and influencing technical change. Spending $1 billion a year on research, NIH is a key agenda-setter. Grant applications for research are already reviewed by a panel of experts divided into study sections by discipline, with advisory councils for each disease institute reviewing the recommendations of scientific experts. The study sections, the advisory councils, the staff of NIH, and the non-voting representatives from other government agencies as well as private agencies for funding research already form a network well suited to conduct constructive technology assessment. This network presently brings into contact many of the groups having an interest in the outcome of medical research. Weijers argues as well that different parts of a more dynamic decision-making structure can be found in the Dutch situation, but they are not related. Thus there is no feedback and interaction between the different institutions. As in the case of Jelsma's discussion of biotechnology there is a clear need for orchestrating a learning process. It may yet be possible to conceive of medical technologies that are cost-effective and user-responsive.

References

Blume, Stuart S. 1992. *Insight and Industry: On the Dynamics of Technological Change in Medicine.* Cambridge, Mass.: MIT Press.

11 Why the Development Process Should Be Part of Medical Technology Assessment: Examples From the Development of Medical Ultrasound

Ellen B. Koch

Quantitative studies of the safety and efficacy of new medical technologies are regularly held up to the medical and health policy communities as the model of technology assessment which should be strived for in rationalizing the use of all new medical technologies. Every consensus conference sponsored by the U.S. National Institutes of Health cites the lack of good quantitative data, especially randomized controlled clinical trials, as the single greatest hindrance in deciding whether, or how, new medical technologies should be used in routine clinical care.[1] In this paper I highlight some of the problems inherent in assessing medical technologies 1) using solely quantitative measurements of the impact of medical technologies and 2) long after the technologies have been introduced into clinical practice.

Medical innovations are often well entrenched in medical practice before they ever draw attention as needing assessment.[2] By assessing only those technologies which reach the stage of commercial development and clinical application, the current policy process acts as a gatekeeper, trying to prevent the further adoption of new technologies rather than encouraging the development of alternatives. Many potentially useful technologies never reach the stage of a formal technology assessment before they are eliminated for a variety of reasons, often having little to do with the technical merits of the instrument, and much to do with the circumstances of the innovator, the availability of funding, or a supportive institutional setting, among other social factors.

If the assessment of medical technology is to become a dynamic process commensurate with the constant change inherent in technology development, it is important to start assessing medical technologies much earlier, at the stages when *qualitative* factors influence their *development*. Characteristics that influence the clinical application of a technology are built into an instrument during the development stages, and research funding decisions based on non-technical factors affect how instruments are constructed and subsequently used in clinical practice. Contrary to most policy assumptions about medical research, social criteria do play an important part in the decisions about funding of biomedical research proposals. Empirical material on the development of medical ultrasonic instruments illustrates these points (see also Koch 1990).

It is precisely because social criteria are already a part of the government funding system for research that I propose that it is a logical place to introduce explicit social criteria into the developmental trajectories of new technologies. This can be accomplished by balancing the influence of scientific experts with experts in other fields and representatives of the public interest, and by equalizing the directional flow of information so that the scientific experts who now control access to research funds have the opportunity to incorporate the perspectives of other interested parties. I am not proposing "targeted research" in the traditional sense, with a government agency setting priorities in isolation. Rather, I am suggesting that research funding institutions such as the National Institutes of Health (NIH) could be adapted to perform early-stage constructive technology assessment (CTA) where it will be most useful, helping to decide which technologies need intensive assessment, what types of patients need to be studied, and what parameters are to be measured in subsequent quantitative studies.[3]

Current Status of Medical Technology Assessment (MTA)

Medical technology assessment (MTA) encompasses a great range of activities, in part because medical technology is such a broad concept. The most common definition of medical technology among health policymakers is the one used by the U.S. Office of Technology Assessment: "all elements of medical practice that are knowledge based....the set of techniques, drugs, equipment, and procedures used by health-care professionals in delivering medical care to individuals and the systems within which such care is delivered" (OTA 1976: 4).[4] Under this definition *all* of medical practice could be construed as technology.

Comprehensive technology assessment involves examining medical practice (which is constantly evolving) for appropriateness, risks, benefits, costs, and long-term social and ethical consequences. Such comprehensive studies are rarely undertaken for even one single technology, and would be an impossible load to conduct for every aspect of medical practice.

Instead, medical technology assessment is currently a piecemeal activity conducted by many different groups for many different purposes. Clinical trials are an attempt to document the status of scientific knowledge at a given point. The testing required by the Food and Drug Administration (FDA) from drug and device manufacturers is intended to help identify ways in which the technology may become risky as it is diffused into practice. Consensus conferences are an attempt to direct the future by influencing practitioners' styles of practice. Current types of MTA offer only a static, snapshot view of a technology at one stage in its development and diffusion. None of these methods of MTA is a dynamic process that can follow a technology as it changes.

An MTA usually begins with the perception -- whether by a professional group, a government agency, a private insurer, or the public -- that a technology seems to be posing a problem of high costs, unacceptable risk, limited access to care, or rationing of scarce resources. In other words, medical technologies are now assessed when they are already established in medical practice, and they are creating a problem. The MTA is meant to address and fix a particular problem, not to do a comprehensive study identifying potential paths of development for the future. The groups conducting MTAs usually have a strong interest in reducing costs (Medicare, Medicaid, private insurance companies), proving safety and efficacy for FDA approval for marketing (pharmaceutical or medical equipment companies), or addressing issues of resource distribution (policy institute studies, Congressional hearings, Office of Technology Assessment studies).

Even though many different types of MTA have been explored, the vast majority of medical technology assessment in the United States, as well as in other countries, is conducted by pharmaceutical companies to gain governmental approval for marketing drugs (Institute of Medicine 1985). In the assessment of drugs, the randomized controlled clinical trial (RCT) has become the gold standard of technology assessment (Marks 1987, 1988). Because the RCT has been very successfully applied to the assessment of drugs, it is often viewed as the best and most conclusive method of assessing all types of medical technologies including devices and techniques of medical practice.

However, even very well conducted RCTs can be fraught with problems. Long-term maintenance of protocol is difficult and expensive, and sample sizes have to be very large to identify rare events. Quantitative MTA is slow and is usually done only once, not taking into account changes in the technology, alternative technologies, or in the patterns and mode of applications. New technology or applications of an existing technology may evolve outside of RCT into a superior version, making the results of the RCT useless or the continued denial of the technique to the control population unethical.[5]

In addition to the difficulties inherent in designing and conducting RCTs, a methodology appropriate for studying drugs is not necessarily appropriate for studying other types of medical technology. With many devices or techniques, it is impossible to perform a truly "blind" study, in which neither doctors nor patients are aware of whether the patient is part of the treated group or the control group. Even in the assessment of drugs, an RCT is very time-consuming, requires a large study population, and is very expensive. In the end, an RCT gives conclusive information only about one particular technology used by particular experienced clinical researchers on a particular patient population. It gives no conclusive information about any modifications of the technology, how effective the technology will be in the hands of general practitioners, or how effective the technology might be in applications to slightly different diseases, conditions, or types of patients (IOM 1985: 32-63).

In a broad analytical study of medical technology assessment, the Institute of Medicine (IOM 1985: 70-175) identified several other techniques of medical technology assessment that are rarely used, but which are more appropriate than the RCT for assessing certain types of technology: sample surveys and case studies for generating hypotheses about the use of medical technology; epidemiological studies to identify rare events such as adverse drug reactions; quantitative synthesis and group judgment to identify the current stage of knowledge; and mathematical modelling to simulate future applications of a technology on the basis of available data. In each case, the IOM (1985: 160) suggested that the assessment technique needs to be improved, in large part by strengthening the quantitative aspects of the assessment.

Even when formal quantitative technology assessments are very well performed for individual medical technologies and many ethical, legal, economic, psychological, and other social aspects of the technology are considered, the conclusions of the assessment often have little impact on common medical practice (Fineberg 1985). Well-designed, scientifically objective methods of assessing medical technologies may produce clear indications for how and when a technology should be used, but they give little guidance on how to encourage adoption of those guidelines in medical practice.

Consensus conferences are the type of assessment closest to the definition of CTA presented in this book. Consensus conferences were originally designed, at least in the United States, to broaden the definition of MTA to include the explicit consideration of all the social factors that affect the safe and appropriate implementation of a medical technology in clinical practice, as well as the assessment of clinical benefits and risk, and the incorporation of majority and minority opinions about the appropriate use of technology. They were intended to provide a forum for intensive discussion between all interested parties, including patients, physicians, and experts representing the positions of the government, insurance companies, ethicists, lawyers, economists, academic researchers, and the community of practicing physicians (Perry and Kalberer 1980).

Yet even consensus conferences are notoriously ineffective in influencing the patterns of use of a medical technology, or in changing the thinking among the medical profession about the relative merits and drawbacks of using particular technologies (Thompson et al. 1981; IOM 1985: 194-5; Kanouse et al. 1989). In fact, consensus conferences often follow changes in physician behavior (Hill et al. 1988). Evaluation of consensus conferences by Martha Hill and Carol Weisman (1991) indicates that they fail to achieve their goals for several reasons. Consensus conferences come too late, after a pattern of practice is already well established. By reporting only the points upon which consensus was reached, they publicize the lowest common denominator of agreement among the interested parties, who often have highly divergent points of view, rather than highlighting the points of disagreement about appropriate use of a technology. Physicians view consensus conferences as biased, and the credibility of consensus statements is linked to the credibility of their sources rather than to the quality of the assessment or the validity of the data used. Government-sponsored conferences are seen as the least reliable, because they are viewed as coercive attempts to interfere in medical practice, while conferences sponsored by medical organizations are seen as more reliable sources of information. Statements coming from government agencies are much less likely to be adopted than those from a professional organization. Information conveyed in journal articles or continuing education courses presented by opinion leaders is tacitly recognized as more influential than other sources (Hill and Weisman 1991).

Two consensus conferences conducted on the use of diagnostic ultrasound in pregnancy in the United States (NIH 1984) and in Norway (Dept. of Social Affairs, 1987) illustrate some of the issues raised above. The two conferences used the same data, yet reached opposite conclusions about the appropriate uses of ultrasound in pregnancy: The American consensus was that ultrasound should be offered only to women when there was some other clinical indication of abnormality with the pregnancy; the Norwegian consensus was that every single pregnant woman should be offered a diagnostic ultrasonic screening test at 17 weeks of gestation.

The starting points for the two conferences were quite different, even though the issue of cost was a major consideration for both countries. The American conference was called in reaction to a marked increase in the use of diagnostic ultrasound in pregnancy during the previous few years, and the proposal by some physicians writing in the medical literature that ultrasound should be used as a routine screening technique in all pregnancies. The NIH consensus panel estimated that between 15 and 40 percent of pregnant women in the United States were exposed to at least one ultrasonic scan in 1983. But as the panel pointed out, these figures were underestimates because they did not include exposure to Doppler devices used in monitoring fetal heart beat at clinic checkups or during delivery. The panel favored restricting access to ultrasound because they had no evidence that ultrasound reduced maternal and infant rates of death, disease, or disability. They also had no data on long-term harmful effects of ultrasonic exposure to the fetus. The panel concluded that "Ultrasound examinations performed solely to satisfy the family's desire to know the fetal sex, to view the fetus, or to obtain a picture of the fetus should be discouraged" (NIH 1984: 594). The panel called for further research on the bioeffects of ultrasound in animals and better data on the effect of ultrasonic diagnosis on birth outcome.

The Norwegian conference undertook an independent evaluation even though no new and decisive findings had emerged since the American conference. There still was no scientific evidence that screening would improve the health of the mother or child at delivery. But the Norwegian conference was less interested in improving mortality or morbidity rates at birth (which already rank as one of the best in the world) than in decreasing the widespread use of ultrasound. In 1986, 94 percent of all pregnant women in Norway were examined with ultrasound, with each woman receiving an average of almost three scans in a pregnancy. The panel found a high variation in the quality of service and the use of ultrasound in different geographic areas. In the

Norwegian case it was no longer a question of whether to use ultrasound routinely in pregnancy, but a question of how to improve the organization of services and reduce over-utilization of the technology. The panel wished to make access to ultrasound more equitable and to maximize psychological, social, and economic benefits of prenatal ultrasound to the women and their families. The Norwegian conference differed markedly from the American group in including less easily measured values. The Norwegians considered as important such factors as the increase in confidence among parents after seeing images of their babies. Since high quality prenatal care is widely available in Norway, the panel recognized that a change in the pattern of ultrasound use would only marginally increase the number of fetal abnormalities discovered, and would have little impact on the mortality rates of pregnant women and their babies.

The effect of the American and the Norwegian conferences on practice patterns has been very difficult to document. Physicians in the United States continue to debate the routine use of ultrasonic diagnosis as a screening technique, with the advocates of routine screening emphasizing the behavioral and psychological benefits in including the father in the pregnancy, relieving parental anxiety, and increasing patient compliance with medical care. Critics emphasize the absence of any concrete evidence that ultrasound improves the quality of care, the unidentified but potential dangers of ultrasound, and the costs of using ultrasound routinely (e.g. Youngblood 1989; Ewigman 1989). In the United States the pattern of ultrasound use in pregnancy continues to be highly variable, with physicians recommending ultrasonic diagnosis for their patients from 10-90% of the time. By contrast, the patterns of use in European countries are more even. In Britain and in Norway, where a single screening sonogram is recommended, and in France and Germany, where two are recommended (in all four cases by an official agency or national organization), the evidence suggests that nearly all physicians now follow these recommendations (Blondel, Ringa, and Breart 1989; Backe 1988; Jackson, 1985).

One article (Ewigman et al. 1991) suggests that the policy statements of national organizations were influential in setting physicians' use of ultrasound, both in the United States, where routine use is explicitly discouraged, and in European countries where it is approved. Given the evidence that statements issued by government agencies are viewed by physicians practicing in the United States as one of the most biased and unreliable of sources for information (Hill and Weisman 1991), it is much more likely that other factors are more influential. For instance, the social health insurance systems in England, France, Norway, and

Germany are in a position to regulate the reimbursement for diagnostic ultrasound through a centralized organization. No such organization exists in the United States, where each insurance company decides which applications of a technology to pay for according to their own community-based model of "usual and reasonable care." Efforts by government or professional organizations to limit the use of ultrasound by publicizing consensus statements to the medical profession are offset by patient demand, physicians' views of ultrasound as a profitable procedure, and physicians' fear of malpractice suits (Perone, Carpenter and Robertson 1984; Jack, Empkie and Kates 1987).

A great deal of research on improving the quality and delivery of health care and the impact of technology on medicine has focused on the diffusion stage of medicine. Medical sociologists and policy analysts have identified such social factors as peer influence, profit motives, incentives and disincentives in the reimbursement structure for medical care, fear of legal liability, pressure from public interest groups, and the association of "high" technology with prestige as influential in the diffusion of medical technologies (Rogers 1981). The diffusion of medical technologies can be highly unpredictable. It is at the diffusion stage that many of the current problems of health care seem to appear and where most policy efforts to influence the patterns of technology use are directed. Overutilization of a technology can create grave cost problems. The inaccessibility of a very expensive technology can create dilemmas of rationing and ethical choices. The use of a technology for a different purpose than the one for which it was developed can radically shift the balance of risks and benefits, making a technology that is in some circumstances risky, but appropriate, become overly risky and inappropriate.

The IOM report concludes that the diffusion of new technologies into medical practice is affected by ten factors, only some of which are subject to policy influence (IOM 1985: 176-210). The attributes of the technology are among the factors the IOM report cites as insensitive to policy. Prevailing theory, features of the clinical situation, and the presence of an advocate for the use of a technology are considered (by the IOM) as relatively insensitive to policy intervention. Subject to policy influence over time are the practice setting, decision-making processes, and characteristics of the adopters. Environmental constraints, conduct and methods of evaluation, and channels of communication are very open to policy influence.

Implicit in the extensive literature on health policy and the diffusion of medical technologies are two assumptions. First, that the scientific and technical content of a medical device is developed in a social and policy vacuum and that social factors only enter the equation at the stage of diffusion. Second, that the diffusion stage of a medical technology is the appropriate place to assess the social factors that play a role in the adoption of medical technologies.

The case studies in the development of diagnostic ultrasound presented below challenge both of these assumptions. The development of the technical and scientific capabilities of diagnostic ultrasound was considerably influenced by social factors. In the very earliest stages of research, investigators working on ultrasound "hardwired" into their equipment two very different assumptions about disease and diagnosis, one based on a visual paradigm and the other on the acoustic behavior of tissues. The present trajectory of the use of ultrasound in medicine took shape when the research funding process selected the visual paradigm over the acoustic paradigm. The visual paradigm is the basis of ultrasound's current extensive use in obstetrics, and it still dominates the patterns of use in other medical specialties. In effect, disciplinary rivalries between biophysicists and physiologists as well as the desire of NIH to promote education in the basic biomedical sciences, resulted in the application of the visual paradigm to diagnosis with ultrasound. Consequently, the use of ultrasound to characterize tissues languished until recently. Since social criteria already play an important role in the funding for developmental research, it is an appropriate place to begin CTA, especially the explicit consideration of social factors in the development and diffusion of medical technologies.

Simultaneous Inventions — Or Hardwired Assumptions

Between 1948 and 1952 Douglass Howry, a radiologist in Denver, and John Wild, a surgeon in Minneapolis, each independently developed a method of producing cross-sectional images of soft tissues using ultrasound, or sound waves above the range of human hearing. Howry and Wild were quite explicit at a conference in 1955 that they were investigating different clinical problems, and as a result were using very different machines. Their respective equipment differed in circuitry, frequencies, levels of picture resolution, ranges of penetration into the body, and methods of image display. Howry and Wild approached ultrasound from very different perspectives, incorporating in their equipment quite distinctive ideas about medical diagnosis and which aspects of ultrasound would be most useful in clinical applications. They were interested in obtaining very different information from the ultrasonic scan (Howry 1957).

In 1957 the research scientists who reviewed Wild's and Howry's grant applications to the National Cancer Institute (part of the National Institutes of Health) declared that Wild's work on ultrasonic instrumentation had become redundant. They suggested that NCI provide Wild with $50,000 (equivalent to two to four years of research support) to purchase equipment from Howry so that Wild could concentrate on clinical studies with the "technically superior" equipment instead of working on building his own equipment (NIH 1957). Four of those six reviewers — Theodor Hueter, William Fry, Otto Schmitt and Russell Morgan — were well equipped to analyse and understand the detailed technical aspects of the two projects since they were themselves working on ultrasound or closely related topics in biophysics.

Howry and Wild were intent on obtaining different information from tissues, largely because of their differences in background. A well-developed tradition of inquiry in radiology shaped Howry's perception of what was useful about ultrasound, and what the new technology should do -- it should be a better X-ray. Howry set out with the express intention of producing two-dimensional images that were clearly recognizable to the clinician as anatomical structures.

In 1957, when Howry and Wild were compared by NCI reviewers, Howry and his colleagues in the departments of medicine, radiology, and electrical engineering at the University of Colorado had recently begun testing on patients a new ultrasonic scanner that produced a beautifully clear and detailed image of the human body in cross-section. This "compound scanner" incorporated features from several previous instruments. In each scanner including the compound scanner, Howry had used a frequency between 1 and 2 megahertz (Mhz), on the basis that this frequency range was the *only* one that could penetrate a sufficient distance in the human body to produce an image. However, this commitment to a particular frequency range created its own set of technical problems. Howry compensated for the lack of resolution obtainable with 1-2 Mhz by narrowly focussing the beam of sound and consequently the aperture of the receiver. This meant that Howry, in striving for good image resolution, had eliminated from his image much of the ultrasound that was reflected, or scattered, by structures inside the body. The compound scanner made a very detailed map of structures inside the body, using ultrasound to measure distance just as radar measured distance in the air by timing returned echoes. Diagnosis of disease with the compound scanner, as with all of Howry's previous instruments, was made on the basis of the measurement of abnormal size or placement of tissues.

Howry's research group had devoted considerable time to standardizing the image produced by the compound scanner. Their goal was to make similar structures appear the same, regardless of their distance from the crystal that produced the ultrasound waves, so that two scans of the same patient would be identical, and so that scans of two different patients could be compared to each other. Standardization of the image required elaborate automatic electronic systems to compensate for different tissue responses to ultrasound and to prevent operator prejudice in producing the image. They had found that the image could be altered considerably by "the wishful thinking and manual control of the operator" (Howry 1955: 19).

Throughout their work, Howry's group strove for standardized, reproducible results, working from the theoretical to the practical, drawing on existing theory and models in acoustics, physics, and electrical engineering, testing each iteration of equipment on artificial laboratory models before moving on to the complexities of working with living organisms.

Wild, trained as a surgeon, turned to ultrasound with a very different clinical goal in mind and a different perception of the clinical capabilities of ultrasound. To the surgical community of the 1950s, the diagnosis and surgical treatment of cancer was a major concern. When Wild first turned to ultrasound in 1949, he intended to use it to measure distances in tissues, just as Howry did. Yet irregularities in his initial tests suggested that ultrasound responded to differences in tissue characteristics. Fifteen-Mhz ultrasound is reflected by irregularities *within* tissues, not just by tissue surfaces, as was the case with the 1-2Mhz ultrasound used by Howry. *In vitro* tests on diseased tissue samples indicated that cancerous tissue both reflected more sound and scattered it more than did the normal tissue. From that point, Wild pursued the use of ultrasound as a method for detecting malignancies that were too small or in such a location that they could not be found by the surgeon's usual method of palpation.

The very characteristic of ultrasound that Howry found a hindrance in producing a "true picture" was one that Wild was using in 1957 to distinguish between malignancies and healthy tissue. For Howry, backscatter caused undesirable aberrations in scans, and he tried many tactics before he managed to eliminate it in the compound scanner. Wild, on the other hand, was intent on capturing as much backscatter as possible, since both the intensity and the degree of scattering indicated tissue character. To Wild, the *relative acoustic behavior* of the tissue, not any direct visualization of abnormal lumps, was the indication of pathology. When Wild and his collaborator Jack Reid, an electrical engineer, adapted their equipment for two-dimensional display in 1951, it was with the primary purpose of speeding up the scanning process, not of producing an image recognizable as anatomical structure. With their two-dimensional display, the brightness and size of a spot on the screen corresponded to the intensity and degree of scatter of the return echo, *not* to the exact size or location of a structure. The diagnostic utility of Wild's two-dimensional echograph still relied on the comparison of acoustic behavior of tissues (not on their spatial relationships, as in Howry's scans).

The equipment discussed by Wild and Howry in their respective grant applications in 1957 differed in all aspects except one: they both produced a two-dimensional image (see Table I). Similarities in their two-dimensional images produced by the two machines were deceptive, however. Even "identical" spots of light embodied different information. Howry's machine was incapable of producing the information about tissue character that Wild wanted, just as Wild's machine was incapable of producing the clarity of image Howry wanted.

Table I: Features of Grant Applications by Howry and Wild

FEATURE	HOWRY	WILD
Technical image	2 dimensional	1 and 2 dimensional
Frequency	1-2 MHz	15 MHz
Penetration	> 6 inches	1.5 inches
Beam width	Narrow aperture	Wide aperture
Backscatter	Eliminated from scan	Captured in scan
Variable measured	Distance from skin to tissue	Intensity of echo and degree of scatter from microscopic structures
Applicability	One instrument for all applications in the body	Different instruments for application to each part of the body
Basis of diagnosis	Examination of one scan to identify abnormal size or location of structures	Comparision of two scans to identify cellular masses with unusual acoustic properties
Disciplinary background	Radiology	Surgery
Affiliation	Medical school and university	Private hospital
Collaborators	Faculty in physiology, radiology, electrical engineering	Non-academic electrical engineers
Approach	Theory to practice	Practice to theory

As Peter Galison has commented in discussing the instruments of experimental physics, "While machines can certify results without constant reference to theory, they can also import assumptions *built into the apparatus itself*. We might do well to call these hardwired assumptions 'technological presuppositions' to remind ourselves that machines are not neutral" (Galison 1987: 152). By 1957 Wild and Howry had both hardwired into their equipment very different assumptions about disease and diagnosis.

The comparison of Wild's and Howry's equipment for two-dimensional imaging illustrates two points about the development of medical technologies that have pertinence to CTA: First, seemingly similar technologies can have hardwired into them quite different assumptions about disease, health, diagnosis, and therapy. (This may be a truism in technology studies, but it is still a relatively unexplored idea in the medical and health policy communities.) Howry and Wild both developed instruments which produced two-dimensional images of the body, but which embodied very different information about the tissues being scanned. Each one of these instruments acquired critical differences in their early stages of development which irrevocably set the parameters of their potential use in clinical practice, as well as their literal design circuitry and instrumentation. Reflecting their respective disciplinary backgrounds and specific clinical goals, each innovator built into his instruments assumptions about the physics of ultrasound and about disease, diagnosis and therapy.

These same assumptions about disease, therapy, and diagnosis are built into instruments, and those assumptions can significantly influence subsequent applications of the instrument. It is important to identify what assumptions are "hardwired" into an emerging technology. What information or results does a technology *really* produce? How flexible or inflexible is the design of the instrument? For instance, Howry's instrument could be applied to any part of the body which could be immersed in a tank of water, so potentially his instrument could be used by any number of medical specialties. By contrast, each of Wild's instruments was designed to detect cancer in a very specific part of the body. His breast scanner and his colon scanner were each site-specific, and could not be used for diagnosis on other body parts without considerable modification in design.

Second, these hardwired assumptions are often invisible to potential adopters, who see only the end-product of the instrument (such as a two-dimensional image) instead of the process of development that produced the instrument. So, for instance, Howry and Wild's instruments appeared to physicians to do the same thing — produce a two-dimensional image of the body — except Howry's images seemed to be much clearer. Unless potential adopters are trained in the disciplines which fostered the innovations, they may have difficulties identifying or appreciating the ways in which hardwired assumptions may bias future applications or modifications of the instrument.

The questions that should be asked of an emerging technology are: What backgrounds do the innovators have and within which disciplinary frameworks were they working? How compatible is that background of the developer with the training of those who will probably use the instrument? In other words, can the "technological presuppositions" of a medical technology be understood and accepted by its potential adopters? Compatibility between the assumptions built into the instrument, and the assumptions made by clinical adopters (in addition to the other factors already identified by other researchers as influential in diffusion) may determine how readily an instrument is adopted in regular clinical practice, how quickly its use diffuses, and how effectively it is used.

Funding Mechanisms As Inadvertent Steering

The biophysicists who in 1957 declared that Howry's and Wild's equipment were comparable, if unequal in merit, were judging the men and their research techniques, not the characteristics of their machines. The reviewers were perfectly capable of understanding the technical differences between the two instruments. In declaring Howry's superior, they were really claiming that: (1) the visual paradigm of diagnosis incorporated in Howry's machine was a better and more adaptable means of diagnosis,[6] and (2) Howry's research methodology in the emerging field of bioengineering and instrumentation was preferable.

Theodor Hueter's and William Fry's judgement of the two grant applications was influenced by their own circumstances as ultrasonics researchers. They were concerned with the difficulties of obtaining funding and scientific respect for their work, when clinical applications of ultrasound already had acquired a tainted reputation as a fraudulent technology in the United States after an over-enthusiastic adoption of ultrasound in physical therapy and inflated claims for its efficacy in treating everything from the treatment of arthritis to diphtheria carriers and bedwetters in Europe during the 1940s (Kremkau 1979; Kobak 1954).

In their own research, both Hueter's and Fry's groups were trying to identify the mechanism of ultrasonic damage in nerve tissues, and to establish a threshold dosage curve for ultrasonic damage in the brain. To address this question of when and how ultrasound damaged nerve tissues, they both concentrated on studying acoustic field characteristics of their instruments and refining their equipment so that it would produce controllable, known quantities of ultrasonic energy in the desired location in the brain.

Hueter's and Fry's work on therapeutic applications of high-intensity ultrasound was regularly confused with two other medical uses of ultrasound, which used low-intensity, unfocussed sound waves: the use of ultrasonic diathermy for deep heating of tissue, developed and widely adopted in Europe during the 1940s; and the use of ultrasound to irradiate the brain in cases of intractable pain, as a non-surgical substitute for lobotomy, a technique recently developed in the United States by a neurosurgeon, Petter Lindstrom.[7] Both focussed, high-intensity sound and unfocussed, low-intensity sound came under the rubric "ultrasound," a general term which applied to a wide range of frequencies, and took no account of the intensity, focus, or method of delivery produced by each instrument (Fry 1958). Fry regularly protested to potential funding groups that there was "essentially no relation between these 'ultrasonic diathermy procedures' and ultrasonic techniques which we are suggesting" (Fry 1956). He was concerned that "very serious misunderstandings might arise on the capabilities of ultrasound in neurology" from the confusion of his and Hueter's work with that of clinical researchers working on therapy under less controlled circumstances than Fry and Hueter arranged in their research laboratories (Fry 1955).

The prevalent confusion between diathermy instruments and high-intensity instruments was a critical hindrance to obtaining funding for lengthy and expensive instrumentation and dosage studies. Fry had troubles convincing biomedically trained reviewers that research on high-intensity, focussed ultrasound was justified and that it required extensive instrumentation when commercial diathermy instruments were readily available, and when some clinicians were already applying the commercial instrument to clinical studies on the brain for such purposes as the control of intractable pain.

In their effort to obtain funding and scientific respect for their own work, Fry and Hueter acted as reformers in both the fields of diagnostic and therapeutic ultrasonics research as they tried to lend scientific credibility to ultrasonics research by imprinting the growing field of biophysics research with the perspectives and methodological approaches of the physics community. Hueter and Fry agreed completely on the proper approach to clinical biophysics research. Hueter's and Fry's groups were each trying to establish a threshold dosage curve for ultrasonic damage to nerve tissue, and to identify the mechanism of ultrasonic damage in nerve tissues. To address this question of when and how ultrasound damaged nerve tissues, both groups concentrated on studying acoustic field characteristics of their instruments and refining their equipment so that it would produce controllable, known quantities of ultrasonic energy in the desired location in the brain. In their own work on ultrasound, Fry and Hueter incorporated precise quantification, control of variables in research design, and reproducibility of results, at a time when clinical research design was itself under much dispute (Marks 1987).

Fry and Hueter did agree that several physical mechanisms were responsible for the action of ultrasound on the brain. Yet their experimental results caused a rift within the ultrasonics community as neurosurgeons, physiologists, and physicists disagreed about the conditions under which thermal action and mechanical action of ultrasound would predominate. Fry insisted that the action of ultrasound on brain cells was non-thermal, and he could use the specific action of ultrasound to impair brain cells without damaging vascular structures. The neurosurgeons working with Hueter at MIT and at Massachusetts General Hospital (MGH) felt their instrument, which thermally coagulated *all* of the tissue in a specific site and did not differentiate between thermal and mechanical effects, was much more reliable *for clinical applications*.

The dispute within the ultrasonics community over the thermal or non-thermal action of ultrasound was at the crux of a more general power struggle between physicists, biologists, and clinical investigators over the control of the field of biophysics. Fry's criticism of the work at MGH/MIT escalated in direct proportion to the control of the project by biomedically trained researchers. Theodor Hueter apparently agreed. He quit MIT in 1957 for a job with Raytheon Corporation, leaving biophysics entirely, with the comment to Fry that: "I am discouraged with the lack of integration in this field and unwilling to figure as a maladjusted sideliner" (Hueter 1957). Fry and Hueter apparently felt that biophysics was becoming a field where biomedical researchers used the tools of physics to answer questions about biology, to the neglect of advancing knowledge about the physical side of the equation. Physically trained biophysicists like Fry and Hueter were in danger of having their research agendas preempted by biomedically trained researchers or of becoming instrument makers working for others rather than researchers in control of their own work. In this context, Howry's emphasis on working from the theoretical to the practical, testing ideas on artificial laboratory models, and expressing his results in standardized, quantifiable terms was more in keeping with his reviewers' efforts to raise the scientific standards of instrumentation research than was Wild's approach. Howry's work contributed to the understanding of the physics of ultrasound in biological tissues in a way which Wild's work did not.

Wild's approach was exemplified by his comment that "a system having a nightmare of complexity to the physicist may be considered simple by the biologist in his blissful ignorance ... increasing exploitation of the observed phenomena does not necessarily need to wait upon whatever fundamental explanations may be forthcoming" (Wild and Reid 1953: 270). This approach directly contradicted everything Hueter and Fry were striving for in increasing the scientific reputation and fundability of ultrasonics research. Wild's work, exploiting physical phenomena for clinical application, without exploring fundamental knowledge about the interaction between sound and tissue, was exactly the sort of biophysics research that Hueter and Fry were trying to avoid having institutionalized.

In favoring Howry's grant application, Hueter and Fry were also conforming to an unpublished but explicit policy in the NCI that gave first priority for funding to research projects which served an educational role, training graduate students or newcomers to biomedical research, especially in critical fields such as biophysics.[8] Howry's research, under the joint sponsorship of a university and its affiliated medical school, included medical or graduate students; Wild's work at an independent private hospital did not.

The idea of judging the men and their methods rather than the specific project was not just a peculiarity of the reviewers for the two projects on ultrasound; it was an idea proposed by a joint committee of all the National Institutes of Health in 1952 and endorsed by the National Cancer Advisory Council (National Cancer Advisory Council 1952). In trying to incorporate the implicit goal of fostering education into the NIH mandate to sponsor the best quality of research, the advisory councils favored assessing the abilities (or "promise") of the investigator more than looking at the scientific merits of a particular grant proposal.

In 1956 a committee on methods of funding research for the National Cancer Advisory Council reiterated the idea, specifically endorsing the policy "*that in the future there should be greater confidence in men and less emphasis on the specific detailed single research studies to be undertaken*"[9] (emphasis in original). In making this statement, the committee members were pressing the idea that methodology was far more important in guaranteeing successful research than the specific idea being researched. In other words, a researcher with the "right" approach was more likely to produce useful and valid information than a well-conceived research goal pursued with the "wrong" methodology. According to the peer reviewers trained in the physical sciences, Howry's group utilized the correct approach to studying ultrasound; Wild's group did not.

The assessment of Howry and Wild was part of a continuing pattern of funding for the two projects. Wild had long-standing problems with sponsoring institutions. Even though both were given similar amounts of funding, Wild's was year by year, with no commitment for renewal, so that he could not make long-term plans, nor offer job security to employees. Moreover, while salaries came out of grant funds for Wild's group, Howry's group had the benefit of academic salaries. In effect, Howry had a far larger percentage of his grant funds to put into the instrument itself. Even though Wild managed to obtain funding for several more years after the 1957 episode discussed here, negative funding decisions forced him to abandon his research. Characterizing tissues according to acoustic properties lay dormant for more than twenty years until a new generation of ultrasonics researchers picked up the idea and built upon it with the much more sophisticated electronics that became available. It has recently found application in clinical practice in such techniques as scanning breasts for cancer and examining the placenta for deterioration in late stages of pregnancy. The visual paradigm that Howry introduced still predominates in current uses of ultrasound, particularly in obstetrics where assessment of the health of the fetus is based on very precise measurement of the size of the head and limbs, and visual examination of such body parts as the heart, kidneys, bladder, and spine for unusual shape, size, or function. The funding decisions made about ultrasonics research considerably influenced the type and capabilities of instruments made available for clinicians for several decades, even though they did not permanently suppress work on ultrasonic tissue characterization.

Thus, the development of medical ultrasound was considerably influenced by factors extrinsic to the state of scientific and technical knowledge. Because Fry and Hueter were concerned about professional respect for biophysics research and the fundability of their own projects, they judged other ultrasonics researchers and their research proposals according to style of research and institutional circumstances, rather than according to the intrinsic clinical merits of their technologies. Fry's and Hueter's efforts to improve the reputation of ultrasonics research and to imprint the emerging discipline of bioengineering with their own research methodology, as well as NIH's policy of giving funding priority to research projects with an educational role, determined the shape of ultrasound technology for an entire generation.

The story of funding for the development of ultrasound is fairly typical. Medical technologies are developed in circumstances different from many other technologies. Those who pay for the development of medical technologies are rarely those who benefit from them, and those who benefit from new medical technologies rarely pay for them. Hueter, Fry, Wild and Howry each received years of funding for their instrumentation work from various government funding agencies. Yet those agencies never intended to use the instruments. The peer reviewers such as Hueter and Fry likewise did not intend to use the instruments produced by grant recipients. What Hueter and Fry would benefit or suffer from, however indirectly, was the contribution of Howry and Wild to the scientific reputation of ultrasonics research as a whole. In the academic research setting -- admittedly not the only source of developmental work on medical technologies, but certainly a major source -- the incentives to innovate and the indications of that success are often measured in terms of *scientific prestige*, rather than in standard market economy terms.[10] It will be important to keep this in mind.

Funding Mechanisms As Overt Steering Devices

Since the 1950s, many things have changed in the environment for research on medical technology in the United States. For instance, research involving animals and humans is much more stringently regulated. Commercializing a new technology is much more costly due to the U.S. Food and Drug Administration (FDA) requirements for proof of safety and efficacy. And government rules prohibit reimbursement through the government-sponsored insurance plans, Medicare (for the elderly) and Medicaid (for the poor), for "experimental" technologies. Medicare and Medicaid reimbursement decisions are often used by private insurance companies to determine which procedures they will reimburse for privately insured patients. In this respect, the insurance structure in the United States provides a strong incentive against innovation.

What has not changed is the mechanism for funding academic research. NIH is still a major source of funds for research, and academe is still a major source of innovation. Grant applications for research are still reviewed by a panel of experts divided into study sections by discipline, with the advisory councils for each disease institute reviewing the recommendations of the scientific experts.[11] In essence, the study sections, the advisory councils, the staff of NIH, and the non-voting representatives from other government agencies and private agencies for funding research form a network already well-suited to conduct constructive technology assessment. This network already brings into contact many of the groups which have an interest in the outcome of medical research. NIH is probably the only existing institution in the United States that brings these groups together on a routine basis (rather than in reaction to such "crises" as bioengineering researchers scheduling the release of altered organisms at a specific site).

Yet NIH has not acted optimally as a mechanism of CTA for at least three reasons: First, the reliance on *scientific* experts completely neglects the contribution that experts from other fields might contribute to shaping perspectives on what research should be funded. Laymen on advisory councils are supposed to represent the interests of the public, but in fact their opinions can be, and are, overridden by the scientific experts (Koch 1991). Even in instances where NIH has recognized complex legal and ethical issues in the developmental stages, as in the human genome project, the experts on the social aspects of the research are carefully kept separate, as though their perspectives have no bearing on the funding and course of the scientific aspects of the project (Stemerding and Jelsma forthcoming). Second, even though techniques for improving the health of the general population have always been the *explicit* purpose of NIH-funding for medical research, scientific knowledge and scientific prestige are the *implicit* goals of NIH-funded research projects. When the current format for reviewing grant applications for research funding was established after World War II, this bias was built into the funding mechanism, with academic researchers designing the research, judging the proposals according to disciplinary criteria, and allocating funds (Fox 1987).[12] Finally, the over-reliance on scientific experts in the funding process masks disciplinary rivalries, ulterior motives in ranking grant applications, and emphasizes the reputations of individual researchers within their own disciplines. The assessment of the technology's characteristics or applicability is secondary to the assessment of the *scientific* validity of the research design, and hence the reliability of the information produced by the researcher.

The NIH system of funding offers a potential policy steering instrument. It *already* combines expert knowledge of a field (in the form of peer reviewers) and social knowledge and patient advocacy (in the form of lay members of advisory boards) with the capacity to tap existing quantitative data. By funding appropriate research NIH can generate good quality quantitative data that is pertinent to CTA. Much of the existing data from clinical trials is of little use in assessing medical technologies at a later stage, either because it is poorly designed research, or the research addresses different questions than are needed for CTA.

The existing structure of funding could be moved significantly toward CTA by increasing the influence of non-experts who currently serve on advisory boards and incorporating experts from other fields (such as law, ethics, economics, and health administration) in funding decisions. There needs to be two-way communication about research grants, so that representatives of the non-scientific aspects of research have the opportunity to inform the scientific panels about their concerns, just as much as the peer reviewers currently inform the advisory councils about their assessments of the scientific validity of research proposals. The hierarchical structure of NIH, originally designed to protect the autonomy of scientific researchers from outside interference in their research designs, also prevents the study sections from learning from and incorporating the perspectives of others, or from discussing with the advisory council their particular reasons for favoring some research over others. There is currently no direct dialog between the advisory councils, where social criteria are supposedly incorporated in the funding decision, and the study sections, where social criteria of a different sort are implicitly incorporated in funding decisions. Social criteria exist and should be made explicit at both stages. The communication between the groups should be facilitated so that a dialogical community evolves, with each group working to convince the others of the validity of their perspectives and, one would hope, learning from each other and adapting their own views in the process.[13]

For the funding mechanism to be a successful contribution to CTA it will be important to: (1) Document criteria for decision making. One problem, as illustrated by the NCI advisory council's internal policy favoring projects with an educational role, is that explicit policies can be circumvented or manipulated to serve the interests of the group making funding decisions. (2) Include a neutral moderator who is not a research scientist, physician, health care administrator, or insurer, all of whom have particular stakes in the outcome of decisions about the funding of research and health care. (3) Make allowances for minority opinion so that funding does not simply follow established, accepted routes. For instance, NIH could make a point of funding research that challenges or contradicts mainstream science.

CTA will succeed in medical technology to the extent it gains the endorsement and active cooperation of opinion leaders from each discipline of medical research, from among technology innovators, from among patients, and from among practitioners. Practitioners are key, after all, since they use technologies on patients and in many ways control access to medical technologies. It may be most difficult to identify and incorporate opinion leaders in areas of innovation into the communication and assessment process, because the participants (like those working on ultrasound) are often from very different backgrounds and have yet to establish common ground. Without adequate representation of all these groups, however, CTA, like current MTA, will be largely ignored by those conducting research or applying technologies in practice.

Conclusions

The current construction of the different types of MTA presupposes that good quantitative clinical data indicating the clinical efficacy and safety of a technology *must* precede any MTA of the social aspects of a technology. I turn this assumption on its head, and propose that an examination of the social factors influencing the development of a medical technology *must* precede the design of clinical studies to establish quantitative data on the safety and efficacy of a medical technology. Before a new medical technology ever reaches the point of an initial clinical application some assumptions have already been built into the hardware. By that stage, completely non-technical or non-scientific factors such as the research style of the investigator, or the reputation of the sponsoring institution, have played a role in determining which research is funded. A key task in CTA is to identify areas which may lead to innovation, and to promote technologies that will be safe, effective, and socially appropriate. CTA must then consider how assumptions are built into medical technologies (drugs, instruments, and techniques) and what factors, including disciplinary disagreements and political pressures, eliminate some technologies in their developmental stages before they ever reach clinical application.

Agencies that fund medical research are in a good position to influence how social criteria are incorporated into the funding of research on medical technologies. They already bring into contact many of the parties with an interest in the outcome of medical research: physicians, patient advocates, researchers, and government representatives. What remains to be done is to make the incorporation of social criteria in decision making explicit, rather than implicit, and to equalize the social learning process so that the academic elite who have so far controlled access to government research funds learn to incorporate the priorities of practitioners and patients.

Even though NIH provides only a portion of the funds for research each year, the close collaboration between NIH and other private funding agencies would make the impact of any change in approach at NIH even more far-reaching. The social learning process at an institutional level already exists at NIH, but it has heretofore been mostly unidirectional, with scientific researchers informing, and other groups learning. The academic research community has traditionally protected its autonomy, on the basis that the nature of the scientific practice and scientific expertise provide adequate justification for the self-regulation of science. Only in cases where the public has perceived a grave risk of externalities, as in genetic engineering research, have the public and experts in other fields demanded and received a participatory role in monitoring the agenda and conduct of research (Dutton 1988).

Why not have the same sort of dialogical forum for all of biomedical research? The early identification and discussion of developing technologies' hardwired assumptions may help alleviate the currently crushing burden of MTA. It could highlight potential problem areas in the diffusion and adoption of new technologies. It could also encourage researchers in the early stages of development to begin negotiating the types of information that will be required, the style of research that will produce convincing evidence to all concerned parties, and the baseline measurements that will be necessary for tracking and comparing multiple applications of technologies as they are adapted later in clinical practice.[14]

NIH is a unique institution, peculiar to the United States. But other countries could adapt the principle proposed here by examining their own research funding structures or institutions for the support of the development of medical technologies (whether academic or industrial) to locate existing areas of common ground between several different interest groups. They could then expand upon the existing nexus to incorporate broader representation and dialog on the social aspects of technology development.

Notes

1. See any NIH consensus conference statement, for instance, National Institutes of Health (1984).

2. John McKinlay (1981) made the point, in a much-cited article on the stages of innovation in medicine, that randomized controlled trials come *after* the medical innovation has been diffused widely into regular clinical practice. The seven stages he identified are: 1) promising report, 2) professional and organizational adaptation, 3) public acceptance and third party endorsement, 4) standard procedure and observational reports, 5) randomized controlled trial, 6) professional denunciation, 7) erosion and descreditation. Not all medical technologies go through all stages.

3. Burtram (1992) shows how the topics for NIH consensus conferences are chosen according to the availability of data from RCTs. Technologies that are directed toward populations underrepresented in NIH funded research (such as treatments for such ailments as migraines and ovarian cancer that disproportionately, or only, afflict women) rarely are the subject of consensus conferences, precisely because the quantitative data available are considered to be an insufficient basis for assessing questions of medical efficacy, much less questions of social, economic, or ethical impact. In effect, choices that are made about the funding of academic research set the range of possible agendas for later MTA by limiting the availability of quantitative data to those subjects deemed important by the discipline-based study sections in NIH, and other funding agencies.

4. Bell (1986: 1-32) argues that there is another, much less common definition of medical technology also used in health policy circles. This is technology as a "product or embodiment of human activity. In this respect, medical technology contains concepts as well as political, social and economic structures .... it embodies the internal dynamics and interactions between the physical and biological sciences, technology (physical engineering and bioengineering), and medical practice."

5. For instance, see Jones (1981) for a discussion of a long-term study on the effects of syphilis in black Americans. Penicillin was developed halfway through the study and shown to be effective on syphilis, but the participants in the study were denied access to penicillin. This case is often cited in discussion of the ethics of clinical trials to illustrate the unethical nature of continuing along one research protocol after the development of alternative and effective therapeutic medical technologies.

6. Edward Yoxen (1987) has also made the point that Howry's image fits better with existing paradigms for visualization of the body in medical practice. See also Blume (1992).

7. In 1952, Petter Lindstrom first proposed using ultrasonic irradiation of the brain for treatment of cortical epilepsy, while Chief of Neurosurgery at Harbor General Hospital, the teaching hospital for the medical school at University of California Los Angeles. The Chief of Surgery at UCLA declined the idea. In 1952, when Lindstrom assumed the position of Chief of Neurosurgery at the V.A. Medical Center in Pittsburgh, then the second-largest V.A. hospital in the country, he gained cooperation of the Addison-Gibson Laboratory, a basic research laboratory at the University of Pittsburgh, for research on the effect of ultrasound on brain tissue in rabbits, cats, dogs, and monkeys. Contrary to Fry's findings in 1953 that the cortex was more sensitive to ultrasonic damage, Lindstrom found that the white matter was far more sensitive. He reported his findings at Scandinavian Neurosurgical Society meetings in 1953, and published in 1954. Lindstrom's results were later confirmed by other researchers, including Fry himself. In 1954 Lindstrom was given permission by the Lobotomy committee of the University of Pittsburgh to try ultrasonic irradiation on patients with intractable pain and depression due to metastatic malignancies.

8. In 1952 President Dwight Eisenhower specifically opposed the use of federal funds to support medical education, as did the American Medical Association. But members of the Advisory Council to the National Cancer Institute, in their council meetings repeatedly, and explicitly,

espoused an internal policy favoring the funding of projects that served educational roles. Thus grant proposals originating from graduate departments in universities and university-affiliated medical schools (as Howry's was) were given higher priority than grants originating from non-university-affiliated hospitals and research institutes or industry (Koch 1991).

9. Memorandum to National Cancer Advisory Council from Planning Committee of the Council, regarding methods of supporting research and training and Council Procedures, June 1, 1956. National Archives R.G. 443, Minutes of Meeting Folder, Box 15.

10. Boelie Elzen (1988) makes a similar point about the development of ultracentrifuges.

11. NIH staff members assign incoming grant proposals to a study section for review. Each study section, say in pharmacology or biophysics, reviews applications for all of the disease institutes, so study section members do have an overview of all of the research being funded in their discipline regardless of the disease institute which received the grant application. Study sections review for scientific validity, and pass their recommendations on to the advisory councils. The advisory councils (the only place where laymen play a role in the funding decision) each rank all the grant proposals under their umbrella disease catagory (cancer, mental health, etc.) according to their perception of the priority of the research. The advisory council communicates back to the study sections solely through memos carrying general policy statements, or by reversing or changing the priority rating given by study sections to individual grant proposals.

12. In the wake of a large controversy over national health insurance in the United States after World War II, the U.S. Congress turned to NIH funding for research as a non-controversial means of improving the health of the general population. After the huge successes of government sponsored research on penicillin, steriods, sulphonamides, and shock, their optimism for the potential of medical research to quickly translate into improved health care was shared by many. What few people realized at the time was the impact that the shift from government-initiated research topics to investigator-initiated research topics would have on the whole complexion of medical research. With that shift, the implicit goal of NIH funding for research became the funding of research according to academic and disciplinary definitions of what constituted important research, rather than according to a definition of important research according to incidence of disease in the general population. Each time Congress introduced a bill for a new disease institute to be added to NIH, it was an effort to bring medical research back to addressing the health needs of the population (Strickland 1971).

13. A dialogical community is a term used by Richard Bernstein to describe a process of "dialogue, conversation, undistorted communication, communal judgement, and the type of rational wooing that can take place when individuals confront each other as equals and participants" (Bernstein 1983:223). Bernstein's concept of a dialogical community is based upon common themes that he identifies in the work of Hans-Georg Gadamer, Jurgen Habermas, Hannah Arendt, and Richard Rorty in political science and philosophy. Bernstein proposes the dialogical community as a solution to the present problems of society stumbling under the burden of instrumental and objectivist notions of rationality. In his dialogical community, technical experts have exactly the same standing as any other interested party who wishes to participate in the discussion. John Dryzek (1990) revises Berstein's concept of dialogical community, applying it to a very practical and local sphere, and trying to fit it within existing political structures. He provides examples of mediation, regulatory negotiation, and social movements to illustrate the mechanisms that lead to and support dialogical communities. A key point in Dryzek's book is the idea that in discursive democracy, "the only remaining authority is that of a good argument, which can be advanced on behalf of the veracity of empirical description, explanation and understanding and, equally important, the validity of normative judgements" (Dryzek 1990:15). In other words, experts, whether scientific or otherwise, have special authority only to the extent that they are capable of convincing others of the legitimacy of their argument. In both of these views of communicative rationality, the *process* of mutual accommodation and learning through discussion is more important than the particulars of any decision, since the dialog always continues and the consensus changes in response to new input.

14. Wilkinson (1989), in a study of the approval process in the FDA for medical devices, documents how the design of clinical research, the choice of data to accumulate and report, and the choice of when a technology is sufficiently studied to merit approval for clinical application, are negotiated from the outset between manufacturers and the FDA. Manufacturers who fail to consult the FDA about what points need to be studied, and what constitutes acceptable results, often find it much more difficult to gain approval. It seems that a similar type of negotiation for academic research could be appropriate. Such a process of communication could identify what different parties view as critical issues, and the research could be designed to address those issues. Perhaps then NIH consensus conference panels would not routinely lament the fact that available data rarely provided answers to the questions that the panel is charged with addressing.

References

Backe, B. 1988. "Quality Assurance in Perinatal and Obstetrical Care: The Norwegian Approach." *Australian Clinical Review* 8(28): 13-18.

Bell, Susan. 1986. "A New Model of Medical Technology Development: A Case Study of DES." *Research in the Sociology of Health Care Volume 4: The Adoption and Social Consequences of Medical Technology*: 1-32.

Bernstein, Richard J. 1983. *Beyond Objectivism and Relativism: Science, Hermeneutics and Praxis*. Philadelphia: University of Pennsylvania Press.

Blondel, Beatrice, Virginie Ringa and Gerard Breart. 1989. "The Use of Ultrasound Examinations, Intrapartum Fetal Heart Rate Monitoring and Beta-mimetic Drugs in France." *British Journal of Obstetrics and Gynaecology* 996: 44-51.

Blume, Stuart. 1992. *Insight and Industry: On the Dynamics of Technological Change in Medicine*. Cambridge, Mass.: MIT Press.

Burtram, Sarah G. 1992. "A Critical Assessment of the Consensus Development Conference of the National Institutes of Health." Ph.D. Dissertation, University of Texas School of Public Health.

Department of Social Affairs, Norwegian Institute of Hospital Research. 1987. "Ultrasound in Pregnancy: Consensus Statement, 1986." *International Journal of Technology Assessment in Health Care* 3: 463-470.

Dryzek, John. 1990. *Discursive Democracy: Politics, Policy and Political Science*. New York: Cambridge University Press.

Dutton, Diana B. 1988. *Worse than the Disease: Pitfalls of Medical Progress*. Cambridge: Cambridge University Press.

Elzen, Boelie. 1988. "Scientists and Rotors: The Development of Biochemical Ultracentrifuges." Ph.D. Dissertation. University of Twente.

Ewigman, Bernard. 1989. "An Opposing View." *Journal of Family Practice* 29: 660-4.

Ewigman, Bernard, Sharon Cornelison, Darla Horman, Michael LeFevre. 1991. "Use of Routine Prenatal Ultrasound by Private Practice Obstetricians in Iowa." *Journal of Ultrasound in Medicine* 10: 427-431.

Fineberg, Harvey. 1985. "Effects of Clinical Evaluation on the Diffusion of Medical Technology," in: *Assessing Medical Technologies*. Institute of Medicine. Washington, D.C.: National Academy Press, 185-95.

Fox, Daniel M. 1987. "The Politics of the NIH Extramural Program, 1947-1950." *Journal of the History of Medicine and the Allied Sciences* 42: 447-466.

Fry, William. 1955. Letter to Percival Bailey, University of Illinois, College of Medicine. May 25, 1955. Bioacoustics Research Laboratory Papers, University of Illinois, Champaign-Urbana, Illinois.

Fry, William. 1956. Letter to C.P. Rhoads, December 18, 1956. Bioacoustics Research Laboratory Papers, University of Illinois, Champaign-Urbana, Illinois.

Fry, William. 1958. Grant application to NSF, 12/10/58. "Facilities for Biophysical Research Laboratory." Bioacoustics Research Laboratory Papers, University of Illinois, Champaign-Urbana, Illinois.

Galison, Peter. 1987. *How Experiments End*. Chicago: University of Chicago Press.

Hill, Martha N. and Carol S. Weisman. 1991. "Physicians' Perceptions of Consensus Reports." *International Journal of Technology Assessment in Health Care* 7: 30-41.

Hill, Martha N., David M. Levine and Paul K. Whelton. 1988. "Awareness, Use and Impact of the 1984 Joint National Committee Consensus Report on High Blood Pressure." *American Journal of Public Health* 78: 1190-4.

Howry, Douglass H. 1957. "Techniques Used in Ultrasonic Visualization of Soft Tissues," in: Elizabeth Kelly (ed.). *Ultrasound in Biology and Medicine*. Washington, D.C.: American Institute of Biological Science, 49-65.

Howry, Douglass, and Joseph Holmes. 1955. Progress Report, Grant C-2423, 31 August 1955. American Institute for Ultrasound in Medicine Archives, Philadelphia, Pennsylvania.

Hueter, Theodor F. 1957. Letter to William Fry. January 15, 1957. Bioacoustics Research Laboratory Papers, University of Illinois, Champaign-Urbana, Illinois.

Institute of Medicine. 1985. *Assessing Medical Technologies*. Washington, D.C.: National Academy Press.

Jack, Brian W., Timothy M. Empkie, Richard B. Kates. 1987. "Routine Obstetrical Ultrasound." *American Family Physician* 35(5): 173-82.

Jackson, R. 1985. "Editorial: The Use of Ultrasound in Obstetrics." *Irish Medical Journal* 78: 149-50.

Jones, James. 1981. *Bad Blood: The Tuskegee Syphilis Experiment*. New York: The Free Press.

Kanouse, David E., John D. Winkler, Jacqueline Kesecoff, et al. 1989. *Changing Medical Practice Through Technology Assessment: An Evaluation of the NIH Consensus Development Process.* Ann Arbor, MI: Health Administration Press.

Kobak, Disraeli. 1954. "Unity and Concord a Desideratum in Clinical Research on Ultrasonics." *American Journal of Physical Medicine* 33: 2-4.

Koch, Ellen. 1990. "The Process of Innovation in Medical Technology: American Research on Ultrasound, 1947-1962." Ph.D. Dissertation, University of Pennsylvania.

Koch, Ellen. 1991. "NIH Extramural Grant Funding in the 1950s." Paper presented at meetings of the American Association for the History of Medicine, Cincinnati, Ohio, May 1991.

Kremkau, F.W. 1979. "Cancer Therapy with Ultrasound: A Historical Review." *Journal of Clinical Ultrasound* 7: 287-300.

Marks, Harry. 1987. "Ideas as Reforms: Therapeutic Experiments and Medical Practice, 1900-1980." Ph.D. Dissertation, Massachusetts Institute of Technology.

Marks, Harry. 1988. "Notes from the Underground: The Social Organization of Therapeutic Research," in: Russell C. Maulitz and Diana E. Long. *Grand Rounds: One Hundred Years of Internal Medicine.* Philadelphia: University of Pennsylvania Press, 297-336.

McKinlay, John B. 1981. "From 'Promising Report' to 'Standard Procedure': Seven Stages in the Career of a Medical Innovation." *Milbank Memorial Fund Quarterly/ Health and Society* 59: 374-411.

National Cancer Advisory Council. 1952. "Report of Inter-Council Committee." Transcript of Proceedings, November 6, 1952. Record Group No. 443, Box 11. National Archives, Washington D.C.

National Cancer Advisory Council. 1956. "Memorandum from Planning Committee of the Council, Regarding Methods of Supporting Research and Training and Council Procedures, June 1, 1956." Record Group 443, Minutes of Meeting Folder, Box 15. National Archives, Washington, D.C.

National Institutes of Health. 1957. Special Study Section for Grant Application C-2125. "Summary Sheet" November 1957. John Wild Papers, Minneapolis, Minnesota.

National Institutes of Health. 1984. *Diagnostic Ultrasound Imaging in Pregnancy. Consensus Development Conference.* Volume 5, Number 1.

Office of Technology Assessment. 1976. *Development of Medical Technology: Opportunities for Assessment.* Washington, D.C.: Government Printing Office.

Perry, Seymour, and J. T. Kalberer. 1980. "The NIH Consensus-Development Program and the Assessment of Health-care Technologies: The First Two Years." *New England Journal of Medicine* 303: 169-72.

Perone, N., R.J. Carpenter, and J. D. Robertson. 1984. "Legal Liability in the Use of Ultrasound by Office-based Obstetricians." *American Journal of Obstetrics and Gynecology* 150: 801-4.

Rogers, Everett. 1981. "Diffusion of Innovations: An Overview," in: Edward B. Roberts, et al. *Biomedical Innovation.* Cambridge, Mass.: MIT Press, 75-97.

Stemerding, Dirk, and Jaap Jelsma. "Ethics and the Human Genome Project." *Science as Culture* (Forthcoming).

Strickland, Stephen P. 1971. "Integration of Medical Research and Health Policies." *Science* 173: 1093-1101.

Thompson, G.E., M.H. Alderman, S. Wassertheill-Smoller, J.F. Rafter, and R. Samet. 1981. "High Blood Pressure Diagnosis and Treatment: Consensus Recommendations vs. Actual Practice." *American Journal of Public Health* 71: 413-6.

Wild, J.J., and John M. Reid. 1953. "The Effects of Biological Tissues on 15 mc Pulsed Ultrasound." *Journal of the Acoustical Society of America* 25: 270-80.

Wilkinson, Susan. 1989. "Negotiating Risk at the Food and Drug Administration: The Evaluation of Medical Devices in the Division of Cardiovascular Devices." M.A. Thesis, University of Texas Health Science Center, School of Public Health.

Youngblood, James P. 1989. "Should Ultrasound be Used Routinely During Pregnancy?: An Affirmative View." *Journal of Family Practice* 29: 657-664.

Yoxen, Edward. 1987. "Seeing with Sound," in: Wiebe E. Bijker, Thomas P. Hughes and Trevor J. Pinch (eds.). *The Social Construction of Technological Systems.* Cambridge, Mass.: MIT Press, 281-303.

12 Social Criteria in the Commercialisation of Human Reproductive Technology

Vivien Walsh

In the case of human reproductive technology, a particularly wide variety of social criteria are important in promoting, delaying and modifying innovation, including ethical as well as safety considerations, social prejudices as well as user needs, ideological stances of political parties as well as the commercial interests of the health care and pharmaceutical businesses. This chapter will focus on the commercialisation of oral contraceptives, introduced first in the early 1960s. This is largely a historical account that analyses the ways in which social criteria influenced the various stages of the innovation process, and the ways in which they were consciously or deliberately taken into account by the various actors involved or, on the other hand, ignored when they might have been taken into account.

The innovation literature has, over many years, focused on the influences of market demand "pull" and technological discovery "push" as major determinants of innovation. More recently attention has shifted to the changing relationship of these and other factors over time. Meanwhile, both the management and innovation literature have been exploring the nature of company strategy and the interactions of various actors within and outside the firm, not only in identifying appropriate market signals and technological opportunities, but in shaping the way in which each individual firm uniquely responds to these, and organises the innovation process.

The importance of understanding user needs was identified in some of the earliest innovation studies (e.g. SAPPHO), but the way in which they are identified and incorporated in the innovation process has not been very thoroughly examined (see chapter by Akrich). Attempts to forecast the impact of technological change on jobs, skills, health and safety at work, the environment, the safety of products and so on, have also been an important area of research since the 1970s. But again, the study of the interaction between forecasts and the *shaping* of technology *in order to avoid or minimise* negative impacts, has also been limited.

The literature on the social shaping of technology has provided studies of the networks of interactions between various actors. These have explored the way that choices are made in creating science and technology and in diffusing and consolidating its results. Recent studies in this tradition (e.g. Callon 1992, MacKenzie 1992, Latour 1988) have not only emphasised the shaping of science and technology by market and other social forces; but also the actions of scientists and other actors in promoting their own interests in certain choices and outcomes, by actively shaping the selection environment.[1] Constructive Technology Assessment, a concept developed in the introduction of this book, refers to the possibility of *societally desirable criteria* being fed into the innovation and design decision-making process, in order to shape technologies as they develop.

The data analysed in this paper were originally collected in order to examine the process of government control of technology and the impact of public interest groups and pressure groups on the process of technological change. It provides an interesting case of the *potential* for constructive technology assessment, rather than a case of deliberate attempts to incorporate CTA, for example as a policy initiative.

Social Forces Inhibiting Innovation

The oral contraceptive, first commercially introduced in the USA in 1960, was the first new method of birth control for 80 years. According to Peel (1963:116): "with the exception of the oral contraceptive, there is not a single method of birth control in existence today which was not already available, and available in greater variety, in 1880." Contraceptive technology was not unusually advanced in 1880, nor were demand or technological opportunities lacking in the intervening period, but little development work aimed at a new form of contraceptive was initiated until the 1950s. Both technological opportunity and demand had long been present when "The Pill" was introduced. But, for a variety of reasons which this section of the paper will explore, few attempts were made to meet (or even admit) the demand, and even fewer attempts were made to exploit (or even recognise) the technological opportunity (at least, not in the direction of a contraceptive drug).

By 1938, various discoveries had been made which might, in principle, have laid the basis for a project aimed at the development of a birth control drug. Thus, the role of hormones in pregnancy had been established (Corner and Allen 1928). Progesterone and other sex hormones had been isolated in pure form simultaneously in Germany, the USA and in the Ciba laboratories in Switzerland (Butenandt et al. 1934; Allen and Wintersteiner 1934; Hartmann and Wettstein 1934). Progesterone had been shown to inhibit ovulation in rabbits and other animals (Makepeace et al. 1937). Ethisterone, a synthetic compound with a similar effect to progesterone, but active orally and at much lower doses unlike the natural hormone, had been prepared by Schering AG in Germany (Inhoffen et al. 1938). These are just a few of the very many developments reported in the literature at the time.

As early as the 1920s (before some of the above events), Wiesner (1929) carried out a study of steroid hormones, with the specific aim of finding an ovulation inhibitor that could be used as a contraceptive pill. However, no further developments of this work appear to have been pursued, and it is not clear whether this was for technical, financial or other reasons. Further references to the deliberate control of fertility more or less disappeared from the literature on the physiology and endocrinology of human reproduction and on steroid chemistry for more than twenty years. There was not even a reference to any project aimed at evaluating ethisterone as a contraceptive drug until the 1950s. Instead, the literature focused on physiology, steroid chemistry and clinical studies with other goals. These included the extraction or chemical synthesis, isolation and structure determination of new analogues of the sex hormones, anabolic and cortico-steroids; the use of these new materials in animal experiments; and the emergence of clinical research with these steroids on human gynaecological disorders, on overcoming human infertility, on increasing muscular strength and on alleviating the symptoms of rheumatoid arthritis. The development, formulation and commercialisation of drugs using these materials for these purposes also took place.

When the goal of a contraceptive drug was once again mentioned in the literature, the active ingredients tested for the purpose (and indeed first commercialised as "the pill") were very slight variants on the compound, ethisterone, first mentioned in the 1930s. It was as though the researchers in the 1950s (mainly English speaking, American) picked up the work exactly where the earlier researchers (mainly German speaking, European) had left it in the 1930s. The earlier publications in German were widely cited by the American researchers. The interesting question is, why the gap? Clearly, the existence of knowledge is not by itself a sufficient reason for attempts to be made to exploit it, or develop it in a particular direction. Exploitation depends on such things as the availability of the knowledge and the skill to develop it, entrepreneurs willing to invest, the belief in at least a potential market demand, and no greater preference for possible alternative projects.

One alternative "project" was the scientific and military effort involved in the Second World War. But this accounted for only some diversion of effort and some of the delay in innovation.[2] I will argue that social and ethical opposition to work on physiology and endocrinology of reproduction, let alone contraceptive applications, had an important part to play in focusing attention on applying scientific knowledge to the commercial developments mentioned above, rather than on a birth control drug. But first, the demand for a birth control drug will be explored.

Demand for products and services in the area of human reproductive technology is ultimately a reflection of the desire of women and men to have, or not to have children. That in itself is profoundly influenced by a wide range of social, economic and cultural factors which have changed substantially over the course of this century.[3] Evidence suggests a huge if clandestine market for contraception and abortion in the 19th and 20th centuries, during the period when innovation might have taken place but in fact did not. Between 1800 and 1960 the average number of births per white[4] woman in the USA fell from 7.04 to 3.52 (Smith 1973). Gordon (1977:53) quotes estimates of up to 100,000 abortions a year in New York City alone in the 1890s and two million a year throughout the USA; by 1921, one in every 1.7 to 2.3 pregnancies ended in abortion, of which at least 50 percent were criminal.[5] These figures indicate the demand for abortion, but also the potential demand for contraception. *Fortune* magazine (1938) reported that American women spent more than $210 million a year on (illegal) contraceptives, while firms spent nearly $1 million a year on contraceptive advertising (also illegal). Various opinion surveys taken between 1936 and 1945 showed that between 61 percent and 84 percent of the respondents were in favour of birth control (Cantril 1951:41-2). Innovation thus failed to take place despite a combination of demand pull and scientific knowledge (if not "discovery push"). According to Kennedy (1970:213) "the demand for contraceptives outstripped the supply ...... and gave rise to a large industry riddled with quackery and dishonesty."

Opposition to Birth Control

Demand for medical advice, treatment and supplies is not usually determined directly by the consumer, but by doctors who apply new medical treatments and prescribe drugs and appliances. Where contraception and abortion were illegal, demand was more directly expressed by the user responding to word of mouth information (e.g. for back-street abortions) or advertisements, carefully worded to avoid prosecution. For example, the American convention was that anything described as "Portuguese" ("Portuguese female pills — not to be used in pregnancy for they will cause miscarriage") was an abortifacient and anything "French" was a contraceptive (Gordon 1977; New York Times 1871).

Even where illegal services are involved, health care markets — and technologies — are at least partly shaped by the approval of the medical profession and the involvement of doctors in clinical testing, or on the other hand, the profession's opposition to, or ignorance of, the techniques of intervention in human reproduction. Their prestige and influence can counteract prejudices and influence the liberalisation of legislation, or contribute to reinforcing the status quo. They can also speed up or slow down the rate of adoption of innovations.

In the case of contraception a majority of doctors took no initiative for a long time in extending the provision of birth control advice (Peel 1964). The leading medical journals were opposed to contraception until well into the 20th century.[6] Far from doing clinical or scientific research themselves (which might have improved quality and removed the alleged side effects), doctors were generally opposed to it being done either by themselves or anyone else (Shryoch 1948).[7] They were, however, the professional group most likely to be using contraceptives themselves (Banks 1954; Peel and Potts 1969).

The opposition of doctors did not prevent the practice of contraception. Indeed contraception was widely practised as part of the folk medicine and culture of the earliest societies, and much of this knowledge survived through the centuries and was further developed and applied with varying degrees of success (Gordon 1977, Riddle 1991). This accounts for the wide variety of methods available in 1880 despite official disapproval and lack of research. However, this climate of opinion did affect the nature of the technology available. Like many things that are illegal or clandestine, prices were high and quality low. In terms of their (lack of) a research tradition, their lack of "complementary assets" (Teece 1986) for drug manufacture, retailing and compliance with regulations, and their lack of expertise in hormone-related technology, the traditional contraceptive manufacturers were in no position to even consider innovation in contraceptive drugs. Even the innovations in rubber technology,[8] which revolutionised birth control as well as road transport, originated elsewhere. In general, there was very little research, even market research or quality control, by the traditional contraceptive manufacturers (Voge 1933) until *after* the Pill was introduced by another industrial sector, the drug industry, and the traditional manufacturers felt the pressure of competition. This could be said to be a negative example of CTA: social disapproval inhibiting the improvement of technology.

Doctors were particularly concerned that techniques that tended to cross the traditional boundaries of medicine represented a threat to their learned authority, their professional status and, no doubt, their incomes. They opposed birth control on the grounds that it was an area of lay medicine and "quackery," an objection that, by definition, could not be removed without their involvement. In the fields of both contraception and abortion, when the profession moved away from total opposition, they opposed the "social" grounds for using either: that is, they adopted the view that it should be available only to women whose life or health were in danger (a position adopted in law, too). In this way they moved from *hostility* to the new technologies to *control* of them, since it would be they who would judge whether life or health were in danger.

Retailing of medical supplies and drugs (with and without prescription) is the province of pharmacists, but for many years they refused to sell contraceptives, or kept them "under the counter." In the 1950s, letters to the *Pharmaceutical Journal* show that many pharmacists were still strongly opposed to selling, advertising and displaying contraceptives, because they were associated with "promiscuity, vice and prostitution." Some did not sell them until the 1970s. In the USA, condoms were retailed in tobacconists and petrol stations. In Britain they were retailed by barbers. Condoms, douches, pessaries, sponges and spermicides were sold in surgical goods stores, especially in Britain, and in both countries they were advertised in coded language (as described above) for mail order purchase via a variety of newspapers and magazines (Peel 1963).

Historical research thus suggests a widespread practice of birth control but official and professional opinion strongly against it until after the Second World War, at least in the English speaking world. In some countries, such as the USA, the sale, display, advertising, and even the use of contraceptives was illegal, although the application and policing of the laws became less and less stringently applied long before they were repealed.[9] In the state of Massachussetts, anti-birth control legislation was not repealed until 1966, although much of the clinical research on the Pill was carried out in that state before its commercial launch in 1960. In the UK, birth control was not specifically prohibited, though such laws as the Obscene Publications Act (1857) was from time to time up to the Second World War interpreted as including contraceptive information and advertising. Local Authorities were officially instructed by the UK Ministry of Health in 1924 not to provide birth control advice at child welfare clinics (Leathard 1980), or they would lose their grants.

The effect on R & D and innovation
To summarise, the state prohibited or at least discouraged birth control, and the medical and pharmaceutical professions disapproved or were positively hostile to it. In this climate of opinion, researchers engaged in basic research on reproductive endocrinology or physiology (let alone the development of birth control methods) believed they risked "both their professional and private reputations" (Gebhard 1968:391-2). Scientific and technical knowledge converged on deeply rooted emotional and moral attitudes (Zuckerman 1970), and diffusion of knowledge met with a "peculiar and stubborn resistance much greater than resistance to the dissemination of knowledge in general" (Sjövall 1970:117). The whole area of reproduction, let alone contraception, was not entirely socially acceptable (Vaughan 1971). According to Parkes (1966:35), "obscurantism reaches its most extreme and dangerous manifestations in the case of human reproduction." J.R.Baker (1926) referred at the time to the secrecy and taboo surrounding the subject, and indeed was in the position of "assembling his reagents and apparatus on a handcart and trundling this from department to department" (Peel 1964:142, quoting a letter from Baker) as a result of being thrown out of his laboratory at Oxford University by the outraged professor of zoology, and having to search for somewhere else to work.

Even in the late 1950s, the strategy of the drug company G.D. Searle, the first to launch the Pill (and to make "Schumpeterian" monopoly profits as a result), was influenced by their perception of the climate of opinion. "No major pharmaceutical manufacturer had dared to put its name on a contraceptive. The individual reaction of a very large religious minority in the United States could not be gauged. The possibility of losing overnight one fourth of all our personnel, a considerable portion of our hospital business and a crippling number of the physical prescribers of our products was not to be dismissed lightly" (Winter 1970).

Many of those commenting on the climate of opinion in the 1920s-50s did so with hindsight, from the viewpoint of the "permissive" 1960s and 1970s. By this time, contraceptives were more respectable as a subject for research and as topics of conversation, and were becoming accepted as products to be displayed in chemists' shops. There was more concern about the "population explosion" (see next section) than "immorality." Between the wars, many of the researchers, directors of research laboratories, and grant-giving bodies might have been more affected by the moral climate in a passive way: it might not have occurred to them to support or carry out research on birth control, especially as there were other opportunities and other goals for exploitation of the scientific knowledge. It would have seemed more obvious to pursue them than to think of challenging social prejudices. Baker and Wiesner (mentioned earlier) were exceptions: they were financed by and were supporters of the birth control movement (discussed in the next section), their work was explicitly on birth control and Baker's comments about social prejudice were made at the time.

As it turned out, G.D. Searle's fears were unfounded and the diffusion of the Pill was very rapid. Within five years 26 percent of married women under 45 in the USA had used the Pill and only 3 percent had not heard of it (Ryder and Westoff 1966). Public opinion was a lot more liberal than articles in the media and the propaganda of moral campaigners suggested. More recently, the Office of Technology Assessment (1982) reported that contraceptive drugs were the most profitable of all pharmaceuticals, the U.S. retail market being worth $1 billion a year.

The drug industry took over from the traditional contraceptive manufacturers and established birth control as primarily a trajectory of pharmaceutical technology, with R&D, clinical trials, approval of the medical profession, remoteness from patients, and other characteristics associated with drugs but not traditional contraceptive manufacturing. Where medical hostility and social prejudice had previously reinforced the clandestine nature of birth control technology and the reluctance of various actors to initiate or facilitate innovation, the situation had now changed. It appeared, at least from the press, popular scientific publications and even medical journals, that the association between "science" and the birth control pill, ironically or paradoxically, had begun to reinforce the acceptability of the technology.

Negative effects of steroids used for contraceptive purpose have since been reported (early examples are Vessey and Inman 1968; Vessey and Doll 1968), as have the unwanted effects of fertility drugs, cortico-steroids used in arthritis, and the abuse of steroids in athletics. In general, however, these were "unexpected" impacts discovered *after* the commercial introduction of the product. In the case of the Pill they were unexpected only because they were not sought. Although clinical trials for efficacy were quite extensive, long-term toxicity tests were much more limited (Mintz 1969).[10] Attempts were not made to modify the technology in advance of commercial launch, by exploring possible long-term effects and establishing from end-users the properties they wanted in a contraceptive. Once again CTA is evident, in retrospect, only by its absence.

Social Forces Promoting Innovation

The Birth Control Movement
Since public policies were responsible for the illegal status of birth control in America and its clandestine nature in Britain, supporters of birth control attempted to change those policies by campaigning in the public domain. The birth control movement[11] tried to break down the complex of moral, religious and cultural values and prejudices, which denied knowledge of contraceptive techniques to the majority and confined contraception to its own clandestine milieu beneath the surface of "respectable" society. They put pressure on the medical profession to take a more positive attitude to birth control and worked with some of the more progressive doctors in evaluating different contraceptive methods and providing supplies and information in the clinics they established. In Britain, for example, the Women's Labour League campaigned for local authorities to provide public birth control clinics, and for central government to allow it. Deputations of Labour women descended on the first Labour Minister of Health, who was a Roman Catholic and opposed to birth control (Leathard 1980). In 1930, the Minister gave in to the "irresistible and irreversible pressure" and issued a typed memorandum stating that local authorities might, if they chose, provide birth control at child welfare clinics — though only to married women for whom further pregnancy would be a health hazard (Ministry of Health 1930). In the USA the focus of activity by Margaret Sanger and other campaigners was de-criminalisation (Kennedy 1970).

These campaigns stimulated both supply and demand, by making people more aware of the methods available, by making the establishment of clinics possible or providing them themselves, and by bulk purchase of supplies for re-issue. They also influenced the climate of opinion to create a more favourable environment for innovation. All of these activities *indirectly* shaped technology; in addition, the birth control movement *directly* affected the nature of the technology available by evaluating different products and putting pressure on the manufacturers to improve their reliability. Most importantly, it raised money for research and funded projects. In Britain, one of the most significant of these was J.R. Baker's work at Oxford, funded by the Birth Control Investigation Committee, which contributed to the fundamental theory of contraception, developed a test for spermicidal activity and hence the first independent assessment of commercial contraceptives, and led to the invention of the spermicide Volpar (Baker 1935). Even more significant from the point of view of birth control drugs was Wiesner's project mentioned above at Edinburgh.

In America the birth control movement funded Pincus' clinical work on synthetic steroids in the 1950s, which resulted in the first commercially successful oral contraceptives. The latter project was also funded by the companies which had produced the active chemicals used in Pincus' trials, but it was the influence of the birth control movement rather than the industry that persuaded Pincus to initiate this work (Pincus 1965). Firms such as G.D. Searle, although they had established the necessary intellectual property rights, were hesitating about commercial development.

Thus a series of organisations and individuals acting as pressure groups (albeit ones with some wealthy backers and far-sighted strategists) played a key role not only in persuading researchers to investigate contraception in a general sense, but in shaping the science and technology that emerged. They did this by identifying the most promising areas for research and development, based on the belief that the most popular existing methods of contraception (condoms and diaphragms) would be much more effective if combined with a spermicide; while in the future a far more satisfactory solution (i.e. from their point of view at the time, far more convenient to users and more effective in preventing conception) would be a drug rather than a barrier method (Blacker 1929).[12] This forecast was based on the practical experience of trying to provide the best possible birth control clinic service using the techniques available at the time, although probably not the result of a detailed survey of their clients' own wishes.

Pressure groups representing the interests of end-users or the community in general, and which include sympathisers who have expert knowledge in fields of relevance to the groups' goals, may play an important part in constructive technology assessment. However, no similar organisations have recently played the kind of extensive role of the early birth control movement; they have tended to be limited to campaigning. In particular, pressure groups concerned with contraception recently have moved from promotion to criticism, focusing on the issues of safety and informed consent in clinical trials and in use. They have presented evidence to licensing authorities (Rakusen et al. 1983, World Health Organisation 1984), attempted to increase awareness among users, and urged family planning services to make sufficient information available so that their clients really do give their informed consent (Phillips and Rakusen 1989).

Post War Economic and Social Changes
The birth control movement did not, however, singlehandedly transform the environment for innovation in contraception or provide the only impetus for change to firms, doctors, legislators and others. It acted as a catalyst against a background of other social and economic changes. The postwar increase in standards of living in most advanced countries influenced attitudes towards birth control and increased demand for it. For the first time home ownership, a car, domestic appliances, fitted carpets and other consumer goods previously only bought by the wealthier sections of society were within the reach of the mass of the population if they budgeted carefully. The advantages of smaller families became more apparent, and a climate of opinion developed in which the demand for birth control became more open. The increased participation of women (especially married women and mothers) in the labour market, the increased numbers of women staying on for education beyond the statutory school leaving age, greater numbers of women pursuing professional careers, and a general increase in the independence and status of women (and women's desire for further such increases) all contributed further to the growth in demand for contraception.

Finally, a quite violent shift in public, medical, and government attitudes was achieved as a result of concern about the "population explosion" (Petersen 1964), at about the time that the Pill was undergoing clinical trials. In 1959, for example, President Eisenhower said of birth control: "I cannot imagine anything more emphatically a subject that is not a proper political or government responsibility." By 1965 he had changed his attitude completely and became a co-chairman of the honorary sponsors of Planned Parenthood — World Population; four years later Lyndon Johnson became the first president to report federal support for population control (*Congress and the Nation* 1969). Regardless of the validity of the consensus opinion about the population explosion, the strength of the newly prevailing attitudes had an important influence on government policy towards contraception and the plans of the drug companies.

Many of the scientists working on the Pill referred to the importance of the population explosion in stimulating their work. Pincus (1965), who carried out clinical trials on the Pill, cited it as an important impetus. Marrian (1971) and Nelson (1956), both working on the biochemistry of reproduction, also referred to its influence. Petrow (1966), a chemist who synthesised the active ingredients of later contraceptive drugs, frequently referred to concern about world population growth in chemical papers. In fact, it was quite common in the 1950s and 1960s to see a genuflection towards demography in papers that otherwise were restricted to experimental details in chemistry and biology. Popular scientific journals had quite sensational headings: "The explosive increase in world population could be squelched by a tiny pill" (Science Newsletter 1951).

The paradox was that the greatest concern in the advanced countries from the "population explosion" was its manifestation in the Third World, while the market for the Pill was the educated, better off, mainly white women in the USA and other advanced countries. Demands from political leaders in the West for measures to limit reproduction in poverty stricken countries had little to do with the desire of the more affluent social groups in the advanced countries to plan their children to suit their new life styles. What the "population explosion" did was make it impossible to sustain generalised moral or social objections to contraception at home. The population "bomb" legitimised scientists' reasons for doing their research and the drug companies' reasons for investing in a potentially profitable innovation (a drug to be sold not to the sick over a limited time period, but to millions of healthy women, potentially over their whole child-bearing lives). The success of hormonal contraception in the wake of concern about world population growth, did then lead to a focus of attention on ways in which innovations could be adapted to the needs of those promoting population control.

User Needs and User Representation

The needs of users of medical treatment are far from being unproblematic. Users are concerned not just with efficacy but with unwanted effects, which each individual experiences differently; with the service as well as the treatment; with the degree of choice offered; and the amount of information (or uncertainty) they want. Many users may not know what their needs are: at least they will rarely know what they feel about the balance of benefits and negative experiences until they do experience them.

Even supposing the wishes of end users were straightforward, what mechanisms are there for taking them into account? Contraceptives, especially contraceptive drugs, are tested on end-users in clinical trials conducted by family planning organisations. The key interaction is between the clinical practitioner and the manufacturer. End-users generally interact with clinical practitioners rather than manufacturers. Even in the most favourable circumstances "user-producer" interaction is thus bound to be limited. In practice it is further impeded by the cultural and historical conventions governing relationships between doctors and their patients.

Although doctors in any case have expert knowledge usually not shared by patients, their training discourages them from communicating thoughts and especially uncertainties to patients, while social learning processes equally discourage patients from questioning doctors or asserting their preferences. This poses a problem for constructive technology assessment in medical treatment. Human reproductive technologies have been the subject of a great deal of discussion by non-professionals, and end-users are more aware of the pros and cons than in other cases of treatment, and more willing to question the services they receive. But doctors still assert their professional independence and are likely to resist inputs from outside, certainly from "non-experts."

Direct communication of user preferences via doctors to manufacturers is thus very limited unless market research surveys are carried out. Indirect information, such as drop-out rates in clinical trials as well as assumptions by doctors and manufacturers that they *know* what users want, is more prevalent. Manufacturers and doctors also use contraceptives, of course, but they are professionals and preponderantly male, and cannot know from experience what women, and working class users of either sex, *really* want. In the 1930s, for example, the diaphragm was very popular among more prosperous women, but less so for those without bathrooms where the diaphragm could be taken out and washed in privacy. At regular intervals since the launch of the Pill in 1960, newspapers have carried reports of clinical trials of a contraceptive drug for men. So far none has appeared on the market, and the reason seems to be that these drugs are associated with negative effects such as nausea, headaches and weight gain, though some effects are more serious.

The male pill could be a case where informal assessment has ensured that a new form of contraceptive is not produced until unwanted effects can be eliminated. But the continued existence of precisely these side effects in the pill taken by women suggests that those who develop new contraceptive drugs, or those who supply them with marketing or clinical information, work on the basis that women will tolerate such effects while men will not. A policy incorporating constructive technology assessment into drug development clearly needs to take on board the role played by doctors as intermediaries between user and supplier, and establish a way in which the end users' as well as the doctors' preferences can be identified, and taken into account in the development of a new form of therapy or a new contraceptive method.

Regulatory Procedures

Manufacturers establish close relationships with networks of medical consultants who conduct toxicity trials of their products and in due course may be expected to prescribe them. They also have a well established system of sending representatives to visit doctors to promote their products, which might be adapted to promote feedback from the doctors to manufacturers. But just as important to the commercial success of their products is the relationship with the government representatives and health authorities who give the firms licences to carry out clinical trials and then to market their drugs, who negotiate the conditions of the licences and in some countries the prices that may be charged. Both types of relationship are examples of the "complementary assets" (Teece 1986) needed to commercialise a new innovation, ownership of which is sometimes more important than development and ownership of intellectual property, when it comes to the receipt of profits from innovation (see Walsh 1993). Complementary assets do not just permit firms to market their output; they enable firms to shape the selection environment, by negotiating the terms of their compliance with regulations and by persuading professionals to approve their products.

The regulatory framework enables government bodies to insist on modifications to drug technology to improve the safety of the product, reduce its price and minimise the impact of its production process on occupational health and the environment. The main mechanism by which the regulatory body's requirements are communicated to the producer of the technology is via the discussions that take place between representatives of the licensing body and the firm when a new product is submitted for approval, that is, *after* the technology has been developed.

Experience of the requirements of regulatory bodies enables established firms to modify their technologies at an earlier stage. However, at least equally important, is the input of the manufacturers in modifying the regulatory requirements, either as a result of lobbying activity aimed at altering the regulatory framework; or as a result of the negotiations between the licensing authority and the manufacturer, concerning what is required in the way of tests and evidence for a particular application. On this basis, the trajectory of contraceptive pill development became established in the 1970s not as a search for radical alternatives that better met the requirements of users, but as a successive reduction of dosage levels, combined with the search for steroid compounds that had the desired biological activity at lower and lower doses.

Lesser (1983) has argued that the "cosy partnership" established in the UK between the pharmaceutical industry, the regulatory authority, and the latter's expert advisors (the Committee on Safety of Medicines or CSM), poses severe problems or even "complete paralysis" in the event of a major crisis (such as the Opren disaster). The appointment of ex-CSM members as drug company directors underlines this relationship (Williams et al. 1982; Coombs et al. 1992: chapter 9). In the U.S., too, Hartmann (1987) provides evidence of the movement of Food and Drug Administration (FDA) officials from public service to private industry, and of pressure exerted on the FDA by industry.[13]

Existing regulatory systems can transmit public policy as spelt out in regulations, and experts in or advising the licensing authorities have some experience in identifying likely unwanted effects of contraceptive drugs. A more effective CTA system needs some mechanism for identifying and incorporating the end-users' priorities and criteria for product development. A body advising the regulators (like the CSM) would be more independent in a real sense, and reflect these requirements more effectively if it had input from patients' organisations and other public-interest groups and from among less-senior health care workers. An explicit policy commitment to establish that CTA is a goal of regulatory and other procedures is needed.

Public Policy

Public policy concerning reproductive technologies has not been restricted to the regulatory framework. Directly or indirectly it has also been influenced by policies concerning the family and population control. At various times governments have adopted programmes aimed at increasing or restricting the number of births because they considered their population growth rates to be too low or too high.

Thus Romania before December 1989 was notorious for its draconian laws against contraception and abortion, whereas China has been equally well known for its attempts to limit couples to a single child. Politics based on coercion could not be said to be in the interests of society, though they may prove to have had an effect on the shaping of technology. Thus the pill Gossypol has been on clinical trial in China, even after the World Health Organisation had withdrawn funds because the drug also causes heart disease and permanent sterility. Market saturation for the Pill in advanced countries by the late 1970s meant that the focus of research was on new forms of administration of contraceptives, to open up markets in developing countries (Hartmann 1987). These applications are closely connected with population control policies. Population control is now a major market for the pharmaceutical industry.

The governments of the advanced countries also have a major interest in the population control policies of developing countries. Without getting into the debates of the population and development economics literature, it seems clear that their interests are not so much in giving the populations of the Third World the freedom to choose the timing and number of their children, or even to prevent starvation, but as a cheaper alternative to investment in economic growth and even a means to avert riot and revolution.[14]

Apart from having an important influence in promoting innovation in birth control in the first place, the stance of governments like that of the U.S. towards population control has played an important part in shaping technology. By 1983, the U.S. government provided 59 percent of world expenditure on reproductive R&D and the evaluation of long term safety of existing methods. U.S. drug companies provided 21 percent (Atkinson et al. 1985). The reason for the drug industry's relatively low contribution was the cost of innovation and risks of product liability claims. Government concern with population policies has directly shaped technology, by focusing R&D on the following (Hartmann 1987):

1. The female reproductive system. Women are the main targets of population control programmes, and reproduction is seen by researchers as mainly a woman's problem.
2. Systemic and surgical methods rather than barrier methods.
3. Long acting methods requiring little initiative by the user and minimal interaction between the user and provider.
4. Efficacy rather than safety. Only 10 percent of expenditure goes on safety tests (OTA 1982).

Hormonal methods are much more profitable than barrier methods and, in the form of injections, much easier to administer, though barrier methods are safer. Researchers also find that projects in sophisticated areas of hormonal technology attract more prestige, recognition, and research grants. However, a combination of consumer pressure, feminist campaigns and the concern about AIDS has increased the resources going to barrier methods in recent years (Atkinson et al. 1985). The mechanism for this has been mainly publicity, and manufacturers' consequent anticipation of consumer demand.

In constructive technology assessment, the interests of end users and of society as a whole are both considered to be important. However, these interests are not always equivalent, and may sometimes conflict. In areas of technology where ethical issues are prominent this is particularly true. The freedom of choice of individual end users (to use technology to determine the sex of their child, for example) is not necessarily the same as the interests of society. On the other hand, governments normally associated with a defence of freedom of choice may be in favour of state regulations in the interests of protecting public "morality," when it comes to exercising choice in relation to human reproductive technologies (such as the right to abortion). In examining the issues of CTA in this area, then, it is necessary to clarify exactly whose interests are being built into the process of shaping technology.

Social and ethical criteria have had a profound, stimulating and retarding, effect on the development of innovations in the field of human reproductive technology. This paper has attempted to make a start at analysing the ways in which these issues were, or were not, built into the innovation process and exploring ways in which they might be taken into account in a more constructive technology assessment on future occasions.

Notes

1. The idea of the selection environment (Nelson and Winter 1977) captures the process of screening, trying out and rejecting or accepting an innovation, and its subsequent diffusion to other firms, institutions and individuals. It is a broader concept than that of the market and includes non-market selection environments, which are particularly important in the case of medical innovations. A new medical technique may be adopted by doctors, based not on market considerations but on their professional judgement, regulated by spending limits set as a result of political judgements. The regulatory system for licensing drugs and clinics providing certain services, may also provide value signals and incentives in a manner analogous to a market. Rip and Van den Belt (1988) have indicated that, in constructive technology assessment, the selection environment interacts with the technological trajectory rather than being independent of it.

2. The pharmaceutical industry in Nazi Germany had other goals than effective forms of contraception before 1945, and it took a while to recover afterwards. German researchers in general would have had other preoccupations. Some of them emigrated to the USA, and continued to work in the field.

3. Examples include the economic means to raise a family, the availability and cost of housing, the existence of welfare benefits and services for the old, the structure of the labour market (the reliance on children to work on farms and in other family businesses is an incentive for having children and the participation of women in the labour market encourages family "planning"), the religious and cultural importance attached to having children, the knowledge about and availability of techniques that have been developed and whether they are available within public or private health care systems, attitudes to birth outside marriage, and women's social status and independence.

4. Figures for black women are incomplete.

5. Gordon (1977) quotes Jefferis and Nichols (1894), which reports a judge's calculation of the number of abortions in New York, and a study from Stanford University (Meyers 1921) which estimates the national abortion rate. She also comments that better statistics collected in later years suggest that earlier estimates were too *low* rather than the reverse.

6. For example, side effects of contraception were said to include "galloping cancer, sterility and nymphomania in women; mental decay, amnesia and cardiac palpitations in men; in both sexes the practice (is) likely to produce mania leading to suicide" (Routh 1878). By the 1920s, "mental degeneration in subsequent offspring" was still one of the alleged effects (*The Practitioner* 1923).

7. In the UK, the first medical school provided a single lecture on the subject in 1936; by 1957 only a third of the medical schools gave a course on contraception, nearly half of those without clinical instruction (Mears 1961).

8. For example, vulcanisation increased elasticity and strength and reduced the cost of condoms, and later the latex process enabled them to be made thinner and even cheaper.

9. The U.S. Federal Comstock Law (1873), named after the senator who sponsored it, and various state laws allowed the opening of mail and various forms of entrapment. Several states had contradictory regulations. For example, Nebraska forbade the sale of contraceptives but also set quality standards for those that were sold. Mississippi both prohibited contraception and offered birth control advice through the state's own health service (Petersen 1964).

10. It took the thalidomide disaster to secure the enactment of the Kefauver-Harris Amendment to the U.S. Food Drug and Cosmetic Act and the UK Medicines Act (Williams et al. 1982), and these did not come into effect until after the Pill was on the market.

11. The birth control movement is an umbrella term I have used to include individuals such as Margaret Sanger and Hannah Stone in the USA and Marie Stopes and Dora Russell in the UK, and organisations such as the American Birth Control League, the New York Birth Control Clinical Research Bureau, the UK Family Planning Association or the UK Birth Control Investigation Committee, which campaigned for legal and other changes, provided information and services, and/or did research.

12. As a result of findings of negative side effects of drugs used in human reproductive technology, plus increased concern about reliance on drugs in general, not to mention concern about AIDS, an appropriate strategy would today seem to be concentration on barrier methods. At the time, however, the idea of a "magic bullet" was very pervasive.

13. In 1974, fourteen FDA officials brought charges against the agency, claiming they had been removed from positions where they were preparing cautionary labelling or holding up drug approvals, as a result of industry pressure.

14. For example, Robert McNamara, president of the World Bank, was quoted in the journal *Science for the People* (1973, 5 (2) 4-8: Survey of the U.S. Use of Population Control in Latin America) as saying "less than $5 invested in population control is worth $100 invested in economic growth" and "The ultimate triumphs of foreign aid are victories of prevention. They are the shots that did not sound, the blood that did not spill, the treasure that did not have to be spent to stamp out the spreading flames of violence."

References

Allen, W., and O. Wintersteiner. 1934. *Science* 80: 190.

Atkinson, L.E., R. Lincoln and C.D. Forrest. 1985. "Worldwide Trends in Funding for Contraceptive Research and Evaluation." *Family Planning Perspectives* 17 (5): 204.

Baker, J.R. 1926. *Sex in Man and Animals.* London: Routledge.

Baker, J. R. 1935. *The Chemical Control of Conception.* London: Chapman Hall.

Banks, J.A. 1954. *Prosperity and Parenthood.* London: Routledge & Kegan Paul.

Blacker, C.P. (ed.). 1929. *International Medical Group for the Investigation of Birth Control*, Second Report.

Burfoot, A. 1990. "Normalisation of Reproductive Technology," in: McNeil (1990).

Butenandt, A., et al. 1934: *Zeitschrift für Physiologische Chemie* 227: 84.

Callon, M. 1992. "The Dynamics of Techno-Economic Networks," in: Coombs et al. (1992), 72-102.

Cantril, H. (ed.). 1951. *Public Opinion 1935-46.* Princeton: Princeton University Press.

Congress and the Nation 1969. A Review of Government and Politics, Vol. II 1965-68, Congressional Quarterly Service. Washington D.C.: Government Printing Office.

Coombs, R., P. Saviotti and V. Walsh. 1992. *Technological Change and Company Strategies: Economic and Sociological Perspectives.* London: Academic Press.

Corner, G.W., and W.M. Allen. 1928. *American Journal of Physiology* 86: 74; also 88 (1929): 326 and 340.

Crowe, C. 1990. "Whose Mind over Whose Matter?" in: McNeil (1990).

Fortune Magazine 1938. (April):154.

Gebhard, P.H. 1968. in: M. Diamond (ed.). *Perspectives on Reproductive and Sexual Behavior.* Bloomington: Indiana University Press.

Gordon, L. 1977. *Woman's Body, Woman's Right: Birth Control in America.* London: Penguin Books.

Hansard 1990. *Parliamentary Debates*. House of Commons Official Report, 171 (92) HC c 31-133 (April 23); 166-271 (April 24); 933-1044 (June 20); 1134-1222 (June 21).

Hartmann, B. 1987. *Reproductive Rights and Wrongs*. New York: Harper & Row.

Hartmann, M., and A. Wettstein. 1934. *Helvetica Chimica Acta* 17: 1365.

Inhoffen, H., W. Hohlweg and A. Serini. 1938. *Chemische Berichte* 71: 1024.

Jefferis, B.G. and J.L. Nichols. 1894. *Light on Dark Corners: A Complete Sexual & Science Guide to Purity*. Quoted in: L. Gordon (1977).

Kennedy, D.M. 1970. *Birth Control in America*. New Haven: Yale University Press.

Latour, B. 1988. *The Pasteurisation of France*. Cambridge, Mass.: Harvard University Press.

Leathard, A. 1980. *The Fight For Family Planning*. London: Macmillan.

Lesser, F. 1983. "Drugs Monitor Needs Sharper Teeth." *New Scientist* (17 March).

MacKenzie, D. 1992. "Economic and Social Explanation of Technical Change," in: Coombs, et al. (1992).

Makepeace, A., Winstein, G., and Friedman, M. 1937. *American Journal of Physiology* 119: 512.

Marrian, G. F. 1971. Transcript of "The Story of the Pill," BBC Radio 3 Programme, produced by Robin Brightwell.

McNeil, M., Varcoe, I., and Yearley, S. 1990. *The New Reproductive Technologies*. London: Macmillan.

Mears, E. 1961. "The Medical Student and Sex Education." Paper presented to the International Planned Parenthood Conference, The Hague, June. Mimeo, deposited in International Planned Parenthood Library, London.

Meyers, A.W. 1921. "The Frequency and Cause of Abortion." *American Journal of Obstetrics and Gynaecology* 2 (2, August).

Ministry of Health (UK) 1930. *Memorandum*, 153/MCW (July).

Mintz, M. 1969. *The Pill*. New York: Fawcett Publications.

Nelson, R. and S. Winter. 1977. "In Search of Useful Theory of Innovation." *Research Policy* 6: 36-76.

Nelson, W. O. 1956: *Endocrinology* 59: 140.

New York Times 1871, August 23: 6.

Office of Technology Assessment. 1982. *World Population and Fertility Planning Technologies*. Washington, DC: US Government Printing Office.

Parkes, A.S. 1966. *Sex, Science and Society*. Newcastle-upon-Tyne: Oriel Press.

Peel, J. 1963. "The Manufacture and Retailing of Contraceptives in England." *Population Studies* 17: 113.

Peel, J. 1964. "Contraception and the Medical Profession." *Population Studies* 18: 133-145.

Peel, J., and M. Potts. 1969. *Textbook of Contraceptive Practice*. Cambridge: Cambridge University Press.

Petersen, W. 1964. *The Politics of Population*. London: Gollancz.

Petrow, V. 1966. "Steroidal Oral Contraceptive Agents." *Essays in Biochemistry* 2: 117.

Phillips, A., and J. Rakusen. 1989. *The New Our Bodies Ourselves*. London: Penguin Books.

Pincus, G. 1965. *The Control of Fertility*. London: Academic Press.

The Practitioner. 1923. Special issue on contraception (July).

Rakusen, J., J. Robinson. and M. Berer. 1983. *Submission to the Public Hearing on Depo-Provera*. London: Co-ordinating Group on Depo-Provera, Women's Health and Reproductive Rights Information Centre.

Riddle, J. M. 1991. "Oral Contraceptives and Early Term Abortifacients During Classical Antiquity and the Middle Ages." *Past and Present* 132 (August): 3-32.

Rip, Arie, and Henk van den Belt. 1988. *Constructive Technology Assessment: Toward a Theory*. Enschede: University of Twente.

Routh, C. H. F. 1878. *Medical Press and Circular* (October).

Ryder, N.B., and C.F. Westoff. 1966. *Science* 153: 1199.

Science Newsletter. 1951. (October 27).

Shryoch, R.H. 1948. *The Development of Modern Medicine: An Interpretation of the Social and Scientific Factors Involved*. London: Gollancz.

Sjövall, T. 1970. "The Development of Contraception: Psychodynamic Considerations," in: K. Elliott. *The Family and Its Future*. Edinburgh: Churchill, 117-137.

Smith, D. S. 1973. "Family Limitation, Sexual Control and Domestic Feminism in Victorian America." *Feminist Studies* 1 (3-4) Winter-Spring.

Stanworth, M. 1987. *Reproductive Technologies*. Cambridge: Polity Press.

Teece, D. 1986. "Profiting from Technological Innovation: Implications for International Collaboration, Licensing and Public Policy." *Research Policy* 15: 285-305.

Vaughan, P. 1971. Typescript of "The Story of the Pill." BBC Radio 3 Programme, produced by Robin Brightwell.

Vessey, M.P., and W.H.W. Inman. 1968. *British Medical Journal* 2: 193.

Vessey, M.P., and R. Doll. 1968. *British Medical Journal* 2: 199.

Voge, C.I.B. 1933. "The Present Status of the Contraceptive Trade." *Manufacturing Chemist* 289.

Walsh, V. 1993. "Biotechnology, Demand and Public Markets." *Science and Public Policy* 20(3) June: 138-156.

Wiesner, B.P. 1929. "Experiments of Controlled Fertility by Means of Hormonic Interference," in: C. P. Blacker (1929).

Williams, R., R. Roy. and V. Walsh. 1982. *Government and Technology*. Milton Keynes: Open University Press.

Winter, I. 1970. *Journal of the American Medical Association* 212: 1067.

World Health Organisation. 1984. "Breast Cancer, Cervical Cancer and Depo Medroxy Progesterone Acetate." Collaborative Study of Neoplasia and Steroid Contraceptives. *The Lancet* (November 24): 1207-8.

Zuckerman, S. 1970. *Beyond the Ivory Tower*. London: Weidenfeld & Nicolson.

13 Decision Structures and Technology Diffusion: Technical and Therapeutic Trajectories for Diabetes Care

Thea Weijers

While most assessments of medical technologies focus narrowly on their cost effectiveness, a more important question for constructive technology assessment involves the decision-making process that accompanies their introduction. In addition, we know all too little about the critical role played by different actors in the development and diffusion of medical technology, such as physicians, hospitals, patient organizations, as well as the financial and regulatory institutions (including health ministries and/or the health care insurers).[1]

Recently, in recognition that the assessment process needed to be expanded, the Policy Development department of the Dutch Ministry of Health commissioned three case studies. Each of the cases — on the extracorporeal shock wave lithotripter (use of ultrasound to "crush" kidney stones), on the use of soft lasers in physiotherapy, and on insulin infusion pumps and other technologies for the treatment of diabetics — involved technologies with a mix of technical components and institutional compensations.[2] Decision making about each of the technologies, moreover, had in the Dutch context become subject to controversy.

Decision making on the insulin infusion pump was especially interesting because it was a new technology whose optimal use pattern was (and remains) unknown. While some physicians and patients stated that the pump's use should be stimulated, other physicians and several insurers maintained that the pump's use should be limited to a select group of patients. At first this controversy seemed to reflect a disagreement over the "medical-technical" qualities of the pump, but we soon found that it also indicated underlying differences in the approach to diabetes care. We therefore extended our research to the broader issues that are important to consider in diabetes care.

Decision making is often limited to the question of efficacy and efficiency of the technology as such, and is based on the technology's state of the art at that moment. Rarely do decision makers take into account the possibility that a technology might change, through research and development, or that new organizations or therapies might change its application. For instance, the Dutch Ministry of Health decided in February 1987 that no limits were needed in the use of the lithotripter, because (at that moment) the machine was too expensive for widespread hospital use. Unfortunately, the Ministry ignored that technologies often become cheaper and more readily accessible, and this one did both. Consequently, the Ministry is now facing the presence of twice as many lithotripters — including some mobile ones owned by groups of hospitals — as the number necessary to treat all Dutch kidney patients.

The market for health care is not a normal one: users are often not the ones who pay (and vice versa, a point developed in Koch's chapter). Still, medical technologies resemble other technologies in that they go through phases of development, from potential through introduction and diffusion, to stabilization or maturity. As we will see, in the case of the infusion pump and diabetes care, the decision-making process did not sufficiently recognize these phases. There was little attention to the changes that occurred during the diffusion process, when interaction between the technology and the medical-social environment led to changes in therapeutical goals and the organization of care. Coordinating these interactions is especially difficult, partly because the decision-making process on therapies is even more diffuse than that on technologies.

The medical and technological developments important for diabetes care can be described as consisting of three waves. The first wave was the discovery of insulin and the recognition that insulin would treat diabetes. The second introduced genetically engineered human insulin. The third led to the so-called intensive therapy. After discussing these three waves, this chapter critically analyses the decision making of private insurers and government health councils on the "intensive therapy" for diabetes, and especially on the insulin infusion pump. The conclusion presents some recommendations for improving decision making on medical technologies and their effective use.

The Discovery and Improvement of Insulin

In 1889 Von Mehring and Minkowski concluded from a number of experiments with dogs that a malfunction of the pancreas played an important role in a disease known as diabetes mellitus. Previously, the disease had been characterized in terms of symptoms including a high amount of glucose (sugar) in the urine, infections, emaciation, and, in the end, coma and death. At that time the dominant idea was that the cause of this disease lay in the nervous system, and consequently many scientists did not take this novel conclusion seriously. Minkowski initiated experiments to prove that the pancreas secreted a substance active in lowering the sugar levels. This hypothetical matter was called insulin.

For two decades researchers who tried to extract this 'insulin' from the pancreas failed. Injections with an extracted matter often led to a deterioration rather than improvement in the health of the tested animals (and persons). Conviction about the existence of insulin had little to go on until the early 1920s, when among others Banting and Best not only managed to extract insulin, but also showed that insulin could be used in treating diabetes patients. They also postulated that earlier experiments had probably failed because too much insulin had produced shock and coma; these fatal symptoms could be counteracted by giving the patient glucose. Insulin was no longer hypothetical but real, and beginning in 1922 several companies, including Eli Lilly and Nordisk, started to produce insulin on a large scale.[3] Until the present time, these companies have remained the most important insulin producers.

Elaboration of this early work showed that diabetes mellitus affects the specific cells in the pancreas responsible for producing insulin. This hormone helps regulate the storage of glucose (sugar) in the liver produced during digestion, and helps transport glucose into the body's cells.[4] Subsequently, medical researchers have discerned three additional distinctive "types" of the disease, beyond the classic Type 1 juvenile onset diabetes that is the concern of this essay. It is now a commonly accepted theory that diabetes is a result of unknown auto-immune processes.

The discovery of insulin transformed diabetes from a fatal disease into a chronic condition, severe but rarely life-threatening. The treatment of diabetes patients consists of daily injections of insulin to ensure that the level of glucose in the blood (the glycaemia or blood sugar level) remains on average between 3 and 10 mmol/litre. A patient with a blood sugar level below 3 mmol/litre (hypoglycaemia or 'hypo') can become comatose and die. A level higher than 10 mmol/litre (hyperglycaemia or 'hyper') has less visible effects, but in the long run can affect the blood vessels and the nerve system. To avoid these extreme levels, the patient is regulated; that is, it is determined how much insulin and how many injections the patient needs. Until recently that regulation had to be static: abrupt changes in lifestyle, diet, and the like had to be avoided since they lead to disturbances in the blood sugar level.

After the discovery of insulin, researchers concentrated on making the industrially produced insulin behave as endogenously produced insulin behaves in a healthy person. One problem was that in a healthy person the amount of insulin will rise during and shortly after meals, but decline during nights. To make the injected insulin behave in this way, the pharmaceutical companies developed and produced medium- and extreme-slow-working insulin. Another problem stemmed from the industrial insulin being made from the pancreases of animals, usually pigs. This porcine insulin was often contaminated with other porcine hormones that could cause severe allergic reactions. Over the years, researchers improved the purification methods for porcine insulin, and by the 1960s and 1970s only a few patients were still allergic to the purified porcine insulin. For this select group, human insulin was produced from human pancreases.

In 1978 an American company named Genetech managed to produce human insulin artificially by using genetically manipulated bacteria. The American company Eli Lilly bought a license from Genetech, and started to market "Humulin" in 1982. Later, European companies like Novo/Nordisk and Hoechst followed with a human insulin produced through enzymatic processing of porcine insulin.

Initially, in 1983, the Dutch Council of Health Care Funds (*Ziekenfondsraad*) advised that genetically engineered human insulin should, just like the naturally derived human insulin, be given only to allergic diabetics. Non-allergic patients, it was reasoned, did not need this very expensive medicine. But the costs of artificial human insulin declined over the years, and accordingly its market share has risen to 80 percent in 1988.[5] Here, the one-time determination of the insulin's "excessive" cost was rendered meaningless by events: specifically, the falling price of genetically engineered human insulin.

Besides the industrial research on the improvement of insulin, making it behave as the normal, endogenous insulin, medical researchers at the medical faculties and hospitals were interested in the long-term consequences of diabetes and also in the ways the insulin could be administered. This clinically oriented research led to the "intensive therapy," to which we now turn.

The Intensive Therapy

Developing an implantable closed-loop infusion pump that would behave more or less like a human pancreas is the ultimate goal of many researchers working on different methods to administer insulin. There are several reasons for their wish to develop this kind of device. One is that many patients suffer from having to inject themselves with a hypodermic syringe each day — for the rest of their lives. The more accessible places for the injections become sore and bruised, and diabetes patients tend to heal well slowly. Moreover, injections in public are difficult: "They think you're a junkie." Another pressing reason is that many diabetics do not respond well to a regulatory regime of one or two injections each day; sometimes, in the case of so-called brittle patients, not even to three or four injections. Further, female diabetics who want to have a baby form a problematic group. Pregnancy with diabetes places additional strain on the body's regulatory systems, often ending in miscarriage or stillbirth. Especially for "brittle" and pregnant diabetics physicians (mostly internists) believe that a more natural delivery of insulin to the body — that is, a more continuous delivery that corresponds with the actual blood sugar level instead of the hour of the day — would help these patients achieve normal and satisfying lives.

Another reason for the development of a closed-loop pump is that older diabetics, even with a good regulatory regime, suffer from long-term complications, especially kidney disease, blindness, and deterioration of the blood vessels and nervous system. It seems possible that a pump — by regulating more precisely the hour-by-hour fluctuations in blood sugar — could ward off these long-term complications. The closed-loop pump should function more or less without direct human interference. It should contain a system for the infusion of insulin, a sensor that could measure the blood sugar level, and a control device. Based on input from the blood-sugar sensor, the control device would calculate the proper dose of insulin. The main obstacle to closed-loop systems is the difficulty in developing a glucose sensor.[6] Attempts to develop an open-loop system have been more successful.

The idea of an open-loop insulin system had already been suggested in the 1930s, but until the advent of microelectronics it was not possible to create a small and portable electronic pump. In 1977 Pickup et al. published an article on the first successful experiments with a portable infusion pump. Like other open-loop systems, it has a reservoir, pump, and control device — but not a glucose sensor. The pump is carried on the body, with a catheter going from the reservoir into the abdomen. The insulin is administered semi-continuously.

The first open-loop pumps were introduced in several academic hospitals[7] in the Netherlands in 1979 as a result of favourable publicity and assessment of the International Diabetes Foundation and the European Association for the Study of Diabetes. These first pumps were used in treating "brittle" and pregnant diabetics. The main purpose of these experiments was to see whether the use of the pump would lead to better regulation, i.e., a steady blood sugar level. The doctors involved in these experiments shared results through the Netherlands Society for Diabetes Research (*Nederlandse Vereniging Diabetes Onderzoek*).[8] Some of the research was co-funded by the Diabetes Fund Netherlands (*Diabetes Fonds Nederland*), which is connected to the Diabetes Society Netherlands (*Diabetes Vereniging Nederland*), a patients' society. Further use of the pumps, especially outside the hospital, was stimulated by the introduction of blood-sugar test strips.

Blood-sugar test strips were introduced in May 1980 by the company Boehringer-Mannheim. These strips made it possible for the diabetic to measure his or her own blood sugar level several times a day by pricking a finger and applying a drop of blood to the strip. The color of the test strip indicates the blood sugar level. Before test strips were available, blood sugar levels were typically measured only at the internist's regular checkup, usually once every six months to a year. This measurement gave no insight into the daily or weekly course of the disease, and so the patient also had to keep urine samples for analysis. Urine tests are not reliable, however, since urine glucose is a poor reflection of blood glucose. The blood-sugar test strips made it possible for diabetics to monitor their own blood glucose during a day as well as over a longer period of time.

The positive reports on the pump, as well as the possibility of using the pump outside the hospital, convinced more doctors that the pump offered significant advantages over conventional injections. Although the academic hospitals were continuing to experiment with the pumps, more internists, even at the non-academic hospitals, started to use the pump; in a period of less than two years about a thousand pumps were brought into use. It seemed that the experimentation phase was rapidly evolving into a phase of regular use.

This period of diffusion ended suddenly in 1982, when the Centers for Disease Control in the USA reported a number of casualties among people who used the pump. Although investigation showed there was no causal connection between these deaths and the use of the pumps, the number of pumps in the Netherlands dropped to about 200.

In the same year, 1982, the American Diabetes Association pointed to the advantages of the intensive conventional insulin therapy (ICIT). The key to intensive therapy was the blood-sugar test strip used for intensive self monitoring. Finally, it was feasible to adjust the number of injections, and the amount of insulin, to the *actual* blood glucose level. Injections could now closely imitate the natural process, in which the amount of insulin secreted by the pancreas depends on the glucose level, and which doctors had tried to achieve through the use of pumps.

In the two years following the ADA's recommendations, many articles appeared comparing the ICIT with the insulin-infusion pump therapy, creating the impression that the two therapies were in competition with each other. The general conclusion was that the pump therapy offered few advantages over the ICIT as far as regulating the patient's blood sugar was concerned. Although many patients reported feeling better using the pump, others said it created a psychological strain because it reminded them constantly of their disease. However, pumps were indicated when patients required *more* than the two to three injections each day that was considered the practical upper limit with ICIT.

In this period, the academic hospitals continued to use the pump for "brittle" patients. Interest in imitating the natural process, aroused by both the pump and the ICIT, spread among the doctors and patients active in the field of diabetes. The development of human insulin further strengthened this goal of a natural situation. Once, preventing low blood sugar level was enough; now, achieving the accurate regulation of blood glucose became the therapeutic goal. Many believe that a persistent normoglycaemia, that is a normal blood sugar level (between 3 and 10 mmol/litre) will prevent the long-term complications that afflict many diabetics. The ICIT was further supported by the introduction of the HbA1 test. The HbA1 test measures the concentration in the blood of haemoglobin molecules that have reacted with glucose. This concentration is not dependent on the amount of glucose at a single moment, but on the average level of glucose over a long period of time (approximately two months). It therefore indicates whether the goal of normoglycaemia is met across a certain period. To achieve both long-term and short-term regulation, some patients needed more than the practical upper limit of three injections a day. For them, the pump was indicated. The intensive therapy and the pump were no longer seen as competitive but complementary.

With either injections or pump, the intensive therapy demanded more from both doctor and patient. Maintaining a normoglycaemia can, in a way, be compared with walking a rope, of keeping exactly the right balance. Doctors need information on the particular reactions of each patient, while patients need education on when to change the dose and number of injections, and on what to do when the blood sugar gets too high or too low.

The need for education and information was expressed in the arrival of a new professional group, the diabetes nurses. In 1986 they organized themselves in the Netherlands as the Association of Diabetes Nurses, under the auspice of the European Association of Diabetes Educators. The introduction of the intensive therapy, sponsored by a relatively small group of doctors, nurses, and patients, was helped by the development of a new injection device: the insulin injector pen.

The insulin pen introduced by Novo in 1985 is an improved replacement of the traditional hypodermic syringe. The pen, containing a thin needle and an insulin ampoule, looks like an ordinary fountain pen and is easy to use, even in public. The injector pen was introduced in the Netherlands in 1986, and within two years approximately 25 percent of all insulin was administered through the pen. The price of pens varies between Dfl. 80 and Dfl. 150 (approx. $50-$94). But the companies usually hand out the pen for free in an attempt to gain customers (most pens can only contain ampoules of one particular company). The market for insulin is dominated by a few established companies, and especially after they have all introduced human insulin, the only way to extend market share is by better services and supplementary products — such as pumps and pens.[9]

The injector pen was marketed when there was growing attention to the intensive therapy and the multiple-injection scheme it required. In 1980 four-fifths of all diabetics injected themselves only once each day. By 1990 this figure had changed radically: 52 percent inject once, 34 percent inject twice, and 14 percent inject three times or more.[10] Because the pen made it possible to have more than three injections, the use of the pump, at first, became a less attractive solution.

According to the Diabetes Society, industry, and several physicians, the popularity of the multiple injection scheme will lead to a growing number of patients who inject themselves three times or more. Quality of life will become more and more important. This demand for higher quality will eventually lead to a growth in the use of the pump as well, because more patients will find that a scheme requiring three or four injections each day does not bring them the quality of life they wish to have. In the Dutch hospitals known for their well organized diabetes care, such as the Weezenlanden in Zwolle, as well as in many German hospitals, 5 to 10 percent of the patients use the infusion pump. If the multiple injection scheme does become the most-preferred therapy, this pattern of use may well spread to other hospitals.

Decision Making and Financing

As mentioned in the introduction, there is no body in the Netherlands that can really decide on the introduction or development of a therapy, and diabetes therapy is no exception. Most decisions focus on financing specific treatments or devices. And therapies, as we have seen, comprise devices, treatments, organizations, and patterns of use. Decisions on the financing of the pump have nonetheless had dramatic impact on the availability of the intensive therapy.

In the first experimental phase the infusion pump was not yet included in the "basic list of services and provisions," which contains all services and provisions covered by the health care insurers.[11] Pumps were used inside hospitals and were paid for and owned by them. When the use of pumps outgrew the experimental situation at the end of 1981, and they were first used outside the hospitals, it was no longer possible to finance their purchase out of the hospital budgets. Patients had to buy their own pumps, but they were still not part of the "basic list of services and provisions" used by health care insurers. For this situation most Health Care Funds have a small budget, called the supplementary fund, which can be used for new provisions and treatments. This gives each fund some room for experimentation.

In the case of new provisions or treatments, a regional Health Care Fund and a doctor agree on the terms of the coverage out of the supplementary fund. The regional Health Care Fund can — but does not have to — inform the Council of Health Care Funds of any agreement. If several regional agreements reach the Council, this is seen as an indication that the provision should be included in the "basic list of services and provisions." Only then will the Council take action. It is not active beforehand; although it could do so, it does not advise the regional funds on the sorts of agreements other funds have reached concerning use of the supplementary fund. In the absence of any coordination, different agreements occur at the same time; furthermore, it is next to impossible for one Health Care Fund to learn from the experiences of another fund. This is an especially pressing problem for technologies used in chronic care, which receive little or no attention in the public press.

Ideally, the period of region-by-region agreement on financial compensation from the supplementary fund should function as a mechanism to achieve consensus during the experimental phase. The Council of Health Care Funds can then follow the emerging common practice. In the case of blood-sugar test strips this process functioned well: an early consensus resulted in a positive recommendation from the Council in 1983.

But decision making about insulin infusion pumps was more difficult, especially after the reported deaths in 1982. There were many differences between the regional agreements, both on the medical indications and on additional conditions. Differences in medical indications are important because they can indicate differences in therapy: can the pump only be used when conventional therapy fails (brittle patients), or for all patients for whom intensive therapy means having more than three injections a day. The same goes for differences in additional conditions, such as the 24 hour service for pump users some Health Care Funds required in case the pump malfunctioned. This service can be seen as part of the required system of health care, or as a warning that use of the pump is potentially dangerous. These differences were not resolved because most regional agreements did not clearly specify what the indications meant or how conditions such as the 24 hour service should be organized. Because differences in indications and conditions reflected more deep-rooted differences in therapy and care, the bottom-up process did not function as well as in the case of the test strips. Perhaps this was an indication that the experimental phase was not yet finished, but unfortunately it also meant that the Council took no action. The Council did not even try to organize some sort of discussion process. The regional funds and doctors were on their own.

This indirection bothered the Netherlands Society for Diabetes Research, which formulated a set of indications for the pump therapy. It also recommended the 24 hour service as well as a provision that only doctors experienced in the treatment of diabetes should be allowed to prescribe the pump. Around the end of 1983 this advice was sent to the Council of Health Care Funds, but not followed. It is not clear why, but some Council members later said that at that time they still had doubts about the safety and necessity of the pump therapy. The financing of the pump was so unclear that it was discussed in the Parliamentary Commission for Petitions, and the Chairman of Parliament formally requested the Minister of Health to come to a decision in this matter. In 1984 the Minister sought advice from the Council of Health Care Funds, but as we will see that advice took a long time coming.

In 1985 a technical committee from the organization of private insurance companies (or KLOZ) studied the application of infusion pumps for the continual administering of medicine. It found that the infusion pumps were already in use for many applications, especially for insulin. Based on this finding the KLOZ advised all members to cover the pump, under certain conditions. Approved candidates for the pump include patients who are not well regulated despite an intensive treatment of two-to-four injections a day. Additionally, the patient and the doctor must be committed to the therapy. The intensive treatment mentioned here refers especially to the exceptional treatment of "brittle" and pregnant diabetics.

Still, the Council of Health Care Funds did not follow this example. In March 1986 the fund's secretariat informed the Minister of Health that the complexity of the issue made it difficult to make a clear decision. This response is rather strange, because by that time many Health Care Funds had set up agreements to cover the pump out of the supplementary budget. To be sure, the agreements differed among the various regional funds and the number of pumps involved was small. Also, there seemed to be a difference in opinion about a more extensive use of the pump between the specialized internists (who could provide the extra facilities) and the general internists in the smaller hospitals. But some of this diversity was the effect — not the cause — of the Council's sitting on the fence. The Council still felt that there was no consensus, and therefore no decision could be taken. The Minister was not satisfied with this answer and in polite but no uncertain terms reminded the Council of his two year old request for advice.

In October 1986 the Council announced changes in the "Decision on artificial and auxiliary devices." Finally, the Council embraced a positive recommendation to cover the infusion pumps, acknowledging that they were the next step (after the introduction of the test strips) in the self regulation of diabetics. Nevertheless the Council still wanted to limit the use through strict medical indications as well as by conditions such as a 24 hour service. The Council emphasized that the use of the pump was not indicated when the patient leads an irregular life. This statement is wholly remarkable, since the pump's value comes partly from relieving patients from the rigid diet and routine that had to be obeyed before. Irregular working hours had been an independent indication in the advice drawn up by the Netherlands Society for Diabetes Research. Six years after its introduction, the financial question about the pump was settled. By now, the injector pen had arrived on the scene, and the intensive therapy was becoming more popular.

After the decision by Health Care Funds and insurers to cover the pump, no other decisions concerning diabetes care specifically were taken. The financing of the treatment of diabetics has not changed. Doctors who apply the intensive therapy, which includes hiring a specialist nurse and scheduling more frequent patient check-ups, get the same amount of money per patient as doctors who do not. The Dutch system of financing does not reward the extra expenses and efforts involved in reaching higher quality. Only doctors who work in big hospitals or large partnerships can afford to offer the intensive therapy. So far the health care insurers have shown no indication that they are prepared to take action on the differences in therapy and in the quality of the treatment of diabetics.

An Analysis of the Technology and the Care

Medical care must be comprehended within a technical system consisting of related artifacts, infrastructure, and policy environment. In the case of diabetes care, it was the interaction of several technologies that made it possible to formulate a new medical regime. The old medical regime focused on the avoidance of short-term complications such as hypo- or hyperglycaemia through a good standard regulation, while the new medical regime focuses on the avoidance of long-term complications through a flexible regulation imitating the natural situation as well as possible.

A necessary component of the new medical regime was the introduction of blood-glucose test strips. At the same time the introduction of infusion pumps focused attention on the serious long-term complications of diabetes, the starting signal for the intensive therapy. The pen, in turn, made the further diffusion of the intensive therapy easier (the relation between the introduction and diffusion of the different technologies is indicated in Figure 1, the given figures are estimates calculated on the basis of information given by some internists, the Diabetes Society, and industry). The introduction of the HbA1 test (because it indicates whether normal blood sugar is maintained across time) has also helped in the diffusion of the new therapy. But it is not only the new technologies that made the intensive therapy feasible. For an optimal use of the new technologies, the social-organisational setting — the care-system — is equally important. In places where the diabetes care is well organized, such as in Germany (the former BRD) and a number of Dutch hospitals, most patients function well on a multiple injection scheme of three-to-four injections each day, and the number of patients using the infusion pump has stabilized at between five and ten percent.

The intensive therapy changed the roles of patient and doctor. Through self monitoring of blood glucose and deciding the dosage of insulin, patients have gained greater responsibilities: continuously maintaining blood glucose between strictly defined upper and lower limits. Doctors need to support and educate their patients so they are capable of making these decisions. The new medical regime requires doctors to be well aware of all the ins-and-outs of diabetes in general and their patients in particular. Accordingly, doctors can no longer be satisfied with one or two check-ups per patient per year. A number of internists have already specialized themselves into "diabetologists."

Figure I: Diffusion Patterns of Trajectories for Diabetes Care

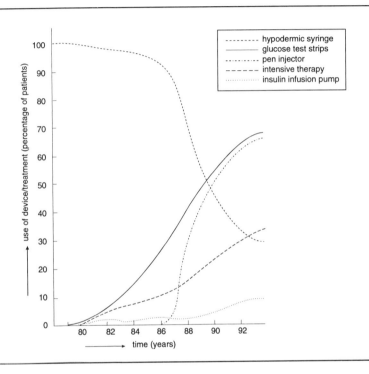

The switch to intensive therapy was combined with a number of organizational innovations. A new professional specialty, the diabetes nurse, arose in hospitals where the new therapy was used. After a period in which they more or less educated themselves, they organized the Association of Diabetes Nurses (under the wings of the European Association of Diabetes Educators). Some hospitals have recently set up special polyclinics and centres for the treatment of diabetic patients.

Although new technologies made the new therapy possible, it will flourish only when diabetes care is reorganized and intensified. Education and information will be important elements. It seems appropriate to speak of a new "care trajectory," consisting of new technologies, new medical regimes, and new organizations. Unfortunately, this new care trajectory has not been sufficiently acknowledged by the financing and regulatory bodies. Even more than the technology trajectory, the care trajectory offers possibilities for an active policy and a constructive technology assessment.

Decision making at the KLOZ and the Council for Health Care Funds was concentrated on the infusion pump. The Council did in the end recognize that the pump was part of the broader goal of optimal self regulation. Perhaps the Council has acknowledged the new care trajectory. But the changing organization of diabetes care did not get much attention from the Council or KLOZ. While KLOZ and the Council have mentioned the organization of care in the context of conditions for the use of the pump, they have said nothing about the importance of availability of doctors, education, and guidance in themselves as part of the intensive therapy. They have issued no guidelines as to how the 24 hour service and the education should be set up or financed. Instead both organizations have focused their decisions on the coverage of the infusion pump, on the technology itself. Both organizations failed to get involved before that decision had to be taken, and after that decision was made both organizations halted their activities.

We recommend that the decision-making process should be better adapted to the different phases of the technology's development: introduction; the first experimental applications; adaptation and modification of both technology and the surrounding care system; and stabilization. The policy as well as decision making should be made in phases as well. It should be directed towards the recognition of new developments (introduction phase) and the stimulation of learning processes (experimental application). Learning for instance could be facilitated by stimulating publications, organizing workshops and consensus conferences. After that, efforts should be made to "tune" the technology and the care system (adaption and modification) and to acknowledge the changed care system as part of the wider health care system (stabilization).

There are several organizations that could play a leading role in this process. For instance, the more independent Health Council and the Council for Medical Research could coordinate the research and discussions necessary in the first phases, especially since they already have some experience in technology-assessment research. The Council for Health Care Funds and the Ministry of Health could be more active in the phases when financing and regulation become more important. The recognition of new developments in this case took place through the supplementary fund, which allows for new developments and experimentation. But, these funds are regional and there is no coordination between regions. Internists and hospitals develop their own plans, which are not attuned to each other. The exchange of information is not organized, so internists at smaller hospitals usually do not know about the newest developments and the experiences of specialist internists. Moreover, the instrument of the supplementary fund leaves the learning process to coincidence. The supplementary funds bottom-up orientation should be complemented with a top-down policy that creates better conditions for learning. For instance, the Council could have stimulated the learning process by publishing the differences in the regional agreements, organizing workshops to identify the real differences in opinion, and working to facilitate the emergence of consensus. Internists and patients, especially those of the smaller hospitals, would have profited.

Most of the organizational innovations, such as the clinics and the nurses, are financed by the hospital's own budget and come about through the enthusiasm of a few doctors and nurses. Therefore the organization of the care and the use of the intensive therapy can vary considerably between hospitals. It is difficult for patients to find out what different therapies there are, what facilities the different hospitals can offer, and what treatments are preferred by the different internists.

Ensuring more transparency in the health care system would materially improve the quality of the health care. Registration of the different facilities offered by hospitals and doctors (the technologies available, the existence of a specialized nursing staff, 24 hour service, etc.) would help create such transparency. This registration could be set up by the insurers, the organization of patients (the Diabetes Society), or the Ministry of Health. The information should be available to all, especially to patients. It might be a good idea to adopt financial rewards for differences in quality, but this would mean a drastic change in the Dutch health care system.

More research on changes in therapy and care is urgently needed. Most health-care research — even that financed by the Diabetes Fund Netherlands — is focused on expanding medical knowledge, as well as devising evermore novel technical solutions. Little research is conducted on the effects of new treatments, on the well being of patients, and other matters vital to a patient's experience. Surprisingly, we do not even know how many Dutch diabetic patients there are, let alone how they are treated.

The lack of research, the lack of information, and the lack of transparency have made it possible for policy makers to ignore the possibilities of the intensive therapy. Recently, in 1991, the Ministry of Health set up a new system for the coverage of medicine, in an attempt to reduce costs. If a cheaper medicine is available the patients must pay for the more expensive one. This is a good idea in itself. But since insulin in bottles (for a hypodermic syringe) is cheaper than in ampoules (for the insulin injector-pen and several pumps), the all-important list of approved items contains only the bottles. Most patients using an injector pen either have to pay for their insulin, or go back to the hypodermic syringe and fewer injections. A new medical regime? A new intensive therapy? The Ministry has never heard of it!

Notes

1. For a review of these issues, see Vrolijk and Straten 1989; Vrolijk 1990; and Weijers and Vrolijk 1990.
2. There are two kinds of "health care insurers" in the Netherlands: the regional co-operative funds, for employees earning less than approximately Dfl. 50,000 before taxes, and private insurance companies, for everybody else. Present policy is directed towards an integration of funds and insurance companies.
3. Wackers 1991.
4. A recent study estimated that approximately 244,000 people suffer from diabetes mellitus in the Netherlands, about 40,000 of whom have the type discussed here (type 1 or juvenile diabetes). Institute for Medical Statistics 1989.
5. The annual turnover of insulin in the Netherlands is more than 30 million guilders.
6. The pumps can be used for the infusion of different fluid medicines, such as cytostatics (chemotherapy) for cancer patients, pain killers, heparin for the treatment of thrombosis, and insulin.
7. Most medical research and experimentation in the Netherlands takes place in academic hospitals.
8. Diabetic patients are usually treated by specialists in internal medicine, also called internists. The Netherlands Society for Diabetes Research consists of internists especially interested in diabetes, many of whom work in academic or large regular hospitals.
9. There are not adequate statistical figures available on diabetes or the use of pumps and pens, but a recent article (Pols 1990) suggests that now about 90 percent of diabetics use the pen.
10. Based on information from a private research and marketing institute, DIMS Institute for Medical Statistics.
11. The Council of Health Care Funds (*Ziekenfondsraad*) makes proposals for that list, which then have to be approved by the Minister of Health. The private insurance companies, organized in the *KLOZ*, use a similar list.

References

"Ambulatory Insulin Infusion Pumps." *Health Devices*, November 1987.

Ballegooie, E. 1984. *Continue Subcutane Insuline Infusie*. Ph.D. Dissertation, University of Groningen.

Durand, T., and T. Gonard. 1986. "Stratégies et ruptures technologiques: le cas de l'industrie de l'insuline." *Revue Française de Gestion* (November-December): 89-99.

European Association of Diabetes Educators. 1989. *Newsletter* (February).

European Association of Diabetes Educators. 1989. *Newsletter* (May).

Institute for Medical Statistics. 1989. "Diabetes Mellitus." *Diagnose Informatie en Medische Statistiek (DIMS)*, special issue no. 7.

Pols, Bram. 1990. "Vergoedingensysteem medicijnen verbijstert diabetici." *NRC Handelsblad* (June 14): 3.

Pickup, J.C., H. Keen, J.A. Parsons and K.G.M.M. Alberti. 1977. "The use of continuous insulin infusion to achieve normoglycemia in diabetic patients." *Diabetologia* 1977 13: 425A.

Rosenberg, S.A., M. Espringate and M.S. Blaiss. 1989. "Nieuwe ontwikkelingen in de behandeling van insulin-afhankelijke diabetes mellitus." *STG-Bulletin*: 19.

Smits, R., and J. Leyten. 1988. "Key Issues in the Institutionalization of Technology Assessment." *Futures* (February): 19-36.

Vrolijk, Hein, and Geert J. Straten. 1989. *Het einde van het Stenen Tijdperk*. Apeldoorn: TNO.

Vrolijk, Hein. 1990. *Hoe soft is de soft laser in de fysiotherapie*. Apeldoorn: TNO.

Wackers, Ger. 1991. "To be or not to be? De identiteitscrisis van insuline." *Wetenschap en Samenleving* 43 (3): 13-19.

Weijers, Thea, and Hein Vrolijk. 1990. *Diabeteszorg, niet bij technologie alleen*. Apeldoorn: TNO.

Ziekenfondsraad. 1986. *Wijziging besluit kunst- en hulpmiddelen ziekenfondsverzekering 1981*. Amstelveen: Ziekenfondsraad.

Part V

Theoretical Analysis of Possibilities for Change

Theoretical Analysis of Possibilities for Change

Introduction

The basic assumption of CTA is the (partial) malleability of technology. The cases in this volume show that there are continual forces which exert some influence, and it is out of such interplay that technology is shaped. This seems to imply that CTA happens all the time. The only problem is that the shaping process sometimes goes in the wrong direction. The challenge is then to be able to have intentional CTA. Some of the essays have already contributed to this aim by providing strategies and instruments. In this section we argue that theory is required for intentional CTA. The authors in this part will assess existing theories and develop new theory by drawing together different sociological and historical models, neoschumpeterian economics, and theory of the firm. In this way authors in this part contribute to the emergent and needed synthesis between the different disciplines.

Drawing on actor-network theory and neoschumpeterian economics, Callon presents a model to address the basic question of how can we avoid the emergence of irreversible or lock-in situations which exclude certain technological options merely because they did not find support at a particular time. Variety is needed (as both Callon and Coombs emphasize) to preserve the desirable properties of our technologies and retain the possibility for political debate about them. The importance of the existence and visibility of alternatives is forcefully demonstrated by Jørgensen and Karnøe and Steward in their case studies. In systematizing these observations, Callon shows that the concept of irreversibility can be analysed in terms of increases and decreases in the convergence of actor-networks. A network is convergent if agreement is reached on the description of its content and on the list of actors and technologies involved. Changes in descriptions and actors and technologies involved depend on the nature of coordination rules present. These rules determine who negotiates with whom, in what order and according to what sequence. It is enough to change the list of actors, or sequences of interactions, for other technologies to develop. Coombs contends that membership in a network will be the most potent means of having effects on technology.

The dynamics of technological development are characterized by alternative processes of convergence and divergence. Creating convergence means forming links between actors and rules of the game. The formation of links will have the structural effect of aggregating previous decisions. Thus degrees of irreversibility depend entirely on the dynamics of networks considered. This draws attention to the importance of monitoring, and the development of means including simulation to model these dynamics.

Network analysis tends to emphasize emerging social structuring and correspondingly neglects the cognitive dimensions of these structuring processes. For example, expectations about the success of following particular technical routes are an important behavioural determinant. Coombs argues that strategies of organization can be perceived as cognitive devices, aiming at making sense of what is happening and creating metaphors. In this view, since organizational learning occurs through storytelling, creating new stories is a forceful strategy to shape the future. CTA must take this cognitive dimension seriously. Cultural engineering, agenda building and participating in discourses about what is constructive are important elements of any CTA effort.

In the epilogue CTA will be presented as a multi-actor decentralized form of control. Not all the action occurs at institutions in the centre. Rather technological changes occur in networks which involve many actors, with nodes that are highly local. In addition it is not necessary to search for one best moment to influence technical change. All case studies in this book make clear that technologies continue to change over their entire lifecycle, and accordingly CTA will involve multiple decision points over time. In generalizing this observation, one could argue that a main strategy for CTA is to nurture the development of places where alignment and interaction occur. Several proposals for CTA action developed in this book illustrate very nicely the potential of the creation of new alignments. For example Coombs suggests the development of new institutes in which debate would take place on requirement for technical change. They would have the capability and scope to act as a "proxy procurement agency."

14 Technological Conception and Adoption Network: Lessons for the CTA Practitioner.

Michel Callon

The Hypotheses Underlying CTA; Definition of Objectives.

CTA may be defined as aiming at broadening the design and implementation of technological systems in order to stimulate the integration of social criteria into the technological development itself (Rip and Van den Belt 1988). In these terms, the CTA concept refers both to a principle of democracy and to a particular interpretation of the technological development process. In this interpretation, three hypotheses can be distinguished:

> 1) Technological development results from a large number of decisions made by numerous heterogeneous actors. These naturally include the scientists and engineers directly involved, but increasingly involve participation by the users, the business and financial worlds, and all levels of government. These partners negotiate the technical options and, in some cases — after what may be a long series of successive approximations — reach mutually satisfactory compromises. The diversity of centres and criteria of decision implies some degree of technical plasticity.
> 2) Technological options can never be reduced to their strictly technical dimension. The design and introduction of a new vehicle, energy production process, or kitchen machine, are indissociable from some degree of social restructuring and role distribution. Hence, the appraisal of technological options is a matter of political debate.
> 3) Technological options bring about irreversible situations, resulting from the gradual disappearance of the margins of choice available to the deciders: as time goes on, their choices are inexorably predetermined by decisions taken earlier. Unlike some decisions which always remain possible to revise, those

that are materialised in technical commitments, such as the capital sunk in the nuclear generation option, lead to durable imbalances and the consequent discarding of options which, with hindsight, might have been thought preferable to those actually taken.

The implementation of a constructive technology assessment, or CTA, must therefore take account of the answers to the following questions: (a) How can we ensure that all the actors involved, especially the non-specialists and the most resourceless, receive a proper hearing during discussion of the technical options, and when it comes to taking decisions? (b) How can several alternative technological options be kept open at all times, bearing in mind that a variety of these must exist if the very notion of choice is not to disappear, and with it all possibility of a political debate? (c) How can we avoid the emergence of irreversible situations which exclude certain technological options merely because they did not find support at a particular time?

If CTA is to become a practical option and not to remain a utopian dream, clarification and further articulation of these problems are necessary. An important element is to model the dynamics of the processes of technological design and adoption, so as to answer three analytical questions: (a) How do we identify the actors who take part in the process of design and of adoption of technologies? In what ways and at what time do they intervene? (b) How do we explain the disappearance of alternative technological options (or of what is called technological variety)? (c) How do we account for the appearance of irreversible (or lock-in) situations?

One starting point is the many studies concerning the dynamics of technological development performed in recent years under the banners of various research disciplines. Technological competition models (Arthur 1989; David 1986 and 1987), analyses in terms of technological systems (Hughes 1983; MacKenzie 1991), of evolution or quasi-evolution (Nelson and Winter 1982; Dosi et al. 1988; Van den Belt and Rip 1987) or of actor networks (Callon 1986a, 1987 and 1989; Latour 1987, 1988 and 1993; Law 1986a and 1987), have in common that they all emphasise the fact that the aggregation of the decisions taken by the actors as of time t-1 contributes heavily to determining the field of choice as of time t. To put it succinctly, both the opportunity of participating in the decision process and the intrinsic nature of the options available, are path dependent.

The other starting point is the network concept which enables us to reconstitute the dynamics of the decisions made by a population of actors confronted with the need to take technological options. It allows us to use the same general categories to analyse both the design and adoption phases. To achieve this, some distance will be taken from the usual definition of this concept. As used here, the term network is not considered as an intermediate coordinating mechanism between an organisation and the market (Thorelli 1986; Powell 1990; Imai and Baba 1991; DeBresson and Amesse 1991; Freeman 1991). It is taken in a more narrow sense to mean the group of unspecified relationships among entities of which the nature itself is undetermined. The picture of a network is thus reduced to a simple graph. The simplicity of this analytical tool is its principal advantage. It offers a minimal framework to describe interactions in all their diversity and richness. This is why it can be applied equally well to design and adoption phases.

This chapter is based on a project (Callon 1993) in which I have identified three relevant models. The first of these, the traditional diffusion model, is applicable to describing the diffusion of one given technological product or process Tj which does not undergo transformation during adoption. It reconstitutes the paths of diffusion by reference to the position of the first adopters and to the form of the networks of interdependence among the actors. It enables us to follow the transformations undergone by these networks in step with the degree of progression of the adoption process: we can thus observe the emergence of new social structures generated by the aggregation of successive decisions.

The second model, competition diffusion, is an extension of the first, placing several substitutable technologies in competition with each other. It refers to the same variable as previously, but is more complicated to analyse, because of the multiplicity of the technologies present and the range of possible trajectories.

Model three, the actor-network model, has the widest scope, integrating the results of the first two with the technological conception, design and development process. This chapter presents — inevitably in summary and abstract form — the third model, which is the one most relevant to CTA purposes.

The diffusion and competition diffusion models have as their main weakness the fact that they assume the technological and human entities (suppliers, adopters, etc.) to be exogenous and immutable variables. The technologies are considered as given: their design is excluded and deemed to have no influence on the adoption process; transformations (adaptation, differentiation) occurring downstream are also neglected.[1] Similarly, the separation between suppliers and adopters, and that between adopters and non-adopters, which existed *ab initio* are not reappraised subsequently. The potential adopter population is circumscribed once and for all, while both suppliers and users are disqualified from influencing the technological options. Economics of technical change and sociology of innovation have made significant contributions to demonstrating the unrealistic nature of these assumptions, such as when drawing attention to the collective and adaptative features of the innovation process. The actor-network model is designed to overcome these limitations. It integrates the technological design and re-design phases of the development process, with the adopters having in principle the same opportunities for taking part in the elaboration of the Tj as do the suppliers. It stresses the co-evolution of technologies and human beings interacting with them, leaving room for variation in the population of potential adopters. It has a relatively indeterministic approach to saying whether a given entity belongs to the technological (adoptable) or the human (designers and/or adopters) universe. This acknowledges the evolutive nature of the conventions (authorities, rules, customs).

Building Blocks of the Actor-Network Model

Actors and Techniques[2]
A distinction can be made between entities which are called 'technical objects' and others which are defined as 'actors.' The first are subject to negotiation concerning their characteristics and production process. They are modified over time as a function of the state of the prevailing forces and compromises reached among actors. These transformations may be more or less significant. This will depend on the degree of convergence or divergence among the viewpoints of the actors participating in the negotiation. Tracy Kidder shows, for example, how within a firm computer scientists, researchers and the sales team negotiate to define the characteristics of a new mini-computer to be put on the market: degree of compatibility with existing models, bit mode to be selected, micro-codes, machine size, micro-processor features, etc. (Kidder 1981). The history of an innovation's conception is mixed up with the discussions and compromises to which it gives rise (MacKenzie 1991). In some cases, at the end of a series of transformations necessary for an agreement acceptable to all the protagonists, the innovation is stabilised and ready to enter the adoption network.

The concept of negotiation leads on to that of actor. An actor is any individual or collective entity who takes part in negotiations and contributes to reach a compromise. Each actor may be simply described in terms of his own view of the technical object to be put into circulation: a member of the marketing team may give more weight to the compatibility of a new model with previous ones, an engineering team may prefer a new microchip promoted by a firm in which they have confidence, the project manager requires that the conception will allow production automatisation, etc.

It is possible to say that the techniques and actors who participate in the conception process evolve together. Take two actors Ai and Aj negotiating technique Tk. They will (sooner or later) compel themselves to find a compromise between their preferred options. This will lead to a redefinition of the Tk whose characteristics will be modified as negotiations progress. But Ai and Aj themselves will also develop because of the simple fact that they are involved in a search for a compromise. Their conceptions, interests, and projects will change as they agree to abandon some of their initial demands to take into consideration those of the other actor. Once agreement is reached, the identities of Ai and Aj are transformed. New actors are created while a new Tk is being defined. This is why I use the idea of the co-evolution of techniques and actors.

The nature of the dynamic which ends with a satisfactory compromise, obviously depends on the distinction established at the beginning of the process between the techniques that are the subject of negotiation and the actors who negotiate. It would be natural to consider that "technical objects" are of necessity material in nature and that actors are human beings, whether individual or collective. This is indeed often the case, but this rule is subject to exceptions, and increasingly so. Techniques are often hybrid encompassing material objects and human beings (for example, an automated factory, a nuclear reactor, etc.) and the subject of design can include the operators, the turbines, the circulating fluids, and the ever moving electrons, etc. The hybrid nature of technical objects is mirrored by the fact the population of potential actors is not limited to human beings. Distributed artificial intelligence is the best example of communities in which robots and expert systems, assisted by human operators, negotiate the production schedule of a new product. In some cases, the separation between what *is negotiated* and *what negotiates*, between the intermediary and the actor, is generally accepted as ambiguous (Callon 1992) and conventions (laws, rules, customs) are introduced to clarify the distinctions. However, nothing may be considered as definitively stable, and ambiguous situations are multiplying under pressure from science and technology. For example, when is an embryo recognised as an actor with rights of expression (through an intermediate spokesperson) rather than a thing on which any experiment or transformation can be conducted? Answers to this type of question are far from fixed and change with the networks within which they are formulated.

Apart from the conventions that delimit the boundaries between actors and techniques, the composition of the set of actors and techniques which are involved at a given time is the second variable which affects the dynamic and the outcome of negotiations. According to the identity of the protagonists and the nature of the technical objects, the end product and its diffusion space may be significantly different.

Socio-Technical Networks: The Negotiation Stakes
There exist sequences during which the conventions governing the division between actors and techniques are stable. In addition, at the beginning of a sequence the list of actors and techniques involved is known (even if, as will be shown, it can evolve during the sequence). Once the identity of the negotiating actors and the techniques to be developed has been defined, it needs to be determined what is at stake in the negotiations. In this context, I would like to recall a crucial finding of innovation sociology (Akrich 1992; Callon 1987; Latour 1988; Law 1991; Rabeharisoa 1992).

Take a technical artefact Tk which is the subject of negotiation by various actors in the conception network. It is possible to show that for each Tk there is a corresponding socio-technical network which describes the world that this technical artefact structures and mobilises. A simple example will illustrate this point. Take an every-day device like an electronic entry barrier to a Paris metro station. This presupposes a human user in possession of a valid ticket sufficiently competent to understand that he should insert the ticket in a slot designed for this and that the ticket should be recuperated, if it is reusable, and who will not panic if the ticket does not re-appear. The most recent versions of this device extend their user model to include a fraudulent and agile user who can jump over the barrier without paying, but will then run into the subsequent gate which has been added to make his life more difficult. The device also presupposes transport tickets with a magnetically recorded strip, a central computer capable of reading this information and identifying and classifying the user (valid or not) and subsequently unlocking the gate, etc, etc...: the description could be developed *ad infinitum*. The entry-barrier-at-the-entrance-to-metro-stations device organises a network of relationships among entities, some human, others technical, of which the roles and links are perfectly defined and are to some degree structured by the device. The consequences of this observation are important for the analysis of the conception process. What is negotiated among actors is not only the characteristics of Tk (e.g. the entry barrier), but also at the same time the socio-technical network linked to Tk. There were deep divergences when defining the project description between engineers at Thomson and those of the sub-contractors and between the design office and the RATP (Paris Public Transport Authority) marketing department. These were not solely concerned with the technical characteristics of the barrier (its size, the additional gate, the speed that the electronic strip was read, the casing which protects the barrier from possible vandalism, etc.). They also involved various definitions of complementary techniques T^{ck} (the design of the ticket, the control algorithm, etc.) but also, and above all, the roles given to the passengers and their ultimate division into distinct sub-populations, e.g. would it be necessary to adopt the entry barriers for disabled travellers or those laden down with heavy luggage. The conceivers of techniques became technicians, sociologists, economists, moralists as they traversed the socio-technical network linked to the object being conceived. Controversies arose on all these issues. In discussing the form of the barrier the various departments of the RATP and Thomson discussed the content and morphology of the socio-technical networks to be built. When compromise was reached, in fact they were agreeing on a particular socio-technical network, which is often the result of tinkering with networks stoutly defended by the various actors (Figure 1). The characteristics of the potential adoption actors and users can be determined in relation to this outcome, and also those of the producers and distributors, etc.

Recruitment
The drawing up of the list of actors and techniques involved depends not only on the contents of the conventions already discussed, but also on the organisational procedures and structures which determine the direction that negotiations concerning techniques should take, for example, concerning the participants, their batting order, their prerogatives, etc. The participating actors may belong to a single organisation, may or may not come from different departments or divisions, or be recruited from different organisations (public laboratories, service firms, lead users, ministries, banks, capital risk companies). The form of relationships and their development over time may range from simple bilateral links, moving relentlessly forward, to ongoing multidirectional interactions. All these rules (formal and informal) have been discussed by sociologists and more recently by the economic theory of organisations. The rules contribute forcefully to the definition of the actors, direct their interventions and co-ordinate their actions (Aoki 1990; Eymard Duvernay 1989; Gaffard 1990).

But recruitment goes beyond the entities physically present during negotiation. To show this, I introduce the concept of representing a network. Each actor Ai exerts force in the negotiation process by positioning itself as a representation of a network, to be denoted as $Ri(Ai)$. Such positioning implies a definition of the actor's own identity, and of the entities Aj brought together by the network. In the case of metro entry barrier, actor Thomson will refer back to its technical and organisational make-up, and a spokesperson for users can refer to her knowledge of (i.e. being representative of) users' habits and preferences.

Figure I: Some elements of the socio-technical network of the entry barrier

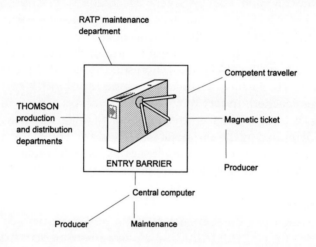

Such representation/identification is itself the outcome of a process, and need not be agreed upon. Entities Aj represented in Ai's version of the network may have a different view, which can be denoted as $Ri(Aj)$. Different divisions within Thomson might well emphasise different aspects of the role of the firm in the Paris metro. When the networks $Ri(Aj)$, $Ri(Ak)$, etc. are similar, it is easier for Ai to position itself as spokesperson for Ri than when there is dissimilarity. In the case of adopters and users, structural equivalence measures can be used in a straightforward way to quantify a similarity measure (Burt 1987). In other cases like Thomson's, it is possible to create a similarity measure by comparing narrative versions of $Ri(Aj)$, $Ri(Ak)$ etc.

The positioning of Ai can be contested during the negotiation process, and Ai may then have to bring in explicitly the entities making up his network, e.g. additional expertise, connections with other technical objects, with suppliers, with ministries. The possibility of doing so is what determines the force of Ai (provided there is enough similarity in network Ri already or can be created). In this way, recruitment of entities outside the immediate scope of the conception process can be analysed.

Another route is anticipatory, when an actor adds new entities to the list, e.g. more sophisticated or more fraudulent users, to support his case for a particular form of the socio-technical network-to-be. This then creates a dependency on these new entities, i.e. a new link in the network, because they should behave (or be made to behave) as set out in the projections. Again, if the new entities have a network definition similar to the projected one (or can be persuaded to have such a one), it becomes relatively easy for the general actor to have them follow the scenario.

The Dynamic of Convergence and Divergence
The conception network's dynamic can be analysed in terms of increases and reductions in convergence. A network is convergent if the three following conditions hold: (a) there is agreement on the initial distinction between actors and techniques; (b) there is agreement on the list of actors and techniques involved in the negotiations; (c) there is agreement on the description of the techniques being conceived, i.e. on the description of the related socio-technical networks. When these agreements are not reached, the conception network is divergent. The concept of agreement was used above, and it was observed that it could be applied to the identification of actors to be recruited (through conventions and organisational rules) but also to the definition of networks that the actors inscribed in the techniques that they wished to negotiate. Thus four configurations can be identified (Figure 2).

Case 1 corresponds to perfect agreement. The consensus applies both to the expected socio-technical networks and the list of participating actors (as well as their rights and duties). The conception network which results is convergent. The networks inscribed in the techniques undergoing conception, i.e. the techniques themselves, are identical. Descriptions no longer give rise to debate or controversy, nor does the list or role of the actors involved in the development of the techniques concerned.

Case 4 is that of total disagreement concerning both the definition of the techniques and the expected socio-technical networks related to them, as well as the list of actors involved and their rights and duties.

Cases 2 and 3 are partial disagreements which give rise to differences of opinion.

Figure II: Conception network dynamics

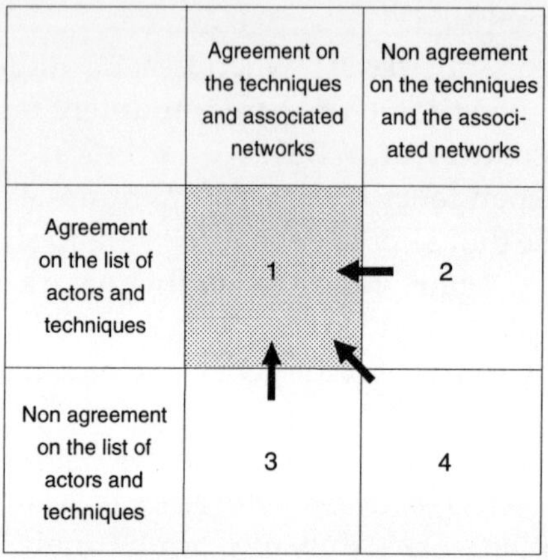

	Agreement on the techniques and associated networks	Non agreement on the techniques and the associated networks
Agreement on the list of actors and techniques	1	2
Non agreement on the list of actors and techniques	3	4

From the above, it can be said that the degree of convergence or divergence is the (always provisional) result of a process which leads to agreement or disagreement on the description of the anticipated network and the actors involved (or involvable). If the starting point is complete disagreement, agreement can only be reached at the end of a negotiation process during which both the definition of techniques and the list of actors eligible to negotiate have been adjusted and modified. To pass from divergence to convergence mutual adaptation occurs. The dynamics of these adaptations depend, to a large extent, on the forms of organisation and the co-ordination rules which prevail in the conception networks, i.e. who negotiates with whom in what order and according to what sequences. The distribution of competence is also crucial: who are the actors and what are their competences, can they be substituted, and to what degree, by other actors through progressively delving more deeply into their associated similarity network. The organisation of these interactions is obviously essential, because the contents of the agreement in particular, i.e. the characteristics of the techniques, depend on their dynamic as does the determination of the adoption network (as will be shown). The network designs itself, as much as it designs technologies (in the sense of artefact). It is enough to change the list of actors authorised to negotiate, the order of their intervention, the morphology of the interactions and also the means by which "represented" actors may be involved (by exploration of the similarity networks) for other techniques to be developed. The dynamic of the conception network corresponds to what economists call "learning." In the present formulation, it is also clear that there are opportunities for interventions as envisaged by CTA. It should be also emphasised that the achievement of workable agreement represents a cost which is not recoverable before subsequent stages come into action.

Conception-Adoption Networks

Adoption
The conception process ends when network convergence is achieved. The adoption phase then begins which goes hand-in-hand with manufacture and distribution, etc. When convergence is obtained within the conception network, agreement is reached on technique Tk. As has been stressed, the agreement is not on the technique in the strict sense. It also applies to the networks inscribed in the technique which will often have been the subject of long negotiations in the conception phase. The identity of the adopters (or more generally, users), the nature of the manufacturing process, the role of the marketing team, and the task of the suppliers, etc., has also been discussed (sometimes, heatedly). The reaching of agreement on the definition of adopters (who they are and their requirements) presupposes the involvement of "spokespersons" for the adopters in the conception process (using the full range of means available to make known the potential users' requirements: market studies, panels, experiments, etc. (Akrich 1992)), and, in certain cases these spokespersons will themselves be the first adopters. What is true for adopters also holds for all other actors in the conception network, some of whom will be progressively mobilised as production increases. Figure 3 shows the conception network and adoption network together with the related sub-contract, manufacturing, research and funding networks, as these are linked up in the sociotechnical network Tk.

When Tk is fixed, the first adopter can be identified at the moment the adoption process begins. In this model the identification of the first adopter is not an exogenous "small event" (Arthur 1989) but the endogenous result of the conception process (Mangematin and Callon 1991). By looking at their similarity networks, the potential adopter population and the morphology of the interdependencies which link them can be determined. This can reduce to the diffusion model which describes the adoption of stable techniques. In general, production, sub-contract and funding networks are also among the networks activated in the second phase. Once the adoption process begins, suppliers supply, producers produce and financiers finance. In this model, supply is as important as demand and Tk's adoption network is matched by the financial and production networks of Tk. The final network operates in several directions simultaneously and, in order to win on the adoption front, must mobilise on the manufacturing and sub-contract front. Again, costs must be attributed to these activities on these several fronts.

Figure III: The negotiation of techniques influences the shape of the concerned networks

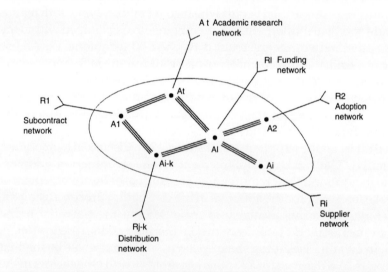

It is also possible to specify situations of technological competition (model 2). It is enough to imagine that other conception networks exist developing Tj which are linked to adoption networks which overlap (partially or more rarely totally) Tk's adoption network. Adoption and competition of Tk and Tj are then described in terms of the techniques (which are stable), the first adopters and their similarity networks .

The Dynamic of Conception-Adoption Networks

The dynamic of the conception/adoption of a technique is a two-stage process with alternating conception and production phases in the strict sense of these terms. First, the techniques take shape and the actors who will participate in their production and diffusion are defined. In the second phase, which is in a manner of speaking the result of the work concluded in the conception networks, a dynamic develops which is that of adoption-conception. It is during conception, i.e. within the network responsible for conception, that the basis for recruitment in the adoption stage is laid. The population of those who will adopt depends on the actors involved, their interests and those they represent and will be determined by the technical choices made. During the conception phase the first adopters are negotiated at the same time as the techniques destined for them. It is on this basis that the adoption networks that will be mobilised emerge. The trajectory of these negotiations determines the starting point of the adoption trajectory. Technology strategy (the conception of Tj) and commercial strategy (identification of the first users who by the interplay of their similarity networks set off the adoption process and the ensuing events) are closely implicated.

Another distinction is useful to describe the trajectory. It is based on the respective duration of the conception and adoption phases. Two extreme configurations can be identified. In the first case, the conception phase lasts a long time and is followed by a short adoption phase: the technique is evolving and probably compatible with small-scale manufacturing. In the other case, the adoption phase lasts a long time in relation to the (re-)conception phase. Techniques are stable and standard. This is compatible with mass production.

A conception-adoption sequence is referred to as the succession of these two phases, and is termed 's'. The sequence begins with the convergence of the conception network which as a result of multiple negotiations stabilises Tj and Ai. This implies drawing in the similarity networks of which the various Ai involved in the conception are the representatives, and continues with the progressive creation and articulation of adoption and production networks.

$Rs(t)$, the socio-technical network which is available at the end of the sequence s, contains the conception, production and adoption network simultaneously. $Rs(t)$ brings together entities which existed in a different form, or which had weaker and different ties, at the beginning of the sequence.

A conception phase always inherits networks formed in former sequences, and in a later sequence the starting point is $Rs(t)$. This means that the conception feeds off pre-formed networks in which techniques and diverse actors (producers, adopters, financiers, marketing personnel and researchers, etc.) are tied together and links may well be stable. In the sequence s_t, the number of actors involved will grow from conception to adoption. But each mobilisation of a new actor (adopter, banker, sub-contractor) opens onto new networks. If the process stops, it is only because the stage has been reached where the probability of entry into what will be Rs(t) falls below the threshold and is too weak to be significant (or to take an economic metaphor: the entry costs are higher than the value of expected benefits). The wider the network extends the more it multiplies the possibilities of connection with other similarity networks. True boundary objects (Hennion 1991; Star and Griesemer 1989) appear which while, of course, belonging to Rs(t) also appear in other similarity networks, distinct from those mobilised by Rs(t). A network profoundly different from that which launched the sequence can, thus, be established. The list of Ai who compose Rs(t) (which is different from the conception network existing at the beginning of the sequence as it has been enlarged by all the actors and all the techniques recruited during the adoption and production phases) forms a reserve of potential participants for the conception network of the following sequence s_{t+1}. In the new sequence the lead users, who participate in the conception network, will be, for example, selected from the last adopters of the preceding sequence; researchers used at the end of the sequence (of whom the similarity networks may be different from those of the beginning of the conception phase), will be chosen as actors in the new conception network. This gives a picture of continuity as well as a transformation principle which rests on the logic of the similarity networks and the permanent space and time disparities involved in them.

The second essential element in the transformation from Rs(t) to Rs(t+1) is linked to the conventions which define access conditions to become an actor or to be a technique. Between t and t+1, it is possible to imagine that these rules can change. Variations in conventions are becoming more common with growing concerns about the environment: hence chlorofluorocarbons were suddenly withdrawn form the list of substances which could be used in the conception of an aerosol. If Rs(t) uses embryos as the material for the development of a medical technique Tk, and if after Tk's adoption phase, the law forbids the use of these materials (which have been given the status of human dignity) the transformation onto Rs(t+1) will be profoundly altered. It is also possible that medical experiments on human beings, forbidden in the past, could be authorised at a later date. There is the bizarre case where chimpanzees were used to test certain pharmaceutical products. It was judged that they were not suitably representative of humans and were replaced, summarily, in derogation of the law by groups of voluntary patients supervised by a panel of experts from various scientific organisations. The chimpanzee, once a legitimate subject for manipulation, was replaced by a group of patients and the researcher-actor by a group of dissenting researchers and experts (Bibard 1991). In any event, the list of actors available to participate in the following conception phase will be modified and it will be necessary to take account of this in explaining $Rs(t+1)$.

Irreversibility and Variety

If the technico-economic trajectory is the series of transformations, $Rs(t)$, $Rs(t+1)$, ..., it is easy to understand under what conditions the trajectory becomes irreversible and produces, or does not produce, variety. In discussing irreversibility, it is sufficient to look at two extreme situations.

The first is the case where $Rs(t+1)$ re-injects into the conception process all the techniques and all the actors present in the conception phase of s_t while adding new actors recruited during the s_t adoption, which opens up possible new adoption networks. Thus, learning is accumulated and networks are extended without loss. Such situations are more common if recruitment has taken place in strongly linked similarity networks. In this case, the latest recruits are quasi-substitutable for the earlier ones. This type of development becomes more and more probable as the sequences succeed one another. As the number of conception-production-adoption sequences increases, the networks $Rs(t)$, $Rs(t+1)$, etc., become more and more connected and stable. Internal links reinforce one another to the detriment of external links. After numerous sequences, the probability of opening up new networks is low, because each actor has become more and more like the other actors in the network and less like the actors who are not in the network but with whom connections remain. The boundaries between the network and its "environment" are reinforced and the network breaks away from its context. Such a situation is usually accompanied by the production of norms and standards which align the behaviour of actors and techniques. The various elements of the network are equivalent and each speaks for the others. Such a development corresponds to a trajectory which is becoming irreversible.

The other situation is that which corresponds to a transition of $Rs(t)$ to $Rs(t+1)$ in which actors and techniques recruited for the new phase of conception are different from those of the conception phase of the previous sequence and belong to the population of s_t's last recruits. As a consequence they may not be too similar to the actors who were involved in the previous phase. Rs(t+1) will have a composition, both of actors and techniques, which places it at some distance from Rs(t). This may happen again in the transition to Rs(t+2) so that there is no convergence over time and very limited irreversibility.

In the case of irreversible trajectories the forms of techniques developed during the successive sequences are relatively stable, whereas in the case of open trajectories they can vary significantly from one sequence to the next.

The fact that a sequence of networks becomes caught up in one trajectory rather than another depends on numerous factors. One of them is the morphology of the networks of the adopters.[3] If they are weakly connected, have long chains of linkages and with regularly falling similarity indices, then open trajectories will probably result. If, on the contrary, they are strongly connected and similarity indices show marked discontinuities (segmentation), an irreversible trajectory will be followed. However neither of the outcomes is automatic nor definitive.

Another determinant of the trajectory stems from the nature and form of the co-ordination mechanisms which play an important role in the recruitment of the conception network's elements in s_{t+1}. If the actors forming the previous conception network are linked by strong bonds their involvement is inevitable, and if the equipment, instrumentation or other resources have not been depreciated or replaced, they form assets which must necessarily be exploited. The trajectory will rapidly become irreversible. If, on the other hand, the actors in the conception network of s_t are co-ordinated through markets or quasi-markets, dissociations and new associations are easier. Simulations can establish the probability of trends in relation to the nature of former transformations.

The generation and maintenance of technical variety can now be addressed in a rigorous manner by giving a dynamic interpretation of technological competition.

Situations of technological competition relate to different configurations. For competition to exist, it is sufficient that adoption networks overlap. But if other actors and/or techniques (researchers, designers, production engineers, experimental techniques, means of production, etc.) in the two competing conception networks are profoundly different from one another, then techniques will not only be dissimilar but also the systems of production, sub-contracting, university co-operation and funding networks will have little in common. Rj(t) and Rk(t) do not share anything other than the same market segment. If, on the other hand, the similarities go so far as to encompass the research actors, the financial circuits, the sub-contractors, etc., the configuration is one of firms or groups who are not only targeting the same users, so all the intermediary configurations of technological competition are possible. To analyse competition, it is thus necessary to change the perspective to a focus on conception networks and the interlinkages which make them more or less dependent.

Similarity of adopters may be the consequence of a multiplicity of factors:

> Strong connectivity of adoption networks which gives rise to a high probability of strong similarity among their spokespersons within the various conception networks (this configuration could be the inheritance of technological competition between highly irreversible trajectories).
>
> Strong links between the conception networks (recourse to identical scientific specialisations, mobilisation of the same research teams, sub-contracting to the same contract research organisations, using the same consultancy partnership, dependence on the same information sources and studies, participation in the same public programmes, joint distribution networks, etc.).

The transition from a competitive to a non-competitive situation occurs when the networks of adopters selected by the conception networks have been (re-)defined in a sufficiently radical manner so as to find themselves without anything in common. Such divergence develops generally over several adoption phases. The development of competition to a non-competitive situation, or vice-versa will be more probable if the trajectories associated with Tj and Tk are themselves open trajectories. In the early sequences, the switch is possible. After many adoption sequences, it is more difficult to escape the situation (of competition or non-competition) which has grown.

Conclusions

The above presentation integrates research findings produced by a variety of disciplines and approaches. It might have been thought that actor-network theory and technological diffusion theory were totally incompatible; that a reconciliation between the technological competition model and an analysis in terms of trajectories was unlikely to succeed. The broader of the models — that of the actor-network — takes on board the design phases (generation of technological variety) in its analysis of the adoption process, creating a unified analytical framework within which specific situations can be specified. The network concept enables us to focus on the recruitment process; it provides the means of following the formation of links between the individual options of each actor and the structural effects of aggregating previous decisions. The actor-network model affirms the importance of the option taken by the first adopters. This option is the outcome of bargaining within the conception network, leading to the identification of the networks of similarities which will subsequently support the diffusion process. The conception networks are also responsible for initiating a situation of competition (or of non-competition) between the technologies in presence.

For CTA, it is important that the actor-network model makes visible all the actors taking part in the development and diffusion processes. Technical options are no longer the exclusive prerogative of the specialists (technicians, scientists or engineers): the marketing executive, the sales agent, the banker and the lead-user are just as active as these when it comes to building up the range of options — as indeed are all the subsequent users making their choices. A collective actor, the actor-network made up of a host of individual actors, comes onto the stage. The dynamics of the actor-network also explain the generation of irreversible situations which illustrate the high degree to which the options taken by the adopters are predetermined: the play between alternating conception-adoption phases, and the types of trajectories followed, are factors to be considered when attempting to explain why and how certain irreversibilities come into being and others are resolved. In the same way, the persistence or creation, or the disappearance, of situations of open competition (i.e. where a range of alternative options is available) have their origins either in the dynamics of the conception networks or in the form taken on by the adoption process.

From the standpoint of a CTA exercise, several lessons can be drawn from the analysis.

One - the importance of the design phase is confirmed:

(i) The shape of the technology and its area of diffusion depend to a great extent on the options of the actors and intermediaries making up the conception network. By the introduction or elimination of a participant, and/or changed attributions of authorship, there can be produced a transformation both of the technical characteristics of the end-product and of the user population.

(ii) The importance of the role of the forms of coordination which govern the negotiations within the conception network is also emphasised. Two networks composed of the same entities, but where their interactions are organised differently, will not give birth to the same technologies or to the same potential adopter populations.

(iii) The degree of variety of the technologies proposed depends on the number of competing design networks and, even more, on the diversity of their composition and on any connections which exist between them.

Two - the distinction made between "technical objects," or better, intermediaries, and actors is a fundamental one, since it draws the borderline between the actants authorised to take options and engage in bargaining procedures, and those destined to play a merely passive role. An actor has a margin of initiative which the intermediary is not allowed. However, the attribution of intermediary status to some particular actant may be of great consequence for the balance of power within the whole network because it will then disappear from the explicit negotiation process. Take, for example, a routine procedure for granting or refusing bank loans to private individuals (such as by applying "scoring" algorithms in an expert system). The decisions made by this system may be accepted without discussion by the banker's agent, leading to the refusal of loan applications from customers not deemed to be credit-worthy. Customers who protest will then be told that the bank has total confidence in its routine system, which blindly applies the instructions fed into it. The user is unable to discuss a decision which has been taken by this "blind" system but attributed to a faceless corporate entity which is the bank in question. The twofold operation which consists in reducing an actant to the status of an intermediary (the knowledge based system), while at the same time allocating that actant-intermediary to one particular actant-actor (the bank), can thus radically upset the balance of power within the network by awarding dominant positions to some actors in relation to the others. From the CTA standpoint, therefore, the way in which the intermediary and actor roles are distributed is of capital importance.

Three - the degrees of irreversibility and convergence depend entirely on the dynamics of the network considered. Identification of the range of options still open at a given time, and assessment of the possibilities of reverting to technologies neglected in the past, are functions that must be sustained by an analysis of the dynamics of the conception and adoption network. This draws attention to the importance of the phenomenon of diffusion as such. The form and development of the networks of similarities play an essential part in the build-up to irreversible situations, as well as in the maintenance or the decrease of technological variety.

Four - given the importance of network dynamics in relation to the decision-making process, it is essential that we possess means of observing the former that are of commensurate efficiency. In this respect, the model-based strategy we advocate, with the possibilities of simulation which it implies, is a valuable additional asset.

For example, it could be used to identify the range of alternative or complementary interventions which emerges as CTA focuses on the conception and adoption network. Questions are:

(i) How can the diversity of actors be maintained, and technological variety be preserved, during an adoption phase? The model provides some of the answers to such questions by defining the crucial factors: identification of the first adopters; introduction of compatibility (common standards) so as to prevent undesirable lock-in situations arising; shaping and transformation of the various networks of similarities (by either encouraging or discouraging particular patterns of association between actors, such as through appropriate tariff, fiscal or regulatory measures, or regional development or information policies).

(ii) The same question can be asked in relation to conception phases, but calls for different answers. Intervention may take the following forms:

introduction of actors to participate in the design process (assistance with hiring scientists and designers; support grants for industrial R&D; development of selected training areas; sub-contracting regulations; worker-participation schemes; regulations opening up access to design network positions for certain categories of actors, and so on);
the way in which the conception network and its negotiating procedures are organised: incentives for joint research programmes involving private firms and public bodies in the environment field, enforcement of official specifications and standards, etc.;
support for emergent conception networks, with a view to maintaining technological variety;
supervision of rules and agreements drawing the line between actors and intermediaries; the aim being to determine those allowed a margin of initiative versus those whose sole function is to comply, and also to influence the distribution of negotiating power as between different actors. Here, we encounter all the major ethical issues arising from the development of hybrid entities, half-way between the human and the non-human, spawned by computer science and biotechnology.

Five - and finally, the approach to CTA presented here creates continuity with regular economic analysis, which is essential to overcome the unproductive contrast between so-called economic forces and societal values to be implemented through CTA. In the present analysis, there has been no separate discussion of the value of, e.g., having lay actors involved in technological development. We have focused on the appropriateness and viability of interventions as these can be derived from a better understanding of the processes of socio-technical development.

Notes

1. Model 2, the more flexible of the two, accommodates only such transformations as do not jeopardise the principle of identical end-use specifications for different versions.
2. This presentation is somewhat stylised. For brevity, not all the hypotheses of translation sociology, on which the following is based, have been discussed here.
3. The morphology of the sub-contracting, funding, research, etc. networks should also be taken into account. For reasons of brevity, only adoption networks are discussed here.

References

Akrich, M. 1992. "The De-scription of Technical Objects," in: Bijker and Law (eds.) (1992), 205-240.

Aoki, M. 1990. "Toward an Economic Model of the Japanese Firm." *Journal of Economic Literature* 28 (March): 1-27.

Arthur, B. 1989. "Competing Technologies, Increasing Returns, and Lock-in by Historical Events." *The Economics Journal* 99 (March): 116-31.

Belt, H. van den, and A. Rip. 1987. "The Nelson-Winter-Dosi Model and Synthetic Dye Industry," in: Bijker, Hughes and Pinch (eds.) (1987), 135-158.

Bibard, L. 1991. *La place et le rôle des sciences dans les innovations techniques: quelques cas en biotechnologie.* Thèse de socio-économie. EHESS.

Bijker, W.E., T.P. Hughes and T. Pinch (eds.). 1987. *The Social Construction of Technological Systems.* Cambridge, Mass.: MIT Press.

Bijker, W.E. and J. Law (eds.). 1992. *Shaping Technology/Building Society: Studies in Sociotechnical Change.* Cambridge, Mass.: MIT Press.

Burt, R. 1987. "Social Contagion and Innovation: Cohesion Versus Structural Equivalence." *American Journal of Sociology* 92 (May): 1287-335.

Callon, M. 1986. "Some Elements of a Sociology of Translation: Domestication of the Scallops and the Fishermen of St. Brieuc Bay," in: Law (ed.) (1986a).

Callon. 1987. "Society in the Making: The Study of Technology as a Tool for Sociological Analysis," in: Bijker, Hughes and Pinch (eds.) (1987), 83-103.

Callon, M. (ed). 1989. *La science et ses réseaux.* La Découverte.

Callon, M. 1992. "The Dynamics of Techno-Economic Networks," in: R. Coombs, P. Saviotti and V. Walsh (eds.). *Technological Change and Company Strategies: Economic and Sociological Perspectives.* London: Academic Press.

Callon, M. 1993. *Modèles de simulations pour la dynamique des réseaux de conception et d'adoption.* Paris: Centre de Sociologie de l'Innovation.

Callon, M., J. Law and A. Rip (eds.). 1986. *Mapping the Dynamics of Science and Technology.* Macmillan.

David, P. 1986. "Understanding the Economics of QWERTY: the Necessity of History," in: W.N. Parker (ed.). *Economic History and the Modern Economist.* Oxford: Basil Blackwell.

David, P. 1987. "New Standards for the Economics of Standardization," in: P. Dasgupta and P. Stoneman (eds.). *Economic Theory and Technology Policy.* Cambridge University Press.

DeBresson, C., and F. Amesse. 1991. "Networks of Innovators: A Review and Introduction to the Issue." *Research Policy* 20 (5): 363-80.

Dosi, G., C. Freeman, R. Nelson, G. Silverberg and L. Soete. 1988. *Technical Change and Economic Theory.* London: Frances Pinter.

Eymard Duvernay, F. 1989. "Conventions de qualité et formes de coordination." *Revue économique* 40.

Freeman, C. 1991. "Networks of Innovators: A Synthesis of Research Issues." *Research Policy* 20 (5): 499-514.

Gaffard, J.L. 1989. "Marché et organisation dans les stratégies économiques des firmes industrielles." *Revue d'Economie Industrielle* 48.

Hennion, A. 1991. *La médiation musicale.* Thèse de Sociologie. EHESS.

Hughes, T. 1983. *Networks of Power: Electrification in Western Society 1880-1930.* Baltimore: Johns Hopkins University Press.

Imai, K., and Y. Baba. 1991. "Systemic Innovation and Cross-Border Networks: Transcending Markets and Hierarchies to Create New Techno-Economic Systems," in: *Technology and Productivity.* ACDE.

Kidder, T. 1981. *The Soul of a New Machine.* London: Allen Lane.

Latour, B. 1987. *Science in Action.* Cambridge, Mass.: Harvard University Press.

Latour, B. 1988. "Mixing Humans and Non-Humans Together: The Sociology of a Door-Closer." *Social Problems* 35 (June): 298-310.

Latour, B. 1993. *Aramis or the Love of Technology.* Cambridge, Mass.: Harvard University Press.

Law, J. (ed.). 1986a. *Power, Action, and Belief: a New Sociology of Knowledge.* Sociological Review Monograph 32. Routledge.

Law, J. 1986b. "On the Methods of Long Distance Control: Vessels, Navigation and the Portugese Route to India," in: J. Law (ed.) (1986).

Law, J. 1987. "Technology and Heterogeneous Engineering: The Case of the Portugese Expansion," in: Bijker, Hughes and Pinch (eds.) (1987), 111-134.

Law, J. (ed.). 1991. *A Sociology of Monsters*. Sociological Review Monograph 38. Routledge.

Law, J. 1992. "The Olympus 320 Engine: A Case Study in Design, Development and Organizational Control." *Technology and Culture* (July): 409-40.

MacKenzie, D. 1991. *Inventing Accuracy: A Historical Sociology of Nuclear Missile Guidance*. Cambridge, Mass.: MIT Press.

Mangematin, V., and M. Callon. 1995. "Technological Competition, Strategies of the Firms and the Choice of the First Users: The Case of Road Guidance Technologies." *Research Policy* (forthcoming).

Nelson, R., and S. Winter. 1982. *An Evolutionary Theory of Economic Change*. Cambridge, Mass.: Harvard University Press.

Powell, W.W. 1990. "Neither Market Nor Hierarchy: Network Forms of Organization." *Research in Organizational Behavior* 12.

Rabeharisoa, V. 1992. *Les indicateurs science-technique-marché*. Paris: Centre de Sociologie de l'Innovation. (PhD).

Rip, A., and H. van den Belt. 1988. *Constructive Technology Assessment: Toward a Theory*. Enschede: University of Twente.

Star, L., and J. Griesemer. 1989. "Institutional Ecology, 'Translations' and Boundary Objects: Amateurs and Professionals in Berkeley's Museum of Vertebrate Zoology, 1907-39." *Social Studies of Science* 19 (August): 387-420.

Thorelli, H.B. 1986. "Networks: Between Markets and Hierarchies." *Strategic Management Journal* 7 (January-February): 37-51.

15 Firm Strategies and Technical Choices

Rod Coombs

The proposition that it is possible to engage in Constructive Technology Assessment (CTA) rests on two fundamental assumptions. The first assumption is that the evolution of technology is neither an autonomous process driven by an inner scientific logic; nor a simple resultant of the operation of market mechanisms; but that there is a space (albeit limited) for social choices to play a part. The second assumption is that this opportunity for social choice can be the occasion for rational collective action based on dialogue and (temporary) consensus formation concerning appropriate normative stances towards technology choices.

In this chapter the first of these assumptions, which we will call "social construction" assumption, is the main focus of attention. The particular question to be addressed is how "social construction" (where "construction" is interpreted as both *active* construction and as construction in the sociological sense of "interpretation") interacts with the position and behaviour of business firms. We focus on the role of business firms because of their self-evident function as the dominant institutional form which organises and deploys technological resources.

Received Interventionist Policy Stances Toward Technology

In order to situate the discussion on CTA, it is useful to consider how stances toward social intervention in technological development have evolved in recent times. In particular, we should also consider how academic research on the mechanisms of technical change has influenced policy styles.

Looking back over the contribution of academic research to policy debates on technological change over the last 30 years or so it is possible to observe two main arenas of policy activity which can be broadly characterised as *promotion* and *control*. (Johnston and Gummett 1979; Coombs, Saviotti and Walsh 1987) The promotion arena centres on the problem of certain nation states or firms not having enough technology of the right kind, at the right time, in order to be economically competitive. The control arena centres on the problems created when technologies have effects, or potential effects, which are regarded as undesirable by some section of the community which is not directly able to influence the creation and use of the technology in question. In the terminology of evolutionary economics (Nelson and Winter 1982) these domains of promotion and control might be called the variety-generating system and the selection environment.

The policy measures which have evolved in the *promotion* arena have tended to be pitched at a fairly aggregate level; dealing essentially with externalities, distributional issues, and various perceived instances of market failure (in the area of basic scientific research for example). The predominant academic input to the policy discourse has come from economic analyses of technical change rather than sociological analyses.

In the *control* arena policy measures have tended to evolve in a pragmatic manner as responses to particular kinds of problem which acquire a high profile at certain times (obvious examples can be found in such areas as environmental pollution, work safety, product testing etc.) By comparison with the reasonably prominent role of economic arguments in the promotion domain, control policy discussions have not been notably influenced by economic arguments, or for that matter by any other social science discipline. Academic involvement with this arena has been restricted to commentary and analysis rather than direct input to the policy debates (risk analysis is perhaps an honourable exception).

The overall context for the development of policy statements on technological development has therefore been somewhat fragmented. Discussions on the control of technology tend to focus on the qualitative features of technology, and its supposed effects on the "quality of life." On the other hand, the promotion agenda is more concerned with the quantitative aspects of technological change and its supposed effects on economic growth. These two discourses have not been able to engage satisfactorily with each other, and have even been seen as to some extent pursuing conflicting objectives. Thus we frequently see specific measures for the control of technology criticised from a "promotion perspective" as having negative effects on investment and growth. Similarly, some promotion policies are criticised from a "control perspective" as being liable to give rise to technological development whose consequences are insufficiently understood and potentially dangerous.

In part, this conflict between promotion and control is an expression of a traditional tension between different political stances of researchers towards the basic features of contemporary market economies. Are such social and economic systems capable in principle of achieving a consensus balance between producing surplus and growth on the one hand, and on the other hand producing patterns of consumption which are consistent with broad political and social values relating to such issues as the environment, health etc.

New Theories Give Rise to New Policy Stances?

But this underlying tension is not the only reason for the failure of the quantitative and qualitative analyses of technology to engage with each other. The other reason lies in the intellectual frameworks which have been developed to understand the generation and economic use of technology. Economic and sociological analyses of technology have proceeded largely independently of each other for several decades. They have had different concerns, and conceptual frameworks, and looked at quite different aspects of technology. Recently however, we have witnessed the beginnings of a more fruitful dialogue between these two approaches. Specifically, the evolutionary economics paradigm has been built on a set of assumptions about firm behaviour which can be inter-related quite successfully with Callon and Latour's actor-network perspective on the creation of technologies.

This emergent synthesis is discussed elsewhere in this volume, (see also Coombs, Saviotti and Walsh 1992), and so can be briefly summarised here. Central concepts such as "decision routines," "technological paradigms" and "technological trajectories" have emerged in evolutionary economics which are used to describe patterns and stabilities in firm behaviour and technological change. In actor-network theory a parallel to routines is provided by the view of decision making as taking place in a context of stabilised networks of actors which socially construct relatively stable interpretations of localised "worlds." Specifically, the actor-network theory acts as a counter-weight to any technological determinist interpretation of such concepts as "technological trajectories" mentioned above. By constituting the trends in the evolution of technologies as themselves social institutions rather than "natural laws" it permits the routines and paradigms within firms to change by means other than a life-or-death survival process achieved through a selection environment.

There is then a degree of commonality amongst these conceptual approaches, in that they all imply a degree of fine structure in technological and organisational features of the world under study. This structure has a short to medium term stability, but it can eventually change by means of transitions to other, different states/ structures. Put at its starkest, we may suggest that where the evolutionary economist sees a stable natural trajectory, the sociologist sees a normalised irreversible techno-economic network: where the evolutionary economist sees a radical innovation, the sociologist sees ruptured networks and the emergence of new networks.

What then is the significance of these new theoretical developments for our policy stances? Can our emergent, more integrated understanding trigger a new era of more effective policy-making? This is the question to be posed next.

In fact, thus far the new intellectual frameworks for understanding technical change have not made a major impact on policy practices. Indeed it is worth asking why, given that we are developing a more and more subtle and convincing account of how technologies are generated and used, there is still a shortfall (or an "agency deficit") in the area of co-ordinated social policies to optimise the "effects" (a term which we can now no longer use so simply) of technology on society?

My view of the reason for the deficit between policy and understanding, put briefly, is this. Our instincts for policy generation, as described earlier in this chapter, are still split between promotion and control. The control agenda in particular is the one which is in need of radical re-thinking in the light of our new intellectual approaches. The received ideas on "control of technology" owe too much to a naive rationalist view of the possibility of controlling anything, let alone technical change, and also to a overly "modernist" hope of clearly defining the normative criteria that are involved in saying what might constitute "good" and "bad" ways of applying technology. By contrast, the network-based understanding of the creation and use of technology forces us to accept that the "interests" of social groups and individuals (from which we might derive normative criteria for CTA) are not completely static and defined by some over-arching social order, but are in fact continuously re-defined and re-constructed, in parallel with the creation and use of technology itself. This is not to reduce the importance of large scale features of the political and economic structure, but simply to assert that they are made present in concrete micro-situations not as fixed and unambiguous "constraints", but rather *through* the interpretations and discourses to be found amongst the actors in a particular network.

It follows from this that the careful analysis of networks, and their role in constituting technologies in society, is a potentially important contributor to the evolution and the outcomes of such networks. Indeed this will be all the more so to the extent that the analysis of the networks, and the increased transparency and visibility which that generates, is transmitted into any given network, or other networks which inter-connect with it. In the limit then, *membership of a network*, (however formal or informal, legitimate or illegitimate) will be the most potent means of having effects within that network, and therefore on the technology.

I would make one major qualification to the argument that active analysis and engagement with networks is the policy corollary of the new intellectual understanding of technical change. The qualification involves making a distinction, which is almost a normative one, between the two domains, identified by evolutionary economics, of *variety generation* and the *selection environment*. If we retain a belief in the "progressive" properties of an economic and social system which generates a high degree of technological variety (however problematic are the choices which variety then faces us with) then we have to be careful that we do not employ policy measures which reduce that variety too much. Indeed, the critique of the "old" control policies was exactly along these lines. As an example we might argue, perhaps provocatively, that the replacement of CFCs with less damaging chemicals depends at least as much on healthy mechanisms for generating technological variety as it does on a socially conditioned selection environment.

Perhaps the benefit of the new intellectual understanding is that it sensitises us to the interface between the variety generation process and the selection environment, and particularly draws our attention to the *internal* selection environment within the firm (the major source of technology). It is here, within the firm, that the networks which are analysed by sociologists of technology act as vectors to concentrate the effects of broader social networks on the emerging technologies and their properties. The challenge for us is to participate in those networks, within the institution of the firm, but from standpoints beyond the firm, in a way which harnesses but defends technological variety. If we accept this challenge, then an intrinsic part of the bargain is that there can be no comfortable certainties about our policies or our actions. Involvement in such networks produces unintentional consequences as well as (if we are lucky), intentional ones.

The Generation of Technology in Firms

The rationale for the project of CTA adopting a clear focus on business firms is clear, and rests on the following points:

> 1. Firms are the major sites of R&D; they are becoming more and more sophisticated in their planning and control of R&D, and R&D in firms is becoming more and more concentrated.
> 2. Firms are the "living proof" that technology does not simply "respond to market forces." Firms' use of technology is frequently designed to disrupt or create markets rather than to respond to them.
> 3. Firms are the source of "technological variety." This variety is not only a source of "problems" for CTA, but also a major source of "solutions."

We therefore need to explore in more detail the ways in which firms generate technology, and the recent changes in these mechanisms. Put another way; what are the networks which create technology in the firm? Answering this question will then enable us to explore the scope for CTA to develop as a social practice through a strategy of systematically amplifying the connections between two categories of social networks. These are firstly, those networks which articulate the perceived contributions of technologies and products to the profitability and survival of business firms, and secondly, those networks which articulate the evaluations of technologies and products made by citizens, consumer groups, and public bodies of various types.

Technology Networks: Trends in the Organisation of Firm R&D

Clearly the first task in this analysis is to review the recent developments in the organisation and management of R&D in firms. This is where a significant part of the intra-firm networks which create and evaluate technology are located. In the last 20 years there has been an extremely marked change in the positioning of R&D functions in large industrial firms, and a marked increase in the integration of technological concerns into the strategic management agendas and processes of those firms. Let us deal with these in turn.

In terms of positioning of R&D, three phases or models can be observed. (see Roussel et al. 1991 and Coombs and Richards 1992 for more detailed accounts.) In the first phase, which probably finished in the early 70s, R&D functions tended to be large concentrated monolithic affairs, funded from the corporate centre rather than by the business units, and with rather vague and long lines of communication to the operational and commercial management structures of business units. Investment in such units was in part an act of faith, premised on a tacit "technology push" theory of innovation. Cultural gaps between the concerns and practices of the R&D personnel and the commercial personnel were relatively large.

In the second phase, which is still not fully concluded, structures and procedures have been put in place which are almost the opposite of the first phase. Central labs have often been broken up or at least reduced in size. R&D units have tended to be located, funded, and controlled at the level of business units or divisions rather than at the corporate level. This approach has grown out of a belief that technology-push approaches had generated too much technology which was "not what the market wanted." The priority became one of making R&D more "market-driven," and ensuring that commercial managers had a direct involvement in decisions on the selection and management of technological projects. This followed through into changes in the culture, expectations and career paths of scientist and engineers employed in firms practising this style of management.

In the third phase, which is now beginning, many firms are beginning to have second thoughts about the severe shift toward market-driven R&D of the 80s. They do not wish to reverse it per se, rather they see a need to correct some new problems which emerge from the severity of the shift. Essentially the issue revolves around the loss of synergy which comes from decentralising technological development responsibility in a large diverse portfolio of business units. Whilst the business units are good at identifying the particular product strategies they wish to follow, and deploying technology to support those strategies, they often fail to use technologies which are being developed in other business units in the same corporate group. Furthermore they tend to under-invest in longer term technical effort; fail to take responsibility for seizing new business opportunities resulting from technical research if it appears to fall outside their current markets; and fail to take responsibility for injecting technological issues into the broader business and corporate strategy discussions taking place beyond the R&D function.

The response to these problems being adopted by a number of firms is to set up "Technology Strategy" units at corporate level, with the specific responsibility to counter all of these dangers. Thus they operate by *"mapping"* the technological competences available across all of the business units; *facilitating* transfer of such competences between business units; and *articulating* an account of the portfolio of technologies held by the firm which can play a part in strategic debates both at business unit and corporate level. Clearly the essence of this third phase is the desire to combine market driven *product development* with strategic control of *technological* competence.

Given that these changes are taking place, we can now identify a clear hierarchy of levels within the large diversified firm at which strategic discussions take place which have a technological component.

> 1. The corporate level: concerned with survival, overall profitability, "shape" (balance between divisions), risk, and finance.
> 2. The technology level: concerned with distinctive ("core") technological assets which may transcend business unit boundaries and contribute to the corporate rationale and shape.
> 3. The business level: concerned with the competitive strategy to be followed in a particular market or related group of markets.
> 4. The product level: concerned with the concrete offerings in a particular market or market segment, the performance levels and characteristics of these products, and the details of the articulation of the product with the changing patterns of demand.

Levels 1 and 3 often receive the most attention in discussions of firm strategies, but it is clear from what has been said so far that levels 2 and 4 may actually offer more favourable terrain for the mission of CTA.

In addition to these changes in the conduct of R&D in large firms, it is of course important to note the parallel rise in importance of small technology-based firms, particularly in areas of emerging technology such as biotechnology. These firms play an important role in shaping technology, both as free-standing firms, and as part of networks with larger firms involving complex patterns of collaboration (see Walsh 1992, for a recent analysis of this phenomenon).

The Social Construction of "Strategies"

Given that these different orders of strategic discourse exist, it is important to be clear about how we should understand the "strategies" articulated in these domains. The literature on strategies in organisations has now thoroughly broken with the rationalist, control-oriented "strategic planning paradigm" of the 60s and 70s. Since the work of Mintzberg (1978), Pettigrew (1987), and others, it has become conventional to recognise the emergent, processual and political character of strategic discussions within organisations. Strategies are at least as much devices for understanding and making sense of where one has been in the past as they are instruments for securing the future. Indeed more recent literature (Knights and Morgan 1990) has gone one step further and argued that the strategic stances adopted by particular subgroups of managers within an organisation can best be interpreted as components of power relationships between themselves and other managers, and between managers and subordinates. Essentially the "authorship" of strategic ideas becomes both a goal and a resource in organisational games and power relationships; success in such endeavours secures both organisational rewards and secures the personal identities of the individuals concerned.

It is clear that this independent trend in the treatment of "strategy" by organisation theorists has some remarkable overlaps and parallels with the Callon-Latour actor-network perspective which we have advocated using for analysis of the emergence and normalisation of technologies. Both approaches draw on the traditions of the social construction of reality. In each case, interpretations of reality strive to achieve the social status of "facts" through the means of alliances, persuasion, dissembling, exploitation of existing powerful actors and so on. In each case, the object of the network, whether it is a technology or a strategy, depends for its existence on the continuing operation and maintenance of the network; a circumstance which depends on the actors in the network, but also on the absence or failure of any "threat" from an alternative. From time to time such threats do emerge and succeed, and there are ruptures in strategies, in technologies, and in networks.

But for us the important point is not only that there is a parallel between the actor-network approach to analysing technology, and the new organisation theory approach to strategies. There is also, in the context of the business firm, a direct empirical connection between *technologies and strategies*, as has been outlined earlier. It follows therefore that the conclusion of this argument, is that it is possible to conceive of the technology and product domains of firm strategy making (levels 2 and 4 in the list above) as sites of actor-networks in Callon's sense. Furthermore these networks have a functional correspondence to decision rules or trajectories in Nelson and Winter's sense. They therefore represent *in principle* sites for the intervention of other actors, whose motivations and values, whilst not being fundamentally opposed to those of the firm, can introduce additional concerns. This then is where the other set of networks mentioned earlier: "those networks which articulate the evaluations of technologies and products made by citizens, consumer groups, and public bodies of various types," might perhaps be brought into engagement with the networks which produce technology in the firm. The question which arises from this observation is how these "public" networks are currently structured, what the possibilities are for them to effectively engage with the firm in this way, and what alternative new forms these networks might take in the future. This is the topic of the next section.

"Public" Networks

There are of course a wide variety of public networks already in existence. These take both official and unofficial forms, and are both formal and informal. Thus we have to recognise the significance of various government agencies for promotion and control; unofficial pressure groups (which deal mostly with control issues, but also sometimes with promotion); and the general expression of public opinion through the media. Examples of all of these types of network could include (in the UK and E.C. context): the Alvey and ESPRIT programmes for the promotion of information technology; the UK Committee on the Safety of Medicines; Greenpeace; The British Computer Society; or the rising public taste for "environment-friendly" products and practices. To a very large extent, all of these types of institution already exercise significant direct and indirect influence on firms' technology programmes; or, in the language of this chapter, these networks engage with the intra-firm networks. Is it not therefore adequate to simply allow this situation to incrementally intensify and become more and more sophisticated and effective? This after all is what has happened steadily over the last 100 years or more of industrial capitalism, beginning with the alkali acts and factory acts of 19th century Britain. Indeed the case for attempting to systematically exploit and develop the links between public networks and intra-firm networks as policy "style" has been made (see for example Schot 1992).

I suspect however, that it is not adequate to rely solely on these received approaches and institutions for the project of CTA. Part of the reasoning for this lies in the view, expressed earlier, that the distinction between promotion and control is no longer helpful and does not sit happily alongside our new intellectual framework for understanding technical change. Furthermore, the ways in which both promotion and control have been exercised have frequently not had a clear view of whether they were targeting the mechanisms for generating technological variety or targeting the selection environment, or both. In so far as they interfered with the variety mechanism there is reason to believe that it has been with, at best, mixed results.

Interfering with the variety mechanism — the way companies generate new technologies — has two major objections associated with it. Firstly, it is difficult to actually accomplish, because it touches on matters so sensitive and close to competitive interests that companies will always resist it. Secondly, it can be argued that it reduces the prospects for CTA, or indeed for any kind of technical and economic progress, because it changes the incentives to firms in such a way as to ultimately reduce the range of technical possibilities and its rate of expansion. Consequently it seems that the best way to strengthen the prospects for CTA is to look again at the networks which shape the selection environment, and at how they can be made to engage more effectively with the networks in the firms which evaluate technologies against firms' perceived interests.

CTA as "Public Procurement"

Imagine a number of large, independent research institutes; each of which is concerned with an area of final consumer demand, such as transport, food, communications, leisure, housing, healthcare, working environment, public services, domestic equipment etc. Each of these institutes could be funded by some combination of public taxation and corporate levy. Each could be staffed by researchers and policy analysts who would be partly recruited in the normal labour market, partly seconded from government departments, and partly seconded from private firms with an interest in the area of responsibility of the institute. Thus the "housing" institute would have a mixture of staff from (in the UK) the Department of Environment, firms with activities in building, building materials, architecture, domestic equipment, furniture, electricity, gas and water, communications etc., as well as "independent researchers."

The governance of the institute would have to also incorporate these private sector and government interests, but would have to be biassed consciously towards other public groups (which might include some of the conventional pressure groups mentioned earlier). This would be vital in order to avoid the institute being "captured" by industrial interests and replicating their own internal firm networks.

What would such institutes do? Their job would be to act as a forum in which debate, underpinned by research, would take place on the issue "what directions do we wish to see housing (or transport or food or whatever) evolve in?" They would have a duty not simply to incorporate the concerns of the relevant government agencies and firms, but to focus directly on the concerns, anxieties, desires and irritations of the vast bulk of consumers. In one sense then, they would be the biggest market research agencies in the history of capitalism! But they would differ from conventional market research in that they would not be working for one commercial interest, or even for all commercial interests, but for commercial interests *and* the "public interest" *and* with due regard for other aspects of public interest as expressed by government policies. These institutes would therefore act as vehicles for the articulation of demand in particular areas of final consumer products and services.

A number of things need to be said immediately about the attractions and limitations of this idea; which is *simply* an idea and not a fully fledged proposal.

1. The function of these bodies is *not* currently met by government departments of transport or housing etc. Their concerns are with how to act given existing constraints of public investment, legal frameworks, and short term political policies. This is not the same thing at all as the articulation of public concerns in a wider sense, although it *does* have a legitimate input to make to these debates.
2. Clearly there would be vast potential for disagreement, lobbying, special pleading and even dissembling and corruption in the interactions between interest groups within these bodies. This is not an argument against their existence (especially since it is already true of all the other bodies which engage in public and private policy making). The difference in this case is that the special interests would be visible, obvious, and transparent. To the extent that groups did no more than defend these interests they would discredit themselves and damage their credibility in the broader sense, and, more importantly, in the market place.
3. In previous decades the system would have had a limited effect on large swathes of industrial technology generation and use because raw materials, components and even sub-assemblies were technically shaped by suppliers rather than by customers. Now the situation is reversed, and final consumer applications "pull through" much of the detail of the technology supplied by firms further back in the supply chain. This amplifies the potential power over technology of interventions directed at final consumer applications.

These bodies would be more legitimate, powerful, and professional than pressure groups. But they would be less constrained by expediency and politics than government bodies. They would have the capability and reach to act like a "proxy procurement agency" for the public in their respective areas.

It is important however to distinguish between two aspects of procurement, namely the "product-specifying function" and the actual *purchasing* function. In the case of these proposed institutes we are focusing only on the product specifying function. The institutes would in effect map out the functional performance specifications which captured the direction of change of public requirements, whilst not specifying the technical means to achieve those requirements, and leaving that task to the variety generating mechanisms of firms (both individually and collectively). Thus they would differ from real procurement agencies in that they would *not actually* make the purchasing decisions. Therefore they would not interfere in the market mechanism directly.

In some senses we can already see some of the principles of this idea paralleled in current developments in the organisation of public services in the UK and other countries. In the UK for example the National Health Service is currently being restructured into purchasing agencies and provider bodies. Agencies which formerly had the responsibility both to fund healthcare *and* manage the hospitals which provided the care are now being divided. Some agencies purchase care, and can purchase from a variety of provider bodies. Providers can sell care to a number of purchasers. Leaving aside the overall merits of such market mechanisms in the specific area of healthcare it can nevertheless readily be seen that the separation and refinement of the purchasing function creates very interesting and new possibilities for the conduct of the relationships between the healthcare sector and suppliers of equipment.

Similarly, in the defence field, the procurement and purchasing function has been considerably developed and professionalised in recent years. Though not without its shortfalls, this area too provides some lessons for how the technical decisions of firms can be influenced from an arms length demand-side position.

Neither the defence example nor the health example map directly onto the main idea being put forward here, since they actually combine the product specifying function with the purchasing function. Nevertheless they have some interesting lessons. In the institutes we are proposing however, it is the product specifying function which is to the fore. If their position, structure, and legitimacy could be correctly balanced with the variety of other public and private bodies with which they would have to interact, then they could provide an exciting and dynamic forum in which constructive detailed appraisal of technologies, products and services could be conducted. Such appraisal would not be defensive reactive regulation to fight fires and ameliorate disasters; it would be prospective identification of public demand in collaboration with firms able to benefit from the information, but with a firm commitment to the maintenance of competition between the firms in the actual products offered to meet that demand.

Conclusion

This chapter has argued that the new frameworks for understanding technical change both demand and stimulate new policy stances. Through a consideration of changes in the sophistication of technology management in firms we have been led to the view that there is a way forward for the idea of CTA as an intervention in the selection environment which preserves the potency of the variety generating mechanism. Thus CTA could develop as a social practice through a strategy of systematically amplifying the connections between two categories of social networks. These are firstly, those networks which articulate the perceived contributions of technologies and products to the profitability and survival of business firms, and secondly, those networks which articulate the evaluations of technologies and products made by citizens, consumer groups, and public bodies of various types. The institutes for the articulation of demand, sketched in the previous section, are a tentative idea as to how this principle might be brought closer to practical reality.

Acknowledgements

1. Some parts of this paper are derived from a paper given at the Conference on Trends in Methodologies in Social Science held at Houthem, the Netherlands, in April 1992. The paper and the conference were supported by the ESRC PICT Programme.
2. I am grateful to my colleague Ken Green for helpful discussions and suggestions which shaped the ideas in the last section of this paper. Deficiencies in the ideas remain my responsibility.

References

Coombs, R., and A. Richards. 1992. "Strategic Control of Technology in Diversified Companies with Decentralised R&D." Paper to the Annual Conference of the British Academy of Management.

Coombs, R., P. Saviotti and V. Walsh. 1987. *Economics and Technological Change*. Macmillan.

Coombs, R., P. Saviotti and V. Walsh. (eds.). 1992. *Technological Change and Company Strategies*. London: Academic Press.

Johnston, R. and P. Gummett. (eds.). 1979. *Directing Technology*. London: Croom Helm.

Knights, D. and G. Morgan 1990. "Corporate Strategy, Organisations and Subjectivity: A Critique." *Organisation Studies* 12 (2): 251-73.

Mintzberg, H. 1978. "Patterns in Strategy Formation." *Management Science* 14: 934-948.

Nelson, R., and S. Winter. 1982. *An Evolutionary Theory of Economic Change*. Cambridge, Mass.: Harvard University Press.

Pettigrew, A. 1987. "Context and Action in the Transformation of the Firm," *Journal of Management Studies* 24(6): 649-669.

Roussel, P., K. Saad and T. Erickson. 1991. *Third Generation R&D*. Boston, Mass.: Harvard Business School Press.

Schot, J. 1992. "The Policy Relevance of the Evolutionary Model: The Case of Stimulating Clean Technologies," in: Coombs, Saviotti and Walsh (eds.) op. cit.

Walsh, V. 1992. "Firm Strategies, Globalisation and New Technological Paradigms: The Case of Biotechnology." Paper given to a colloquium on "Coping with Globalisation." St. Malo, June 1992.

Epilogue

Arie Rip, Thomas J. Misa and Johan Schot

Constructive Technology Assessment identifies the present sequential approach of optimizing the technical before considering uptake, use and effects, as the key area to improve the management of technology in society. Broadening design criteria and/or broadening the kind of actors involved is important to overcome the evident limitations of technology-focussed design. With the increasing recognition of this basic point, and given some exemplary achievements, we can speak of a paradigm of Constructive TA.

The paradigma of CTA is still under construction and a lot more work needs to be done: development of specific tools, analysis of cases, designing of experiments and other policies. With all its reliance on the new perspectives of technology studies (and other areas like sociology, political science and cultural studies) the spirit of CTA is to avoid the temptation to find a substitute for action in thinking and epistemological virtuosity.

Thus we offer five clusters of suggestions for doing CTA. These suggestions are a mixture of perspectives, strategies and tools. Their formulation is rather broad, but is unavoidable since tools for doing CTA will take a different shape for different actors and situations. For industrial firms, for example, CTA can be seen as an extension of concurrent engineering, to cover adoption, diffusion and impacts as well. For governments, on the other hand, it could be a matter of creating the right conditions for so-called social experiments with the introduction of new technology.

Suggestion 1:
Sociotechnical foresight exercises and mapping tools, rather than trend extrapolations should be used.

A key element of CTA, given its upstream focus, is that anticipation of future developments and impacts is a task for all actors. The contingencies of technological development and its embedding in society make the development and application of simple tools fraught with risk. Linear extrapolations, and model-based extrapolations like S-curves are of little use because of the non-linear character of developments. There is branching and cross-linking, there are path dependencies, and there are the effects of sociotechnical linkages between technologies (e.g. the emergence of an all-electric house). Technological development is not continuous progress along dimensions of increasing functionality. On the contrary it is more like a patchwork quilt, or if one prefers a different metaphor, the way yeast cells grow: developments branch off in different directions, cross-connections and interactions occur. The eventual shape of a technology, its usage and the way it is embedded in society can be very different from what it looked like in the beginning. What is needed, and what is also possible, is prospective technological mapping.

It is not only producers of technologies who determine their final fate, as the case-studies in this volume have indicated. For example citizen and consumer action can play a crucial role. In a way they can invent new artefacts (Carlson 1992). Since impacts of technologies are co-produced — they develop during the process — it is clear that it is not possible to fully anticipate, and again, linear trend extrapolations will not work. But the possibility to say something about future developments is not closed altogether. This can be done through sociotechnical and structural mapping: mapping the web of technological and social constraints and opportunities, including the regulatory and cultural landscape, in their dynamic rather than momentary appearances. Remmen and Koch have, therefore, rightfully argued that traditional TA methods still have to play a (restricted) role. In addition, Callon has discussed a more formal way of modelling sociotechnical dynamics.

Suggestion 2:
Forceful visions of the future and forceful demonstrations, pilots, etc., are important in their own right, and are a way to induce debate and learning.

Prospective mapping is an important, but essentially passive activity done by an observer, and without necessarily any implication for action. From cases and theoretical arguments in this book and elsewhere (Coombs, Jørgensen and Karnøe, see also Van Lente 1993), we know now that expectations and visions play a prominent role by providing coordination and guiding choices. Thus, developing visions on the future-state-of-the-world is not futile if these visions are translated into requirements for technical change and hence become forceful.

Several authors in this book have pointed out the importance of experimentation. While Soete's and Herbold's treatments of this theme are directly related to the non-linear character of technical change and hence the unpredictability of impacts and opportunities, there is a second reason to support the production of alternative technologies and experimentation. Technologies can be seen as embodiments of visions (see Jørgensen and Karnøe) and in this way can generate productive technical fixes which include new social practices (for a similar argument see Downey, Jelsma).

Suggestion 3:
Stimulate interactive demand and acceptance articulation.

A number of economists (Clark 1985, Green 1992) have emphasized the importance of articulation of demand. Consumers do not yet have precise requirements. Demand will be articulated only in interaction with technological supply. In analysing the automobile industry in the United States Abernathy et al. (1983) emphasize how technological products and markets were not pre-given, but emergent:

> "Producers gradually learn to distinguish the relevant product attributes for which they must supply technical solutions acceptable to the market. (...) Taken together, these attributes constitute an industry's basis of competition - that is, they define the arena within which different producers stake out their distinctive positions. [For such an arena to emerge some stabilization has to occur of ..] a process of iterative searching and learning (...) among producers and buyers alike" (Abernathy et al. 1983, p. 26, 25).

Akrich has added to this that the supply side already contains user (or demand) "representations", i.e. views of the user that lead to design specifications and configurations that 'script' a particular kind of user. Demand articulation in relation to technical change always involves articulation of cultural and political acceptance as well. The economic aspect is itself embedded in a culture in which goods are valued in particular ways. In addition, users of technology relate to the technology also in terms of their (positional) identity (see Downey).

Articulation of preferences (if preferences is the right concept at all) cannot occur in isolation. Adopters, users, their spokespersons — all must have something to interact with, and also know it is for real. Anticipatory articulation and coordination processes, are a necessary condition to reduce the costs of trying out, often by error, and creating undesirable irreversibilities on the way.

What is not clear, and perhaps can never be known *a priori*, is when articulation has progressed sufficiently. Similarly, learning about handling technology does not have an immediate criterion in which direction the learning should occur, and when it has gone far enough (Wynne, Walsh). What kind of criteria are available to decide if a learning process goes well? A principle answer to this question would be that such criteria emerge in the ongoing practices, and that the criteria of quality refer to process rather than outcome. For example, whether there is a preparedness to listen to views and arguments of other actors.

In real life, learning is an agonistic process, and actors may accuse each other of being interested, or emotional, or naive, and thus not eligible for a hearing. Socio-cognitive power play is part of the process.

A variety of methods have been developed to mitigate these problems, mediation being one recent example. While they have a lot to offer, their focus is on conflict resolution rather than learning. Resistance might well be very productive, because it forces actors to articulate, develop new strategies, including the strategy of adapting the technology. The strong opposition of environmental groups to herbicide-resistant transgene plants has been seen, by technology promoters, as misguided, rigid, and emotional — and perhaps they were right. But it has also forced a move towards disease-resistant varieties, and herbicide resistance is now only taken up if clear environmental advantages can be articulated.

To put the point emphatically: for real, and symmetrical, articulation of supply and demand (to use the economic terms as a shorthand) technology actors might well be advised to arrange for criticism. Competition is recognized as healthy for the quality of products. Similarly, too easy acceptance is unstable, and opposition is healthy for the quality of the technology.

Suggestion 4:
Modulation of technical change processes must recognize the multi-actor, decentralized character of such processes.

We advocated creative incrementalism before, and this is not (just) a matter of expediency. It relates to an analytic and normative diagnosis of the situation, especially in processes of technical change and embedding of technology in society. *De facto* there are many actors involved, and so there are no central institutions which can exert control, but there is mutual dependency and interplay of actor strategies. This is not something to be deplored, and restructured to more hierarchical structures. Multi-actor, decentralized processes require multi-actor, decentralized forms of control.

In political science and policy analysis, the concept of 'policy networks' has become popular to capture this situation (Mayntz 1993). Governments become modulators of policy networks, rather than top-down, authoritative directors of change. For technology in society, the specifics of the processes and structures have to be taken into account; especially the fact that most of the processes occur without a direct relation with a government initiative or a government concern. Conversely, attempts at intervention and modulation are not limited to government actors, even if other actors often try to mobilize government actors for their purposes.

In such a situation of multi-actor decentralized forms of control, indirect interventions rather than direct intervention would be in order. This can most effectively be done through third parties. In the case studies (Jelsma, Steward, Walsh), pressure groups appeared to function as carriers of credibility pressure and as actors who put and maintain a certain issue on the agenda. Thus, pressure groups can act as lead articulators, in the same way that economists have identified the role of lead users. They help articulate a certain demand or mode of acceptance and in this way influence technical change. So influencing relevant pressure groups is a relevant CTA strategy, even if there is little control over the eventual outcomes.

Another opportunity for intervention through third parties is offered by insurance companies. Because of their financial interest in risk reduction, in occupational health and safety, and more recently also in environmental risks, they exert a lot of *de facto* influence on firms and other organizations (including public authorities). Occasionally, this has implications for technologies to be developed and used. A small shift of the insurance companies may lead to large shifts in industry and technology. So this is an example where the route through 'third parties' creates leverage.

For non-governmental CTA agents for change, the implication is that they can be relatively powerless, but may profit from judicious interaction with more powerful actors. For governmental actors, the implication is that they may focus on infrastructural intervention, which ranges from providing the conditions for CTA interaction — the creative game regulator role — to creating new kinds of third parties — including empowerment of citizens, and/or increase learning capacity.

Suggestion 5:
Social learning processes can be facilitated by alignment of actors (sometimes by creating new actors or nodes) and creation of specific loci for societal learning processes.

Embedding new technology in society is a process in which all sorts of actors actively try to exert influence, and/or passively shape what happens by not doing something. It is through the interaction of such explicit and implicit actor strategies that the success of desired sociotechnical transformations is determined. Alignment can be used as a concept that indicates the mutual and well-functioning adjustments between different actors involved: designers, users and other people affected. There is often no actor with special responsibility for alignment. The lack of such alignment actors is part of the modern regime of managing technology in society as it developed since the eighteenth century.

Recently, alignment is recognized to be important, and sought after consciously. Alignment actors are created, at the promotion side in the form of directorates or programme committees for technology stimulation programmes, at the adoption and diffusion side in the form of consortia and platforms (as for High Definition TV and for tele-work/tele-commuting). Such new composite actors are interested in societal learning (on their own terms, but these terms are often shaped by overall societal goals).

Earlier, and entrenched, alignments may be a constraint on societal learning. The medical sector is a clear example where pre-existing alignments and decision-making structures hamper societal learning. In this case learning is obstructed by the presumption of go/no-go decision moments that are forced upon participants by lack of coordination. Remmen's essay describes another example of such constraints. It argues forcefully that government regulation stimulates end-of-pipe technologies instead of more preventive clean technology options which would entail a broader interaction process between user-firms, suppliers, management and workers (see Schot 1992 as well).

So at the end of this epilogue we must also reflect on actions of the CTA agents: as mobilizers, as advisers, as mappers of co-production dynamics, as change agents. To actually exert influence entails becoming an actor oneself, or at least a forceful visitor. CTA agents have interests themselves and will be acted upon by other actors and already in this way be drawn into the games. Strategic behaviour is unavoidable.

At the same time, actors are reflexive, and CTA agents should be reflexive as well. They make judgements about their aims and the situation. Based on such reflections, a theory of normative orientation of CTA action and of the role of CTA agents must be developed. In fact, implicit in the individual chapters in this book one can already find reference to it.

One important question relates to social and political philosophy: if CTA includes value assessments of impacts, *who* represents actual or potential users or impactees — and *how*? There are principle questions here (compare the issue of how future generations should be taken into account) and institutional questions (what sort of accountability can and must be organized for CTA agents).

For a CTA agent, who is, in a sense, an intermediary between the present situation and a future world, the question of 'representation' appears to be easy in cases of broad consensus. This often is a consensus about diffuse goals, like "environment", and more recently, "sustainable development." In these cases the CTA agent can as it were assume their validity and proceed to implement them. If there is no such consensus, a CTA agent has to take responsibility for the goals he serves. In the case of modern biotechnology, to mention one example, this requires hard choices: how the risks of technological progress should be faced, how legitimate public concerns and their spokespersons are.

So value questions cannot be evaded. Moral entrepreneurship for particular causes is of course widespread, in CTA the attitude of being anticipatory and relating this to action or at least, articulation of strategy and the identification of future impacts, independent of specific values, is characteristic. And agents carry the moral thrust, and moral justification of being active in CTA, and so each becomes a bit of a moral entrepreneur. This explains (and contains an argument) why open-ended societal learning processes are foregrounded by CTA agents.

References

Abernathy William, J., Kim B. Clark and Alan M. Kantrow. 1983. *Industrial Renaissance*. New York: Basic Books.

Carlson, W. Bernard. 1992. "Artifacts and Frames of Meaning: Thomas A. Edison, His Managers, and the Cultural Construction of Motion Pictures," in: Wiebe E. Bijker and John Law (eds.). *Shaping Technology/Building Society: Studies in Sociotechnical Change*. Cambridge, Mass.: MIT Press.

Clark, Kim, B. 1985. "The Interaction of Design Hierarchies and Market Concepts in Technological Evolution," *Research Policy* 14: 235-251.

Green, Kenneth. 1992. "Creating Demand for Biotechnology: Shaping Technologies and Markets," in: Rod Coombs, Paolo Saviotti and Vivien Walsh. *Technological Change and Company Strategies. Economic and Sociological Perspectives*. London: Academic, 164-183.

Mayntz, R. 1993. "Policy-Netzwerke und die Logik von Verhandlungssystemen," in: Adrienne Héritier. *Policy-Analyse. Kritik und Neuorientierung*. Opladen: Westdeutscher Verlag.

Rip, Arie. 1992. "A Quasi-evolutionary Model of Technological Development and a Cognitive Approach to Technology Policy." *Rivista Di Studi Epistemologici E Sociali Sulla Scienza E La Technologia* (2): 69-102.

Schot, Johan. 1992. "Constructive Technology Assessment and Technology Dynamics: The Case of Clean Technologies." *Science, Technology and Human Values* 17: 36-56.

Van Lente, Harro. 1993. *Promising Technology. The Dynamics of Expectations in Technological Developments*. Enschede: University of Twente. Ph.D. Thesis.

Index

ACSYNT (AirCraftSYNThesis) 93-98
ACSYNT Institute 94-98
actor world 138, 333
actor-network theory (see also socio-technical network, network alignment, convergence, conception-adoption network) 32, 86, 160, 305-306, 308-327, 333, 339
Advanced Genetic Sciences 151
Agene 111, 113-119
aircraft industry 93-107
alternative energy movement 57-59
alternative technology 4
Alvey programme 340
American Diabetes Association 290-291
American Society of Microbiology 147
Anders, Günther 1
ascorbic acid (vitamin C) 117
Asilomar conference 144, 160
Askov Folk High School (Denmark) 59, 61
Association of Danish Wind Mill Manufacturers 67
Association of Danish Wind Power Owners 67
Association of Diabetes Nurses (The Netherlands) 292, 297
Atomic Energy Commission (Denmark) 63
Atomic Energy Commission (U.S.) 2
Aylesbury Dairy Company 127

Baarder (German fish-processing equipment manufacturer) 205-206
Baker, J. C. 113
Baker, J. R. 267, 270
Barham, George 121-123, 126
Berg, Paul 143-144, 157, 162n7

Bernstein, Richard 255n13
Biarritz pilot experiment (optical-fibre cable network) 174
Bielefeld (Germany) 188-191
Bielefeld model 191-192
biological containment 144, 149
biophysics 244-249
biotechnology (see also recombinant DNA) 23-25, 141-161
  criticism of 141-143
  ethical issues 144
  human insulin 288
Biotechnology Science Coordinating Committee (U.S.) 154, 163n18
Birth Control Investigation Committee (UK) 270
birth control movement 269-271
birth control pill (see oral contraceptive)
Board of Industrialists (Denmark) 64
Boehringer-Mannheim 290
Boeing 94
Bohr, Nils 63
Borden, Gail 123
boron compounds 122, 125, 127
Boron Syndicate 126
boundary disputes 225-226
boundary objects 320
Boyles, Richard 93
Bread and Flour Regulations of 1963 (UK) 131
Brenner, Sidney 161n3
British Baking Industries Research Association 118
British Computer Society 340
British Diary Farmers Association 121, 127
Brundtland Commission 199
Buckley, Wilfred 128-129

CAD/CAM (computer-aided design/ computer-aided manufacturing 83-107
canine hysteria 116
care trajectory 297-300
Centers for Disease Control (U.S.) 290
chlorine dioxide 117-120
cleaner technologies 208-212
  defined 199-200
  in fish-processing industry 206-208
Cleaner Technology Program (Denmark) 209-212
coffret d'abonne (CA) unit 169-181
Cohen, Stanley 146
Collingridge, David 7-8
Collins, Ken 159
Committee on Safety of Medicines (UK) 275, 340
complementary assets 167, 265, 274
comprehensive assessment 203-204, 213-214, 218
Computervision Corporation 89
conception network 315-317
conception-adoption network 317-323
concurrent engineering 8, 349
conflict resolution 3, 350
consensus conferences 6, 233, 235-237
consequence assessment 203-204, 213-214, 218
constructive assessment 203
constructive technology assessment (CTA)
  actors and aspects 214
  and democratic process (see also technological development--public participation in) 307
  and design activities 105-107
  and experiments 46-47, 193
  and localizing strategies 85-86
  and pollution prevention 213-215
  and research funding 232
  and societal integration of technology 44
  and technical fixes 83-84
  and technology policy 15-17
  and user representations 168, 182-183
  as paradigm 5-7, 347
  as public procurement 341-344
  contrasted with traditional TA 15-17, 216, 218
  ethical and normative aspects 5, 353
  goals 3, 7-10, 354-356
  in Denmark 201-203
  in private sector 336
  existing models for 249-253
  need for reflexivity in 32-33
  suggestions for doing 347-352
consumer testing 170-172
Contact Ambience telephone 169-181
contraceptive technology 262-267
control (see multi-actor decentralized control of technology)
control arena 332-333
controversies 16, 157, 188-191
  as informal TA 20
controversist space 86, 100
convergence of network 315-317, 325
Cooperative Wholesale Society (UK) 113
Coppock, J.B.M. 118
Corn Laws (UK) 112
Council for Medical Research (The Netherlands) 298
Council of Health Care Funds (The Netherlands) 288, 293-299
Cow & Gate 121
credibility problems 2
cultural anthropology (see also positional identity) 86
cultural theory of risk 29-30
Curtiss, Roy 161n3, 162n7

dairy industry 121-130
Dairy Show (UK) 121, 127
Danfoss (Danish metal producer) 207
Danish Board of Technology 6-7
Danish Wind Energy Company 60
Davis, Bernard 161n3, 162n7, 162n8
decision routines 333
decision-making processes 22, 192-193, 227-228, 285-286, 288, 293-300
demand (see also user representations)
  and social factors 264
  articulation of 349-350
Denmark and CTA paradigm 5-6, 57-80, 199-219
Department of Defense (U.S.) 94
Department of Energy (U.S.) 2
Department of Social Affairs (Norway) 236-237
design for impacts 8
design processes 4, 84-85, 90-99, 324-325, 347
  and CAD/CAM (computer-aided design computer-aided manufacturing 90-93
  and user representations 169-183
Diabetes Fund Netherlands 290, 299

Index

diabetes mellitus 287-300
Diabetes Society Netherlands 290
diagnostic ultrasound (see also medical ultrasound) 236
dialogic community 228, 251, 255n13
dialogue workshop 139, 213, 217
diffusion model 309
diffusion of technology 208-212, 238-239
drug industry 268, 276
Dryzek, John 255n13
Dybkjær, Lone 199

ECO-labelling 212
ECO-management 212
economics of technical change (see also evolutionary economics, neoschumpeterian economics) 37-46, 167-168, 310
Eisenhower, Dwight 271
Eli Lilly 287-288
Ellul, Jacques 1
ELSAM (Danish power producer) 63
end-of-pipe technologies 200
Energy Information Board (Denmark) 65
Energy Plan 2000 (Denmark) 66
engineering science 92
entrenchment 8-9, 142
environmental audits 210
Environmental Protection Agency (Denmark) 205-206, 211-212
Environmental Protection Agency (U.S.) 152-154, 158, 219n3
environmental strategies 211
ESPRIT programme 340
ethisterone 263
European Association for the Study of Diabetes 290
European Association of Diabetics Educators 292, 297
European Commission 23-25
evaluation research 219n1
evolutionary economics (see also neoschumpeterian economics, variation, selection, technological trajectory) 74, 86, 332-335
Express Dairy Company 121-123, 129
extended peer review 160
externalities of technical change 44-45

F. L. Smith (Danish cement producer) 60, 67
feedback on experience 172-173

fish-processing industry 199-208, 214-215
flour-milling industry 112-121
Food and Drug Administration (U.S.) 163n18, 233, 249, 275
Foundation on Economic Trends 151
Fourastie, Jean 40
France Telecom 169, 181
French Institute 122
Frost-ban (ice-minus bacteria) 151
Fry, William 240, 244-249
future workshops 217
futures research 201, 204

G.D. Searle 267-268, 270
Galison, Peter 243
General Electric 94
generic technologies 42
Genetech 288
genetic fingerprinting 38
genetically modified organisms (GMOs) 152-154
German Federal Waste Disposal Act (1972) 188
Glasgow University 34n1
Great Western Railway 121
Green Party (Germany) 189
Greenpeace 340
Grimwade, F. S. 123
Grindsted Products 207
Grumman Aerospace 94

Health Council (The Netherlands) 298
health insurance system 237-238, 288, 293-300, 301n2
Hehner, Otto 130
High Speed Civil Transport (HSCT) 85, 96, 100-105
Hoechst 288
Hopkins, Frederick Gowland 130
Housewives League (UK) 117-118
Howry, Douglass 239-249
Hueter, Theodor 240, 244-249
human genome project 250
human reproductive technology 261-277
hybrid corn 41

impact constituency 5
Industrial Biotechnology Association (U.S.) 153
Industrial Health and Safety Units (Denmark) 206
infant mortality 125

innovation as recursive process 185
innovation literature 261
innovation stages 254n2, 286, 298
Institute of Medicine (U.S.) 234-235, 238-239
insulin 287-288
integral chain management 8
intensive conventional insulin therapy (ICIT) 291-300
interest theory 88
International Atomic Energy Agency 27
International Diabetes Foundation 290
interorganizational network 138, 209, 211, 214
irreversibility 305-308, 321-323, 325

Jameson, William 117
Johnson, Lyndon 271
Joseph Rank, Ltd. 113, 118-119
Juul, Johannes 60-62

Kent-Jones, D. W. 114
King, Jonathan 162n7
KLOZ (Dutch organization of private insurance companies) 294-300

La Cour, Paul 59-60
Lahl, Uwe 189
Lancet, The 124
learning (see social learning processes)
learning-by-doing 188
learning-by-experience 68-69
learning-by-using 68
Lederberg, Joshua 162n7
Lindstrom, Petter 245, 254n7
Lister Institute 125
Lister, Joseph 125
Lockheed 94
Loombe, C. A. 118

managing technology in society 1-10, 347-356
mapping 350
market research 273
market surveys 169-170
Marshall Plan 61
McDonnell Aircraft 94
medical research 247-253
Medical Research Council 117, 119
medical technology 225-302
medical technology assessment 227, 231-238, 285

medical ultrasound 231-253
Milk Advisory Committee 130
Ministry of Agriculture (The Netherlands) 158
Ministry of Health (The Netherlands) 285-286, 294-300
Ministry of Health (UK) 115-117, 120, 267, 269
Ministry of the Environment (Denmark) 200, 207, 210-212
Minitel system 174-175, 178
modelling 46, 85, 308, 326, 348
modernity and technology 1, 9-10
Monsanto 154, 159
moral entrepreneurship 356
Morgan, Russell 240
multi-actor decentralized control 202, 306, 349-356
Mumford, Lewis 1

National Academy of Science (U.S.) 147
National Aeronautical and Space Administration
   Ames Research Center 94, 101-102
   Langley Research Center 101-102
   Lewis Research Center 102
National Association of British and Irish Millers (NABIM) 112, 114-121
National Association of Flour Importers (UK) 114
National Association of Master Bakers, Confectioners and Caterers (UK) 115
National Cancer Advisory Council 247
National Cancer Institute (U.S.) 240
National Clean Milk Society (UK) 128
National Dairymen's Association 121
National Farmers Union (UK) 115
National Health Service (UK) 343
National Institute for Research in Dairying (UK) 128
National Institutes of Health (U.S.) (see also Recombinant DNA Advisory Committee) 144-148, 158, 228-229, 231-232, 247
   and ultrasound 236-237, 247-249
   as model for CTA 249-253
neoschumpeterian economics 116, 305
Netherlands Organization of Technology Assessment (NOTA) 6-7, 87
Netherlands Society for Diabetes Research 290, 294, 295, 301n8
network alignment 177-182, 352

Index

network externalities 45
network management 160-161
nitrogen trichloride (see also Agene) 113
non-linear learning (see also technical change—non-linear character) 185
Nordisk 287
North American Rockwell 94
Northrop 94
Northwest Jutland People's Centre for Renewable Energy 71, 75-76
Novick, Richard 162n7
Novo/Nordisk 288, 292
nuclear fuel cycle 22
Nuclear Regulatory Commission (U.S.) 2

Obscene Publications Act of 1857 (UK) 267
OECD 6, 40, 39
OEEC 61
Office of Science and Technology Policy (U.S.) 100, 154, 157
Office of Technology Assessment (U.S.) 232-233, 268
open planning process 191-193, 194n15
oral contraceptives (see also birth control pill) 38, 261-277
Organization for Information about Atomic Power (OOA) 63-66, 71
Organization for Renewable Energy (OVE) 66, 71, 75
organizational theory 336-338, 355

Paris Public Transport Authority (RATP) 313
Pasteur, Louis 122
pasteurization 122-124, 128-129
patent law 159
path dependencies 45, 54, 308
Pincus, Gregory 270, 272
Planned Parenthood 271
policy networks 353
polychlorinated biphenyls (PCBs) 188
population control 276
population explosion 271-272
positional identity 55, 86-88, 107, 225
post-modernity 23, 30
pre-competitive R&D support 42
PREPARE 208, 219n3
PRISMA 6
productivity 89
productivity puzzle 39
professions 55

promotion arena 332-333
Public Health (Milk and Cream) Regulations of 1912 (UK) 123
Public Health (Preservatives in Food) Regulations of 1925 (UK) 123, 131
public networks (see also policy networks) 340-341

radiology 240
Raging Diggers 141-142
Raging Potatoes 141
randomized controlled clinical trial (RCT) 227, 233-234
Raytheon Corporation 246
recombinant DNA 143-149
Recombinant DNA Advisory Committee (RAC) 145-152
reflexivity 20-21, 34n1, 193
refrigeration 122, 126-127
regulation 274-275
 biotechnology 151-154
 diary industry 123
 environmental 204-205, 212
 recombinant DNA research 147-149
 two-track approach 2-5, 332-334
regulatory regime 149-151, 156
Reinschmidt, Kenneth 92
Reliable Energy Information (Denmark) 65
Research Association of British Flour Millers 114, 117
Rifkin, Jeremy 151
risk acceptance 192
risk analysis 25-26, 193, 332
risk assessment 2, 22, 155, 159
 and technological choice 112, 120, 127
 animal growth hormones 24
risk perception 26-29
risk problem 143-148, 161
risk society 2
risk studies 20-21
Royal Agricultural Society 127
Royal Society of Arts 126-127
rubber technology 266

Sanger, Margaret 269
Schmitt, Ott 240
Science for the People 146
SEAS (Danish power producer) 61
Securiscan 169-181
selection environment 29, 74, 335-344
sensitivity analysis 99

Simon, Henry 112
simulation (see modelling)
Sinsheimer, Robert 147, 162n8
Skandinavisk Aeroindustri 60
Skandinavisk Aeroteknik 67
social construction of strategies 338-340
social constructivism 86
Social Democratic Party (Germany) 189
social experiments 16, 46, 138, 185-193, 202-203, 213
social learning processes (see also dialogic community) 16, 185, 253
   conditions for 137-139, 155-161
   criteria for 28-33, 34n2
   defined 142
social rule systems theory 162n14
Society of Public Analysts (UK) 124
socio-technical network 312-315
sociology of innovation 310, 312
sociology of translation 327n2
Solow paradox 39
sonic boom 99-105
Spillers, Ltd. 113
SR-71 99-100
Staudenmaier, John 5
strategic management 355
strategic planning paradigm 338
Strauss, Nathan 123, 128
summer diarrhoea 125
sustainable development 199
sustainable production 213

technical change
   measurement of 38
   non-linear character 350
   users in (see also user-producer interactions) 167-168
technical fixes 138, 144, 149, 154, 161, 349
technical systems 28
technological determinism 84, 178-179
technological momentum 53-54
technological paradigms 333
technological presuppositions 243-244
technological taxonomies 38
technological trajectory (see also care trajectory) 53-54, 62, 111-112, 130-131, 333
technology
   and democratic process 202
   and modernity 1, 9-10
   as "black box" 37
   as social process 40
   co-production of effects 3-4
   control of 7-8, 334
   criticism of 2
   ethical issues 277, 326
   tacit social models of viability conditions 30-33
Technology and Resources (TOR) project 204-215
technology assessment (see also medical technology assessment) 38, 120, 348
   and technology policy 8, 37
   conventional 2-6, 200-201, 216
   criticism of 19-20, 34n4
   in Denmark (see also consequence assessment, comprehensive assessment, constructive assessment) 201-203
technology development
   as social experiment 186-187
   bottom-up strategy 68-70
   public participation in 191-193
   top-down strategy 69-70
technology dynamics 7-8, 85-86, 333, 347
technology forcing 4, 212
technology networks 336-338
technology policy
   and "market failure" 15, 41-43
   and societal integration of technology 16, 37, 43-46
   and technology assessment 8
technology strategy units 337
Test Station for Smaller Turbines (Denmark) 67
theory of acceptance 86-88, 107
theory of the firm 305, 336-340
Thomson 313-314
Three Mile Island accident 65
Toxic Substances Control Act (U.S.) 153
Tvind Folk High School (Jutland) 66-67
two-track approach (see managing technology in society, regulation--two-track approach)

ultrasonic diathermy 245
ultrasound (see medical ultrasound)
ultrasound research 239-249
United Dairies 121, 129
United Dairies Equipment Company 129
University of California 150-151
user needs 261, 272-274
user representations 138, 167-183, 227-228, 272-274, 349
user-producer interactions 167, 226, 273

values learning (see also social learning processes) 33, 192
Vanderplaats, Garret 96-97
variation 74, 335-338, 341
Virginia Polytechnic Institute 94-95
Virulent Viruses 141
VMK (Swedish fish-processing equipment manufacturer) 205-206
Volpar (spermicide) 270

Wallace and Tiernan 113-114, 117
waste management 185-193
waste sciences 187-188
Watson, Jim 162n7
Wild, John 239-249
wind turbine industry (Denmark) 57-78
Windpower Commission (Denmark) 61
Windscale Public Inquiry (1977) 21-23
Winner, Langdon 1
Women's Labour League (UK) 269
working conditions 205-208

Zoutendijk, G. 96